THE AMATEUR ENTOMOLOGISTS' SOCIETY

A COLEOPTERIST'S HANDBOOK

(3rd Edition)

by Jonathan Cooter
et alii

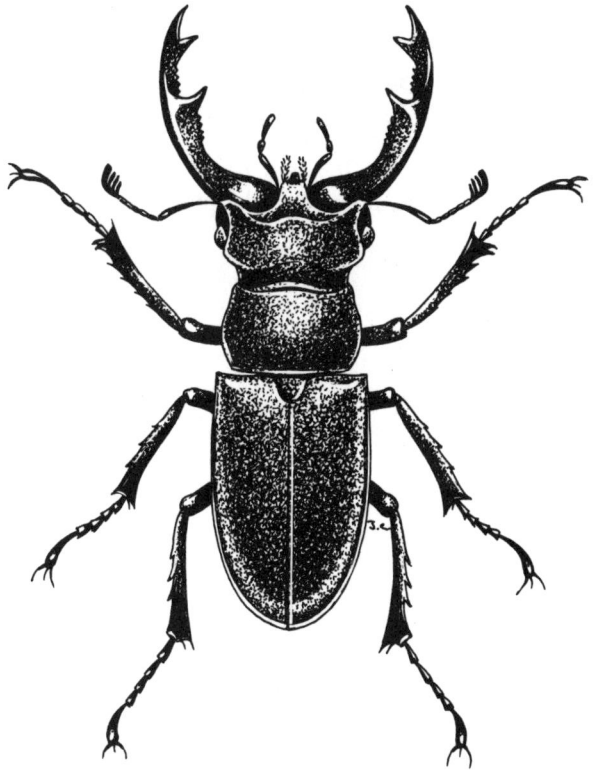

General Editor: P.W. Cribb

1991

Dedicated to my parents
John and Anne Cooter
and to
Joseph Cribb
Sculptor & Coleopterist
1892–1967

©1991. The Amateur Entomologists' Society

ISBN 0 900054 53 0

Published by the Amateur Entomologists' Society
22 Salisbury Road, Feltham, Middlesex TW13 5OP

Printed by Grosvenor Press (Portsmouth) Ltd

CONTENTS

		Page
PREFACE		vii
PART I	PRACTICAL ASPECTS	
	Collecting Equipment	1
	Collecting – Sample Habitats	12
	Starting and Building up a Collection	30
	Indoor Work	33
	Identification	47
	Examination of Coleopterous genitalia	52
	Diaries and Notebooks	66
PART II	THE BEETLE FAMILIES	68
	Carabidae	69
	Water beetles	73
	Histeroidea	79
	Ptillidae	81
	Leiodidae	84
	Silphidae	89
	Scydmaenidae	90
	Scaphidiidae	91
	Staphylinidae	92
	Pselaphidae	95
	Scarabaeoidea	97
	Clambidae	103
	Scirtidae	103
	Byrrhidae	104
	Heteroceridae	106
	Buprestidae	106
	Elateridae	108
	Throscidae	111
	Eucnemidae	113
	Cantharidae	114
	Lampyridae	116
	Dermestidae	117
	Anobiidae	117
	Ptinidae	118
	Cleroidea	119
	Kateretidae	121
	Nitidulidae	123
	Rhizophagidae	129
	Cryptophagidae	129
	Phalacridae	131
	Corylophidae	132
	Coccinellidae	133

	Lathridiidae	137
	Cisidae	138
	Heteromera	139
	Scraptiidae	140
	Mordellidae	141
	Cerambycidae	145
	Chrysomelidae	156
	Curculionidae	171
	Scolytidae	150
PART III	BEETLE ASSOCIATIONS	
	Beetles associated with Vascular plants	183
	Beetles associated with stored food products	199
	Beetles associated with ants and ant nests	210
PART IV	BEETLE LARVAE	
	Beetle larvae, their morphology and identification	217
	Rearing beetle larvae	239
PART V	BIOLOGICAL RECORDING	251
	CONSERVATION AND THE COLEOPTERIST	256
APPEN-DICES	GLOSSARY	275
	INDEX OF GENERA REFERRED TO IN TEXT	287

PREFACE

During January 1974 the second edition of this *Handbook* was published. Finances dictated its format, with new material added as appendices. The result was adequate but not entirely satisfactory and it was obvious that, once the second edition became out of print, a completely revised edition would be required. With stocks low, I was again approached by Peter Cribb and set about the work. A new format was devised and the help of many specialists became a required necessity; my own knowledge, possibly in common with many other coleopterists, is selectively gleaned to suit personal interests and bias. The whole exercise has taught me four lessons:

My own knowledge of British Coleoptera is patchy.

Use of a word-processor would have made life easier and seen an earlier publication of this book.

To think hard before volunteering for anything!

Dead-lines really are quite flexible.

With regard to the latter I must point out that very often the flexibility was my own and not that of (the bulk of) the individual contributors. With this in mind I tender my apologies to those assiduous specialists who promptly prepared their valued contributions. In all, delays included, I hope this edition meets with approval and will be a useful text for the novice and more experienced coleopterist alike. In revising the *Handbook* it soon became evident that some of the original material (in the 1954 and 1974 editions) needed so little alteration, often only the revising of nomenclature, that to re-write would be not only a waste of time but something akin to plagiarism.

With so many individual contributors, there may be some inconsistency in terminology and style. Efforts have been made to bring a degree of uniformity, but doubtless some instances remain.

I offer my thanks to all those persons that have contributed and are named in the text. My sincere thanks are also offered to Dr Lena Ward and David Spalding (Institute of Terrestrial Ecology) and Dr. M. Majerus. Howard Mendel and Alex Williams, with a naïvety born of friendship, kindly offered to read through the assembled manuscript; both made valuable comments and suggestions, the majority of which have been incorporated. I am pleased to report that burdensome task had no adverse effect on our friendship. Peter Cribb once again has taken on the onerous job of not only giving continual encouragement and not complaining too strongly about my missed dead-lines but for editing the manuscript and seeing it through the publication stages. John Read very kindly produced the excellent drawings of beetles featured in the *Handbook*; to all four gentlemen I tender my heartfelt thanks.

In a book that proposes to give advice and guidance perhaps the "Preface" is the place to mention the best piece of advice that was ever given to me; it has helped me to maintain a sense of perspective over the years. At my very first "Verrall Supper" when about 18 years old, I found myself in conversation with Horace Last and Rev. E.J. Pearce, two authorities with whom I had corresponded for several years, but never previously met in person. Towards the end of our chat Horace (now a close family friend and personal mentor) took me to one side and confided to me "Now remember, Jonathan, there's more to life than beetles".

Jonathan Cooter
1990

Part I – Practical Aspects

COLLECTING EQUIPMENT
By J. Cooter

The novice can do no better than learn from the experience of others, either by asking advice or by careful field observation. Personal experience is a great teacher and one's methods and equipment become adapted to those habitats most often worked or those beetles most avidly sought. One lesson is very quickly learned – collecting equipment should be light but robust and minimal. My own sweep net, pooter, box of 43 $2\frac{1}{4}'' \times \frac{5}{8}''$ tubes, sheet, tins and chisel fit into an ex-Army gas-mask pouch which is carried over one shoulder. The sieve fits over the pouch leaving $1\frac{1}{2}$ arms free for sweeping, turning over logs or stones, grubbing and using the pooter whilst walking to or from the collecting ground.

Most of the popular items of equipment can be purchased from suppliers. These are designed for general entomological use and their efficiency for collecting beetles can be enhanced by simple modifications if desired. In the following notes various modifications have been shown and it is hoped the diagrams will serve to make these self-explanatory.

Collecting equipment falls into two categories:
1. That needed for finding beetles.
2. That needed for securing their capture.

Pooter

This is perhaps the most essential item of equipment; it is a simple device for the rapid collection of small insects. Coleopterists tend to prefer the tube-type pooter (Fig. 1b) rather than the cylinder model (Fig. 1a) much favoured by our near relatives, the dipterists. The tube-type does indeed have advantages when used for capturing insects that walk rather than fly. The head can be removed and quickly replaced with a cork, the head being fitted to a fresh empty tube. Emptying a cylinder-type pooter on a hot sunny day requires a swiftness of hand more associated with illusionists; remember dipterists kill or anaesthetise their catch in the pooter before emptying. A spider inadvertently caught should be removed at once as invariably its silken thread causes problems.

The shop-bought tube pooter can be easily modified (see Fig. 1b) or can be built from scratch, based upon this diagram. The main points of modification are: replacing the mouth-piece tube with a shorter length of larger diameter tube; replacing the bulky cloth filter with a neat disc of metal or nylon gauze; replacing the inlet tube with a longer length of metal tube (aluminium, easier to bend cold, or brass) available from model shops; replacing the rubber tube with a length of a slightly larger diameter, a little longer than one's arm. The reasons for these

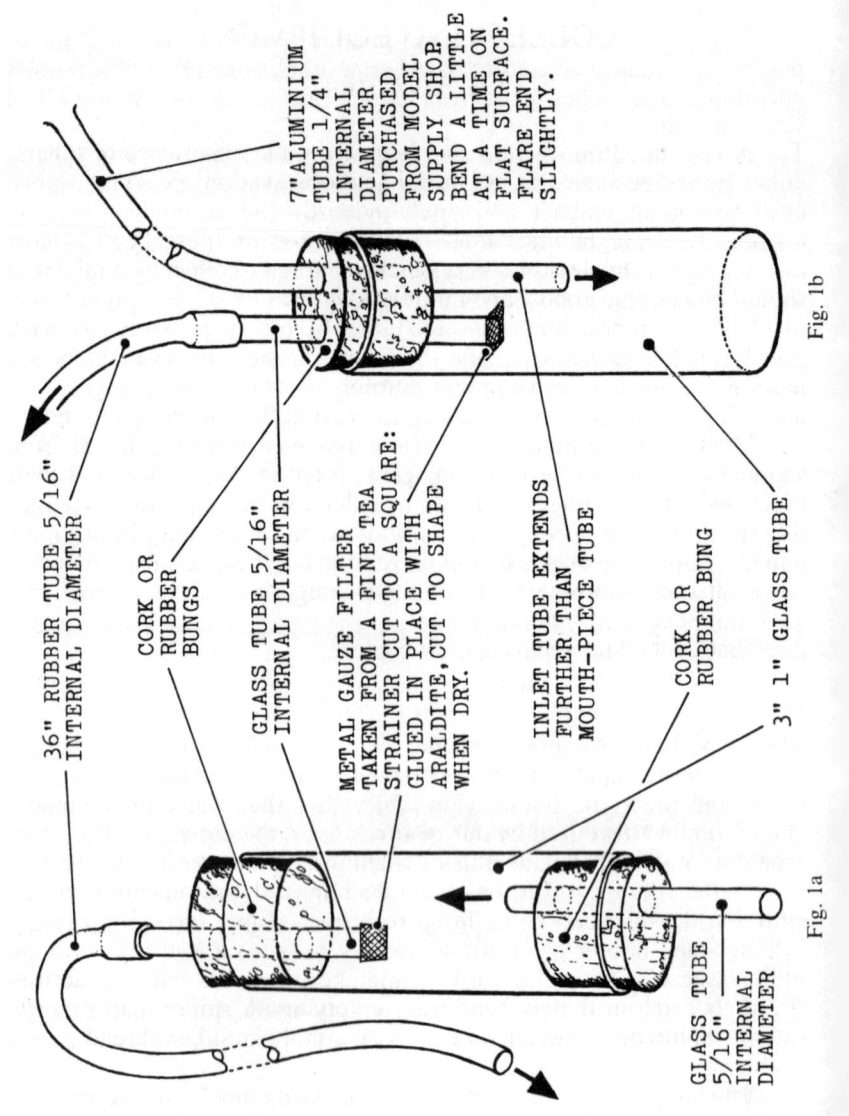

CYLINDER AND TUBE POOTERS

changes are: the cloth filter traps beetles, the inlet tube is stronger and projects longer into the pooter, the larger inlet arrangement permits a more powerful "suck" to be generated and the whole can be used at arm's length.

Many Coleopterists studying smaller beetles make a "micro-pooter" of similar design, but based on a tube about $1\frac{1}{2}'' \times \frac{5}{8}''$ with inlet and mouth-piece tubes made from lengths of electrical insulation tubing stripped from flex.

A variant of the suction pooter relies upon a strong blow but has never gained popular use. The various filters help to weaken the blow effort and prolonged use is exhausting. However, if its efficiency can be improved it would be a useful device for collecting at carrion, dung or in dusty places. It should go without saying that the use of a suction pooter in such places is fraught with health hazards.

The "spider pooter" is a simple tube expanded at one end beyond the filter. It is used for holding a specimen whilst it is examined in the field. It would seem to have minimal application for working Coleoptera.

Tubes

Nowadays the most widely available glass tubes have push-on polypropylene caps, the corked tube going out of fashion. The new type have many advantages – the caps do not work loose or shrink. They are, however, apparently not as thick walled as the older cork stoppered ones and can break when the cap is pushed in. Some beetles used to eat their way out of the cork, or burrow into it; cork does however, permit the tube to "breathe", the plastic has an air-tight fit.

Tubes are available in many sizes, the most popular being perhaps $1\frac{1}{2}''$–2" long $\times \frac{3}{8}''$ to $\frac{5}{8}''$ diameter. A few of larger size are useful, say, 3" long $\times$ 1" diameter for larger beetles, and double up for replacement pooter tubes.

Some form of box should be made or used to hold the tubes as they are easily broken if left loose in the collecting bag, and a box will also protect the beetles against sunlight and the consequent formation of moisture inside the tube.

Tins

A tin, box or even polythene bag is very useful for bringing home various oddments, especially larvae with a supply of their pabulum.

Sweep-net

The main requirement for a good sweep-net is a strong bag deep enough to fold over the frame to prevent any active beetle taking flight and escaping.

My own is a Lepidopterist's cane-framed kite net with linen bag. It has stood the rigours of twenty or more years' sweeping and I find a handle unnecessary. My experience of the specially designed shop-bought sweep-net has not been very good. The bag is too shallow and the centre-tube is a confounded nuisance. They weigh considerably, more than the cane-framed job and their continued use on evening sweeping is quite tiring. However, this might be personal prejudice as they are widely used either by preference or because they are sold as sweep-nets.

Beating tray

A sweep-net can double as a beating tray, but the tray cannot be used for sweeping; beating is after all a vertical variation of sweeping. The traditional "Bignall Pattern" tray is readily available with white, or upon request, black cloth. The white cloth tends to give a high degree of glare on a sunny day.

To dislodge insects, one or two sharp taps with a stout stick or vigorous shaking will suffice; most of those dislodged should fall onto the tray, some will take flight quickly, others will feign death and remain still for a long time.

Knife

A stout "sheath-knife" or woodworker's firmer chisel of about 1″ to 1½″ width is often required for a variety of tasks, most usually lifting bark and working rotten wood, removing bracket fungi etc. It can even be used for digging under dung in the absence of a trowel. Many coleopterists are now purchasing a tiling hammer – this has a pick-like point at one end of the head and a chisel-shaped point at the other. For working rotten and hard dead wood this tool is superior to a knife. As an example, whilst preparing this book, I found *Lymexylon navale* in its characteristic deep burrows in hard oak. My 10″ knife made little impression, but a colleague's tiling hammer produced the beetle with considerable ease – the beetle retreated a good 4″ into the hard wood upon being attacked.

Water nets

For collecting water beetles a suitable net is essential. The net must have a strong enough head to deal with the weight of water and material that enters it, a strong handle and the net must be of a material which will allow the water to drain through quickly without releasing the smallest beetle being sought. One can use a fishing landing net which has a rigid steel head. Aluminium-ring type are unsuitable as they bend too easily. The net should be sewn onto curtain rings of brass and these passed onto the net ring. This is to avoid abrasion of the net material when working the bottom of the pond or stream. However it is better to buy or make a purpose-built net which should give years of service and not let you down while on a trip.

The Freshwater Biological Association has developed a net that has stood the test of time but unfortunately no longer makes them for sale. The pole is of ash, 1″ in diameter. A metal rod or spigot, square in section, and threaded to take a butterfly-nut at one end is inserted into the end of the pole which is then strengthened by a metal ferrule about 3″ long. A hole is drilled through the ferrule, pole and spigot and a locking pin inserted (see sketch).

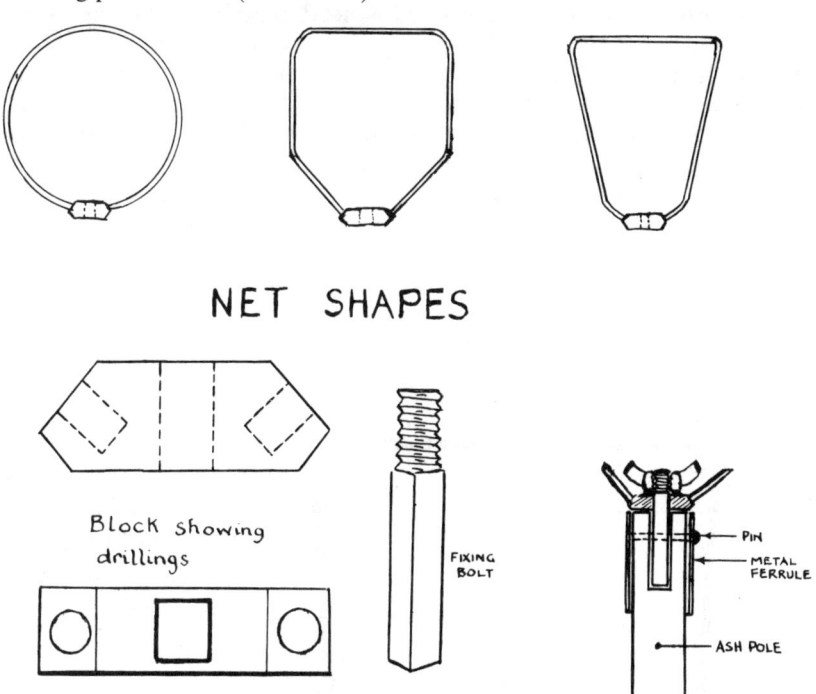

Fig 2 Details of water-net.

The net-head, of copper or aluminium alloy, comprises a boss and a tubular ring. A length of the tube is bent into a circle or the desired shape and the ends are inserted into the drillings in the sides of the boss and soldered or pinned securely. The centre of the boss is pierced by a square hole to fit the spigot. To assemble, the spigot is passed through the square hole of the boss and the head is secured to the pole by means of the butterfly-nut.

The net, about 12″ deep, is of bolting silk (obtainable from John Stannier & Co., Manchester Wire Works, Sherborne Street, Manchester M3 1FD). The top of the net is sewn onto a strip of strong linen, folded to give a double thickness. Between the two layers of linen is passed a length of brass wire bent into a circle of a diameter a little less than that of the main ring. Small square holes are then cut in the linen at suitable intervals, six is usually sufficient, and the brass ring is bound to the main net-ring with copper wire. This arrangement ensures that wear and tear is taken by the main ring and not by the linen which would otherwise abrade rapidly.

A circular net-head is the most useful shape for general purposes but in some instances a square or triangular one may have advantages.

Another method of constructing a water net with a robust head is to use 1" wide × $\frac{1}{8}$" thick aluminium or copper strip. Two strips are needed, one longer than the other with the length depending on the size of the net-head required. A bending board is needed on which bending pins are located to the final net-head shape (see sketch). Bend the shorter strip to shape and then place the second strip outside it and bend again to shape. The two frames are then drilled at suitable intervals so that they can be bolted together using small rust-proof nuts and bolts. Two sets of larger holes are drilled on the end tangs to take two bolts which will bolt the head onto the handle. This can be a metal or wooden handle, a stout broomstick with flats filed on the sides will do. Matching holes are drilled in this handle. The net bag is clamped between the two strips, so protecting it from abrasion. Nets of this design can be purchased complete with extendable handles from GB Nets, Linden Mill, Hebden Bridge, W. Yorks HX7 7DP.

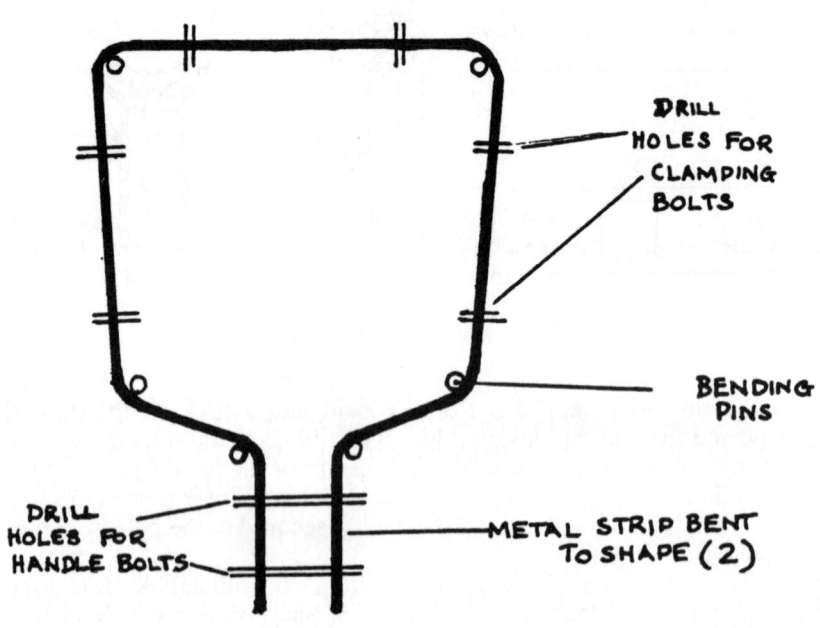

Fig 3 An alternative water-net.

Nets can also be purchased from Philip Harris Biological, Oldmixon, Weston-super-Mare, Somerset.

The net contents are washed into a white dish and the object of the hunt transferred to suitable containers. Some prefer to suck then up into a glass tube with a rubber bulb at one end; others prefer fine flexible forceps.

Sieve and tray/sheet

A sieve and something to sift on to are inseparable. Many coleopterists make their own sieves either from scratch or by adapting or cannibalising a garden or coarse meshed kitchen sieve. The ideal mesh is about $\frac{1}{4}''$.

A plastic tray of the type used by butchers for displaying meat (available from caterer's supply shops or your local butcher) are very good. Not too large, strong white plastic with raised sides preventing escape. A good alternative is a photographic developing dish, available in a wide range of sizes. Being rigid a tray will not wrinkle and can be held standing while examining the siftings.

Opaque white plastic sheet never lies flat in the field and the accumulated pockets of debris will need careful examination. A sheet can fold for stowing in the collecting bag and is useful in that larger quantities of material can be examined – an uprooted tussock or pile of fungi can be placed on the side of the sheet while small amounts are being sifted. Examination of the sheet must be carried out on the ground, not always convenient in boggy areas.

Material can be rough sifted in the field and brought home in a cloth bag (ideally, as this will prevent condensation damaging the contained occupants). It can be sifted piece-meal at leisure or placed in an extraction device. (see below)

Extraction funnel

This can be made by anyone with an interest in DIY. All that is needed is a large funnel – this can be purchased from a laboratory supply company or home-made from thin sheet aluminium. If purchased, the spout has to be cut short. A mesh grid is placed inside over the smallest diameter hole, this will prevent excessive amounts of debris entering the collection jar. The whole will need support in a home-made cradle. A conical lid or cover with light bulb is also required.

A large tube or glass jar is placed under the extractor and should, ideally, contain alcohol (industrial spirit or iso-propyl alcohol) thus all the creatures extracted will be killed. Often a predator will damage a number of its fellow occupants and a spider or two will cause havoc by spinning threads. Extraction may take several days.

Debris is placed into the funnel, the lid placed on and the light bulb turned on. This gives a gentle heat causing the debris to dry out from the top, driving the inhabitants to the lower more moist regions, and eventually into the collecting jar.

It is a good idea to examine the jar regularly and so deal frequently with small batches, rather than sort one large mass.

Winkler extractor

This extractor is nowadays becoming popular in Britain after many years of use by our neighbours abroad. They can be purchased from Hildegard Winkler (see list of suppliers) but are quite expensive. Instructions for making your own complete with cutting diagram have been given by Owen (1987). Basically the extractor consists of an oblong box-like cage of cloth supported by wires, tied at the top to prevent escapes and at the bottom tied to a collecting jar. Inside are a number of net bags. The whole is suspended from a rafter in the garage or similar support.

Debris is placed in the net bags, top tied, jar fitted. As the material dries out, the beetles and other inhabitants will crawl out and fall or otherwise make their way into the collecting jar. With the larger surface areas exposed, this extraction method is much quicker than the funnel. As no heat source is employed, the extraction tends to be more thorough. Generally, the bulk of the inhabitants will have been extracted within a couple of days.

Traps

It is a good idea to encourage beetles to catch themselves by using a range of traps. These can be quite simple and offer the beetles a concentration of a biotope otherwise scarce in that locality. In literature one can read of bunches of *Heracleum* being placed in Scottish woodlands attracting a range of scarce Highland beetles – a very hit and miss method but it would seem to produce good results if weather and other conditions are favourable.

Piles of fungi and decaying fruit/vegetables can be left and regularly sifted over the weeks. I once had good results by placing a dead cat on top of a large biscuit tin filled with sand. As the corpse decayed a succession of beetles came and went. By sifting through the sand a very rich fauna was sampled which, if on bare earth, would have been scattered in a much larger volume.

Lawn mowings can be placed in paper sacks or plastic bags for a few days to rot a little. This is then taken into woodland and placed in piles at bases of trees. After a week or so and at intervals thereafter, the pile can be sifted. Such rotting vegetation attracts a wide variety of uncommon Staphylinidae, Pselaphidae, Ptiliidae and other small-fry.

Faggot piles – bundles of twigs made up in woodlands and stacked to dry out, although largely a product of past woodland management, can be made. They are untied and beaten over a sheet or beating tray. Often uncommon species have been produced in this way.

A "bird-nest" trap was described by J.A. Owen (1976) who used them with success in many parts of Britian. They can be left and inspected at long intervals or more regularly if time and travel permits.

Some types of trap can be used with success throughout the year, especially unbaited pit-fall traps which can produce good results through November to March. With a glycol-based fluid (car antifreeze) in the trap the contents will not freeze, but the traps should still be inspected regularly, ideally each week.

Although interception traps seem to be most productive from spring to autumn when the bulk of beetles are active, they continue to function well throughout the winter.

Pit-fall traps

This type of trap can be used with a bait or without. When used with a bait, the vast majority of the beetles captured will of course have been attracted to the bait. Without bait, the catch is more varied and often some quite rare species might be found.

As with all trapping methods some places give better results than others, often a "likely" spot will give disappointing results. Fortunately traps can be moved from place to place quite easily and pit-fall traps can be set out in some numbers in various places. Be warned though, it takes a lot of time to service traps and to sort the accumulated catch; use only sufficient that can be dealt with.

A basic pit-fall trap is a plastic container of the type and size of a yoghurt pot. It is sunk in the ground so that its top is level with the surface. If inspected regularly no fluid or simply water plus a few drops of detergent need be added. If inspected at weekly intervals, use a glycol-based car antifreeze solution. Fill the pot about $\frac{3}{8}$–$\frac{1}{4}$ full. The trap should be placed away from the curious passer-by or vandal (often the two are synonymous) but with regular inspection, there will be no need to fix up a "roof" to keep rain out (a flat stone or piece of bark propped up). Some wastage will occur as a result of attention from foxes and other scavenging animals. A series of traps is best set out in any one locality, either randomly or in a line. Remember their location, as vegetation grows during the year the precise sites might become difficult to locate.

Baited traps have been used with great effect for special purposes. Welch (1964) describes a method for collecting insects from rabbit burrows. He sunk a jam-jar at arms length inside the burrow and used dead roach (handily available at the time) as bait. From March to late June 1963 he thus captured beetles belonging to seventeen families, including 1,182 Leiodidae from 13 traps in one warren. Another remarkable capture of Leiodidae was achieved by Ashby (see Cooter, 1989).

Flight interception traps

This simple trap has been used with great success for a number of years by American and Canadian entomologists, especially when collecting in the tropics. It is slowly finding use in Britain and seems capable of trapping a range of beetles not normally encountered, other than in odd specimens, by other collecting methods.

Manser and Goulet (1981) describe a trap used with success for collecting hymenoptera. Coleopterists prefer a screen 6' long by 3' high and in Britain these traps are generally used without the rain-deflecting roof.

The material used should be black (sprayed with a shoe dye if necessary) polyester marquisette (Dacron or Terylene) – any material that affords a foot-hold for insects should be avoided. The rectangle is bound with tape around its edges except at the bottom and loops fitted at the sides through which the supporting poles are passed. The trap is

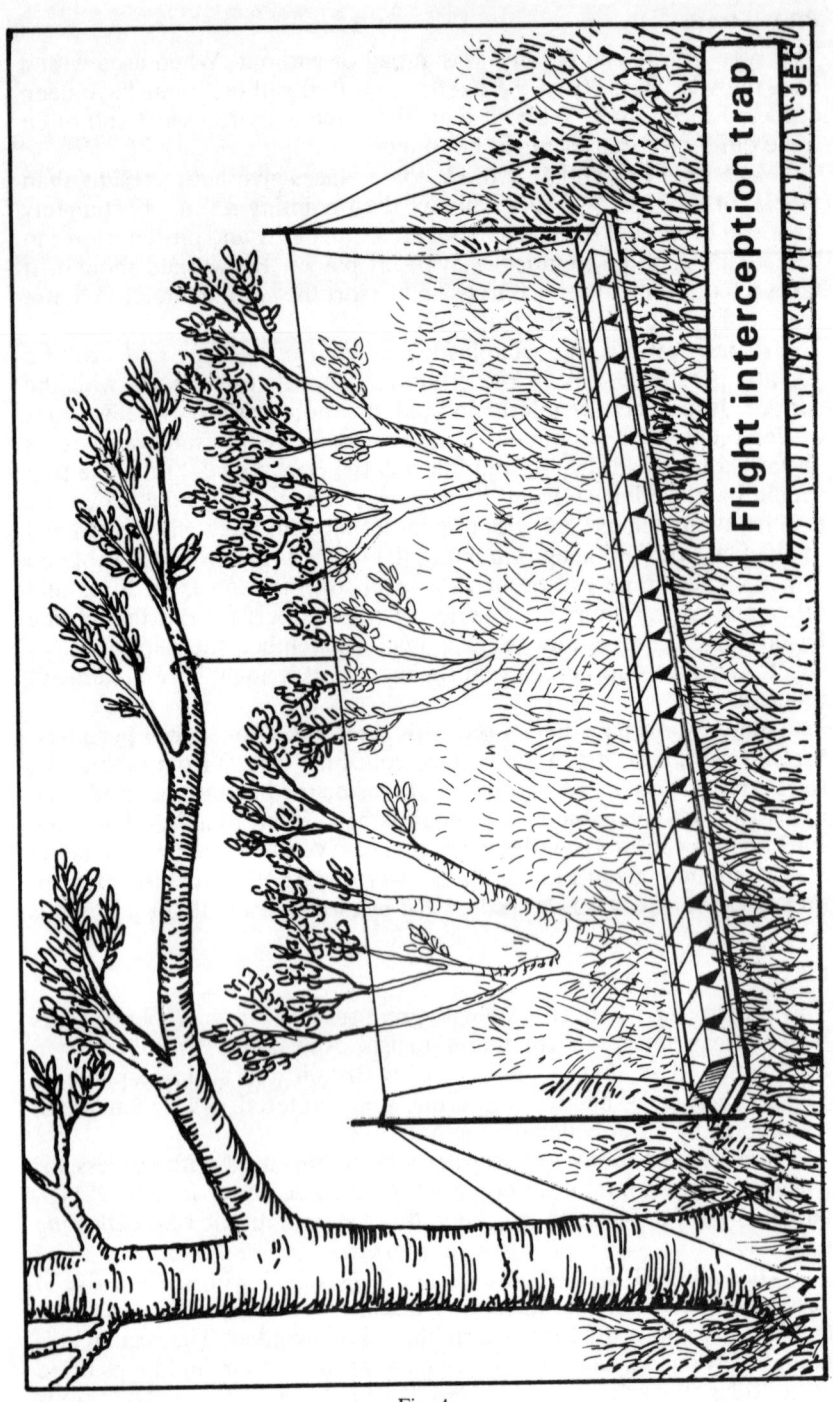

Fig. 4

set up vertically and the poles supported with guy ropes, the whole being taut and to offer minimum movement in wind.

Under the bottom, place a line of aluminium food containers (of the type take-away foods are sold in) or a length of gutter. Guttering is more tiresome as it will need some digging and packing to be horizontal over its 6' length as well as being unwieldy to empty. If inspected daily there is no need to put fluid in the receptacles, simply add strips of corrugated card to offer the catch protection. If inspected every few days, water with a few drops of detergent will suffice, but if left a week or longer a glycol-based car antifreeze should be used.

The traps are best set up on private land away from public gaze and vandalism. Always get the owner's permission and take time to select a likely site – a woodland ride or forest edge or some natural corridor seem likely places, but make sure these are not frequented by deer or other large animals.

Marris House Nets (address in list of suppliers) have made interception traps for the British Museum (Natural History) and for private individuals and now list this item for general sale.

REFERENCES

Cooter, J., 1989. Some notes on the British *Leiodes* Latreille (Col., Leiodidae) *Entomologist's Gazette* **40(4)**:

Manser & Goulet, 1981. A new Model of flight interception trap for some Hymenopterous insects *Entomological News* **92(5)**: pp. 199–202.

Owen, J.A., 1976. An Artificial nest for the study of bird nest beetles. *Proc. Trans. Br. Ent. Nat. Hist. Soc.*, **9**: pp. 34–35.

Owen, J.A., 1987. The "Winkler Extractor" *Proc. Trans. Br. Ent. Nat. Hist. Soc.*, **20**: pp. 129–132.

Welch, R.C., 1964. A simple method of collecting insects from rabbit burrows. *Entomologist's mon. Mag.*, **100**: pp. 99–100.

WATER NETS: These are discussed at length in the *Balfour-Browne Club Newsletter* number **13** (July 1979).

Suppliers

General equipment
  Watkins & Doncaster, Four Throws, Hawkhurst, Kent.
  Hildegard Winkler, A–1180 WIEN, Dittesgasse 11, Austria.
  Sciences Nat, 2 rue André MELLENNE, Venette, F–60200 COMPIEGNE, France.

Forceps etc.
  Southern Watch and Clock Suppliers Ltd., Precista House, 48–56 High Street, Orpington, Kent BR6 0JH.

Nets, Malaise and Interception Traps
  Marris House Nets, 54 Richmond Park Avenue, Queen's Park, Bournemouth, Dorset BH8 9DR.

Water Nets
  Philip Harris Biological, Oldmixon, Weston-super-Mare, Avon.
  G.B. Nets, Linden Mill, Hebden Bridge, W. Yorks HX7 7DP.

COLLECTING – SAMPLE HABITATS
By G. B. Walsh – Revised by J. Cooter

1. Introduction

Initially the beginner will find plenty of interesting beetles by turning over stones and logs (remember to examine the crevices in the bark of the over-turned log) and by general use of the beating tray, sweep- and water-net. The more experienced coleopterist will also find these methods productive on the first visit to a new area. After a while more specialised collecting methods will have to be adopted and it becomes a good idea to make a search for a particular, usually a rare, species. During the course of searching for the desired quarry, a wide range of other species will be encountered as a matter of course.

Nearly all coleopterists will be able to recount "beginner's luck" stories, and the occasional rarity will continue to crop up unexpectedly throughout your collecting. However, to try for consistently good results should be the aim and, to achieve this, an amount of simple research is necessary. Advice from other coleopterists, reading species notes in the journals, the study of maps are all essential preparation. If a beetle is phytophagous, its precise requirements might be known – does it feed on the leaf, in the stem, at roots for example. If its biology is unknown, be careful to note its habits when you eventually find it. Quite often the coleopterist embarks upon a spot of "trophy hunting" – visiting a well-known locality with the sole aim of securing the rarity which that place is famous for. Such collection often involves a long journey from home followed by an assiduous search in the precise place worked by others. The upshot, while the desired species may well be secured, is that the rest of the site is unworked and species lists for such places (often SSSI's or Nature Reserves) are small and strongly biased towards the rarities. Try to spend some time working other parts of these sites – such time is rarely wasted and often pleasant surprises are made.

Many people have an interest in more than one insect Order; often Hemiptera and certain families of Diptera are paired with the Coleoptera. It is generally best to embark upon a day's collecting with one of these Orders in mind, but keeping an eye open for desirable species of the other Order which might be turned up during the day.

As stated in the first edition of this Handbook, patience or better PATIENCE is needed and hard work (which is in itself enjoyable and productive) has to be put in. Many beetles are restricted to very small areas of what to us looks like suitable and identical habitat. Courage is often required – I recall the late Eric Gardner telling me of a search he made in the New Forest for *Velleius dilatatus* (F.) in a hollow tree with a hornet's nest. After accidentally disturbing the hornets and receiving so many stings on his bald head that he had to visit hospital, he returned to the site, continued working the debris under the nest and "eventually got my series".

The following notes are arranged seasonally at times when optimum results can be expected. They can of course be followed at any time with varying results, a different range of species can be expected at different seasons.

2. Winter

Winter collecting is usually very disappointing if the temperature is at or below freezing-point or if the night has been very cold.

(a) *Flood-refuse*: This is a most productive method of collecting at any time of the year and will often yield many thousands of specimens. Long continued rain in the winter (or a few days' heavy fall in summer) will cause rivers to overflow their banks and bring down quantities of refuse. With luck, one may find a bridge that spans the stream and fish out with a net the material that is being carried down with the flood water, or we may search the banks to find a place where the refuse has been piled up in rows or in little heaps. In the winter, beetles will remain in the refuse for days or even weeks; but in the summer most of them will have flown in 24 hours.

The best flood-refuse is the fine material formed of small pieces of grass, stubble, horse-dung and general refuse of the banks; twigs, large grass-tufts and coarser material generally contain few specimens. The material is lifted up from the bottom and put on the waterproof sheet and is then rough-sieved through the fingers to remove the bigger useless material. The result is put in a bag or sack and carried home to await examination there. We then sit down at the table, first having got permission from our womenfolk and warned them that small fry may creep or fly. Then, with the following articles before use – a waterproof sheet on the table, on it a white tray, a fairly fine garden sieve, small dry specimen tubes, killing tubes and a fine camel hair brush – a small handful of the flood-refuse is put in the sieve and shaken over the dish so as to leave a scattered deposit. The bigger beetles are put into the killing tubes and the smaller specimens are picked up with the brush and put into the dry tubes. An hour's work like this will give us enough material to last for the rest of the evening.

Before the material left in the sieve is rejected, it should be turned out on the dish to see if we have missed any large specimens which have not passed through the meshes.

The sack is fastened up at the top and put into a cool place to be examined again later. If we can identify the beetles we have got, it will help us to make a selection when next we sieve the material. The unused material – both dead and alive – should ideally be returned to or near the place from where it was collected.

Some authorities recommend warming the sack of refuse in front of the fire, but the beetles may then become so active that it is difficult to deal with them.

Some species, e.g. *Pseudopsis, Micropeplus* and *Monotoma*, do not move for some time and a special look-out should be kept for them;

their presence is best recognised when they run about. A fairly large reading-glass is of great help here.

In summer, the same method is used, but the beetles are much more active and perhaps numerous, and the refuse cannot be kept so long.

(**b**) *Damp refuse* from the bottom of hay-ricks and straw-stacks can be treated in the same way. We are particularly fortunate if we can find a hay-stack that has been standing neglected for some years and is beginning to ferment at the bottom. Here we shall find "hot-pockets" and the material round these will repay examination. Here we may find *Pseudopsis* and species of *Cryptophagus, Atomaria, Acrotrichis, Euplectus*, etc.

(**c**) *Big tufts of feathery moss* should be squeezed free of much of their water, roughly sifted and the residue packed into cloth bags and brought home for examination as before. The best results are obtained from mosses such as *Mnium* and *Hypnum*; close growing mosses such as *Leucobryum, Polytrichum* and similar types are only poor. In November *Mniophila muscorum* (Koch) is to be found in this way, as well as swarms of other small beetles, e.g., *Tachyporus, Sepedophilus*, etc.

(**d**) *Tussocks of coarse grass*: The Rev. E.J. Pearce ["The Invertebrate Fauna of Grass-Tussocks", 1948, *Ent. mon. Mag.*, **84**: 169–174] gives the following directions: – "The working of grass-tussocks for their fauna . . . requires the following apparatus – (a) a small fine-toothed hack-saw about 1′ long. This is greatly preferable to a sharp knife, which too easily and quickly blunts and rusts, and so loses its efficiency. In the summer it is necessary to remove the tussock quickly or some of the larger and more active inhabitants will escape. If the handle of the meat-saw is set at an angle to the blade, this is a refinement which increases its efficiency and avoids the hand being hindered by pressure against the ground. (b) A stout waterproof (or at least damp-proof) sheet not less than 4′ by 3′ and better larger still. If the edges are so constructed as to turn over and form a pocket all the way round, this will prevent the escape of the larger and active light-shunning species. The stronger, smoother and stiffer the material, the better it will serve, and it should be light in shade. When one wishes to make a quick general population estimate, it is an advantage to have the sheet squared in some convenient manner. (c) The usual supply of tubes, phials, bottles, etc., for the reception of specimens. If it be desired to bring the material home, it can be placed in a bag, either before or after sieving.

"Having selected the tussock, clear away the surrounding vegetation and with the meat-saw cut it across its base as quickly as possible about one inch (deeper if desired) below the level of the ground. Immediately place it on the spread out sheet, invert it and gently shake until sufficient of its contents have been dislodged for the first examination to be made. While this is proceeding stand the tussock the right way up on a smaller sheet nearby; in summer it may be necessary to cover it over as well. If too much earth is adhering to the base, it is well to remove this earth first unless its contents are being specially studied.

When the first shower of creatures has been dealt with, the shaking operation is repeated as often as may be necessary. Sharp taps with the flat palm of the hand are also effective in dislodging the contents. When further methods fail, tear the tussocks into portions and shake more vigourously until the whole process is completed".

Isolated tufts at the edges of fields and woods yield the best results. This is a fine haunt for species of *Sepedophilus, Cypha* and *Tachyporus*.

(e) *Bark* can be torn off with the chisel or the ripper. Under the bark of living trees such as sycamore and willow we may find beetles that have gone into hibernation there, as well as truly subcortical species, e.g. spp. of *Dromius. Carpophilus sexpustulatus* (E) occurs under the bark of freshly felled oaks. The loose bark of palings can be removed as well as that of dead trees. The best results are achieved when the bark peels off fairly easily, and underneath it the wood is still sappy; here we may hope to find species of *Rhizophagus* and *Cerylon, Agathidium, Acrulia inflata* (Gyll.), *Glischrochilus, Pityophagus, Rhinosimus, Xylotenus, Atheta, Hylastes, Triplax,* etc. Very old bark with woodlice, the hyphae of fungi, and worms, usually has very little underneath it except perhaps hibernating Carabids. Both moss and bark are usually less productive after a night's frost.

(f) *Bracket fungi* on trees, especially birch, may be ripped off for *Cis, Triplax, Tetratoma,* etc., and if larvae are present in the fungus, it can be taken home and kept in a tin for breeding purposes. In this way *Tetratoma fungorum* F. can be bred in plenty.

(g) *Stems*: In the early spring the dead leaves of reedmace (*Typha*), carefully pulled apart, reveal hibernating beetles, e.g. *Alianta incana* (Er.), *Telmatophilus typhae* (Fall.). Good results are often found during the summer in stems which are bored by the moth *Nonagria typhae* (Thnb.).

(h) *Carrion traps* can be used during mild spells in the winter, but better results are obtained in the spring and summer.

Unbaited pit-fall traps can be used with success throughout the winter. Less frequently encountered species of Leiodidae can thus be caught, often in numbers from Autumn to mid-January.

(i) *Herbaria, etc*: In old and perhaps neglected herbaria, especially in museums, there may occasionally be found the slender Lathridiid *Dienerella filum* (Aubé); and in stored tobacco *Lasioderma serricorne* (F.) may sometimes be found, occasionally becoming almost a pest.

(j) *Beating:* Some species are at large during the winter. Beating suitable trees during mild spells in late winter may produce species of *Anthonomus*. Old oaks should be beaten for *Phloiophilus edwardsi* Stph. in late Autumn through to mid-Winter, its host fungus being *Phloebia* spp.

3. Spring

With the coming of Spring, more and more outdoor work may be done. The winter methods may still be followed, with the following new ones.

(a) *Ground beetles*: This is an excellent time of the year for turning over loose stones, clods and fallen trees – every type of habitat should be tried – edges of fields, the coast, moorlands, hillsides, woods especially forest glades, edges of streams. On moorlands, pieces of peat can be turned over; on the edges of fields, tufts of grass and clods of earth uprooted by the plough; on the coast the jetsam thrown up by the tide.

Here, again, the stone, etc., should always be replaced, so that it can be utilised again.

(b) *Moorland beetles*: Good results can be obtained in the spring by lying down near old patches of heather and searching at their roots, or the material can be put on a sheet and examined there.

(c) *Nests of mammals and birds*: Dr. N.H. Joy first pointed out this method of collecting from mammals' nests. The easiest mammals' nest to find is that of the mole to which at least nine species of beetles are specially attracted. Among a number of the smaller mole heaps there is frequently a much larger one, possibly containing about a barrow-load of soil. The top soil is cleared away, and then the trowel is used to dig down through the middle. The nest usually lies about four inches below the surface of the ground, and is about as big as a man's head. The hole is enlarged sufficiently to allow the nest to be lifted out entire with both hands, and placed on the sheet. It is rough-sieved here through the fingers, any captures being tubed, and the residue is then placed in a bag for more careful examination at home. Dr. Joy says "Moles' nests are made of grass, leaves or sedge, the structure depending on the situation in which the nest is placed. I have found those made of leaves as a rule most profitable, and the grass ones least so". Apparently there are more beetles to be found in nests in the south than in the north, and in the winter and early spring than in the summer.

The nests of badgers, rabbits, mice, etc., can also be examined, if we are lucky enough to find them.

Dry birds' nests are too dry for beetles, but nests in holes in trees and other damp nests are often very productive. Starlings' nests, owls' nests, buzzards' nests, all yield beetles of particular interest: these are best examined just after the young have left. It is well to bring home the refuse which contains beetle larvae; it can be put in a tin and kept moist, and so we may perhaps breed the species.

The following are some of the species that may be found:
Atheta nigricornis (Thoms.) – general.
Haploglossa pulla (Gyll.) – starlings, tits, sandmartins and other small birds.
Haploglossa nidicola (Fairm.) – sandmartins.
Haploglossa picipennis (Gyll.) – buzzards.
Gnathoncus schmidti Rtt. Joy (owls).
Atomaria morio Kol. (jackdaws).
Ptinus fur (L.).
Ptinus sexpunctatus Panz.
Quedius puncticollis (Thom.), *Q. nigrocoeruleus* Fauv.
Leptinus testaceus Muell., P.W.J., *Peranus bimaculatus* (L.) in moles' nests.

Leptinus testaceus Muell., P.W.J., in nests of mice.

Aleochara cuniculorum Kraatz at the entrance to the burrows and in the burrows of the rabbit.

Gnathoncus spp. in birds' nests, especially of owls.

Commdr. J.J. Walker (1896, *Ent. mon. mag.*, 260) records from an owl's nest in a hollow tree, *Carcinops pumilio* (Er.), *Dendrophilus punctatus* (Herbst), *Gnathoncus nannetensis* (Marseul) *Trox scaber* (L.), *Mycetaea hirta* (Marsh.).

H. Donisthorpe records *Ocyusa maura* (Er.), from nests of moorhen and swan, and *Deubelia picina* (Aubé) from a moorhen's nest.

(d) *Catkins*. These can be beaten for species of *Dorytomus, Ellescus bipunctatus*, (L.) etc.

(e) *Sub-littoral species*: These are best sought for on a coast with rocks which can be easily split with a cold chisel and a hammer. The Liassic rocks and estuarine series are very suitable. In this way, we may hope to get *Aëpus, Micralymma*, etc. The edges of the rocks are to be worked along to about mid-tide depth, looking for places where it will be reasonably easy to prise off slabs of rock, not necessarily of large size. We are likely to find great numbers of the Collembolan *Anurida maritima*, on which the beetles probably feed, and it is probable that we shall find larvae, pupae and imagines of the beetles.

(f) *Carrion traps*: Carrion traps do well in the spring, species appearing then that later seem to disappear. Fish seems to form the best bait; a cod's head, deposited under leaves in a wood, will speedily have a large population of sarcophagous beetles.

Traps, as described in the foregoing chapter on Field Equipment, should be laid in open woodland, on the moors, in the open country, in lonely places on the coast where the sea cannot touch them. Fish waste from fish-manure works is good, and if we have access to a bone-mill, so much the better.

Keepers' trees should be examined whenever possible and the corpses beaten over a beating tray; there is a succession of species until everything edible has been consumed. Carrion traps can be used all through the year. It is well to keep the specimens alive in moss for a day or two so as to get them clean, and to enable them to empty their intestines; except for tomentose species such as *Oeceoptoma thoracicum* (L.), these are probably best killed with boiling water which removes much of the foreign matter from their bodies; it is well, too, to use a brush to wash their bodies.

(g) *Damp fallen leaves* in woods can be rough-sieved in the open and the fine material brought home for further examination.

(h) *Moss* may yield us *Bradycellus* spp., *Trichocellus* spp., *Otiorrhynchus ovatus* (L.), *O. desertus* Ros., *Byrrhus* spp., and many others. Any moss on old walls may be turned over for *Bembidion quinquestriatum* Gyll. In mid-May moss at the roots of trees, especially ash, in damp places may yield *Caenoscelis ferruginea* (Sahl., C.R.) and *Atomaria fimetarii* (Herbst).

(i) *Moss in waterfalls*: The thick moss, even under small waterfalls, has its own special beetle fauna. The moss is pulled off and squeezed, and then examined on the sheet. Some of the species likely to be found are *Stenus* spp., especially the red-spotted species *S. biguttatus* (Linn.), *S. comma* Lec., *S. guttula* Müll., *S. bimaculatus* Gyll.,: also *S. guynemeri* du Val., *Dianous coerulescens* (Gyll.), *Quedius umbrinus* Er., *Q. maurorufus* (Grav.), *Q. auricomus* Kies., etc.

(j) *Spring flights*: On favourable days, Coleoptera fly in the spring in great numbers of both species and individuals, so that it is possible to use the voile net. The beetles seem to prefer to alight on light objects, such as new fences, white paint, etc. When spring bursts suddenly with a few particularly warm days after a long stretch of cold weather, the swarms of flying beetles (and other winged insects) are particularly noticeable.

(k) *Beach collecting*: During early spring, often around April 14th but sometimes earlier, only very rarely later than late April, sandhills near the sea are amongst the most productive places for beetles. Many species, responding to the increased temperatures, become active and, whether flying or crawling, blunder into objects and become trapped in small depressions in the bare sand. When dry, the sand is so loose that the beetles cannot negotiate the slope and remain trapped. After rain, however, the sand is more consolidated and the beetles can escape. The ideal weather then is bright and sunny with moderate breeze (a stiff wind makes conditions for the coleopterist unpleasant as well as causing the dry sand to blow freely, covering the depressions and their entrapped fauna). The sand must be dry underfoot. A search should be made of the small hollows left by passing walkers as well as the large expanses of breached dune, especially the basal halves of such features. By getting down on hands and knees and crawling along the paths even the smallest sand-coloured beetle will not escape detection. However, with experience, one needs only to walk slightly stooped concentrating on the nature and texture of the sand. Most beetles will not be moving and a few might even be dead (possibly from heat or dehydration). This method is often highly productive and can produce beetles which otherwise are not easily encountered by other collecting methods. For example, in two visits to adjacent sand-dune areas on the Welsh coast in early April, I captured (amongst a lot of more common species) *Apion seniculus* Kirby (many), *atomarium* Kirby (many), *Hypera dauci* (Ol.) (3), *Ceutorhynchus punctiger* Gyll. (many), *atomus* Boh. (3), *erysimi* (F.) (2), *pilosellus* Gyll. (9), *Psammobius asper* (F.) (1), *Dicronychus equeseti* (Hb.) (l.), *Aphanisticus pusillus* (Ol.) (1). Some beetles must be sought in these places by other methods and in other habitats – *Cicindela maritima* likes bare sand, areas of dune not yet fixed by vegetation, rather than the dune slopes. *Hypocaccus rugifrons* (Pk.) can be found at rest on the dune slopes, *Licinus depressus* (Pk.) under refuse, wood, etc., *Phytosus balticus* Kr. under drift wood on damp sand above the high tide mark – look on the underside of the wood as well as minutely examining the sand.

On a return visit to the same Welsh dunes in May, none of the above mentioned beetles was evident, and the hollows in sand were virtually barren.

Possibly bunkers on golf links might act to trap beetles in a similar way.

(l) *Grubbing and Searching*: Many species of phytophagous beetle appear early in spring, often as early as March, and frequent the tender young growth of their host plants. By April or May at the latest, these species are no longer evident. Grubbing at roots as well as painstaking searching the foliage will produce good results. Another advantage from collecting early in the year is that many of the plants are not crowded out by the more varied and vigorous vegetation of summer. Several species of *Longitarsus* can thus be found and with this genus it is most important to record the host plant. *Longitarsus fowleri* Allen occurs in spring on teasle plants. It seems logical that such a diminutive species would attack this plant only when it is young and tender.

(m) *Searching*: After a time, indiscriminate beating and sweeping cease to add new species to the collection, though there will always be the odd newcomer, even in the most thoroughly worked locality. Physically searching plants for beetles is often very rewarding and, at the same time, often frustrating if care is not taken to prevent beetles falling from the plant and being lost in the mass of roots and young growth – weevils are particularly infamous for this, while most Halticines will readily jump and avoid capture. With experience, however, escapes will become less frequent.

A search of the literature, including this Handbook will provide much information as to the host plants and more precise habits of particular species. However, it must be borne in mind that the biology of many species is still not known and there is plenty of scope for the coleopterist to make a valuable contribution to our knowledge. Some beetles will frequent leaves, others the stem, roots, flower or seed. Sweeping therefore will probably not produce a range of root-feeders, other than by capturing the odd specimen straying from its normal biotope. Grubbing at roots, looking for mines and galls and feeding holes in leaves all help in our quest. Difficult genera such as *Longitarsus* and certain sections of *Apion* (species around *seniculus*, for example) become less problematic if the host plant is accurately recorded. Do remember though that a species might be on a particular plant by pure chance, especially isolated specimens.

Many species of *Ceutorhynchus* will be over by the end of June, but others can be found later in the year.

(n) *Grubbing*: When grubbing, it is good practice to lift carefully the leaves and growth resting on the ground; it may be tapped first, ideally over the collecting tray. The surface is then carefully examined. Next the roots and dead leaves, etc. are parted and the ground surface carefully examined. This job is greatly facilitated by the use of a dinner fork, both the pronged end and handle can be used. It has the advantage of being very much smaller than a hand, so does not get in the way so much and can get under small roots more easily and pull them to one side. It can of course be stuck in the soil to hold vegetation apart while the pooter is employed. After this, some soil can be scraped away and roots examined. *Hylobius transversovittatus* (Gz.) makes galls in the roots of Purple Loostrife and is invariably collected by grubbing in spring/early summer. It can also be found in autumn by examining roots for galls. The adult is found either on the soil surface or at the roots.

4. Summer

The activities of Spring and Winter can be carried out during the summer months and new methods can be brought into practice as conditions permit. Generally, early summer is the most productive time. By the end of July or early August most coleopterists are of the opinion that the season is in decline until the autumn species become evident, usually during September. However, there will be plenty to do throughout.

(a) *Beating*: This is often extremely productive of common species. Certain trees such as oak, hazel, birch, sallow, pine are good; others such as horse-chestnut, elder, yield very little. Specimens from the same species of tree should be kept separate for ease of identification. During the second week in August sallows may be beaten for *Oberea oculata* (L.) in suitable localities. Over-mature trees, especially oaks, are particularly productive and will produce may species that inhabit the rotten mould of the interior. Ancient woodlands/parklands harbour some very rare species: *Hypebaeus flavipes* (F.), *Trixagus* spp., *Anaspis schilskyana* Ciski, *Scraptia* spp., *Aderus* spp., anobiids, etc.

(b) *Sweeping*: This, too, gives plenty of material, and once again we should sweep in as great a variety of places as possible. Very often in the heat of the day, from 12 to 4 G.M.T., there is little to get; but as evening draws on, species come out of shelter again and work once more becomes worthwhile. The best places are low, swampy meadows, meadows surrounded by woodlands or on the slopes of hills, undergrowth in open woodland, and neglected fields. Well-eaten pastures are very unproductive. In the daytime we may obtain Cantharidae, Chrysomelidae, Curculionidae, and in the late afternoon and more especially in the short interval from just before sunset to dark, rare Pselaphidae and Scydmaenidae. *Colon* and *Leiodes* may be swept from the tips of grasses, *Colon* spp. apparently preferring the grass Melic (*Melica uniflora* Retz.).

(c) *Evening sweeping*: This can be carried out from sundown until it is too dark to see. In fenland, swarms of *Stenus* can be obtained in this way. Fryer says (1941, *Ent. mon. Mag.*, 280) that *Agriotes* spp. are most active at this time, and can be swept in numbers from the grass along dyke sides, often when the herbage is very wet. The last captures can be put into a calico bag for examination in daylight.

(d) *Searching*: After a time indiscriminate beating and sweeping cease to add new species to the collection, and there is a tendency for the tyro to think that he has exhausted the district. A notebook now comes in useful.

From Fowler or from the later chapter herein upon beetles and their food plants, we note the food plants of the various plant-feeding beetles and the insects we are likely to find on them. We then obtain from a local botanical friend the places where we can find these plants in the district. The plants are then examined where the beetles are likely to be found, e.g. on flowers (for weevils, pollen-feeders, etc.), on the leaves, under the leaves, at the roots. In some cases we split open the stems of plants, e.g. for *Baris picicornis* (Marsh), on *Reseda lutea* L., or pull apart the leave as in *Iris* and reedmace (*Typha*).

In the case of *Meligethes* spp., we may refer to a list of the food plants and their associated beetles, and then search in the corollas is far more likely to give us our desiderata than sweeping; the beetles often get so far down the corolla-tube that the net fails to dislodge them.

(e) *Flowers*: Certain plants when in flower are very attractive to beetles, e.g. hawthorn, guelder-rose, sycamore, hogweed; the beating tray may be used, or in the case of the umbellifers and very low growing plants, search is far better. The best results are obtained in a year when there is little blossom or where the necessary flowers are few and far between, for then the beetles are congregated in greater abundance in one place.

Special flowers sometimes attract certain definite beetles, e.g. *Rhinomacer attelaboides* F., may best be found on male flowers of Scots pine, and species of *Donacia* must be sought on water plants; *D. obscura* Gyll., may be found in June on the Norfolk Broads, on the flowers of *Carex rostrata* Stokes; the flower-heads may be moved and the water beneath examined. In the last week in July and the first week in August, flower-heads of Bog Asphodel (*Narthecium*) may be examined for *Phalacrus substriatus* Gyll. In May hawthorn flowers may be beaten for *Acrulia inflata* (Gyll.), and crab apple blossom for *Anthonomus* spp.

(f) *Dead or dying trees*: These well repay beating over the tray; hawthorn, spruce and pine are very good. These give good results until the wood becomes thoroughly dry, which usually takes place in large trees in two or three years after the death of the tree, and in less time with smaller ones. The bark of such trees is the best collecting place for Cucujidae, Colydiidae, Scolytidae, Histeridae, etc., and the shady side is more productive than the sides exposed to the sun.

Some species must be cut out, others may be found crawling over or resting on the bark in the bright sunshine, while the crepuscular and nocturnal species may be found on the tree towards evening or after dusk, when a torch is needed.

A sharp lookout must be kept for beetles which take to flight almost at once, e.g. *Ernobius nigrinus* (Sturm). Particularly good results may be obtained by beating the lopped branches of pine where the needles are just beginning to die. We may thus obtain *Pogonocherus fasciculatus* (Dg.), *Salpingus castaneus* (Panz.), *Hylobius abietis* (L.), *Pissodes pini* (L.), *P. castaneus* (Dg.).

If the tree is at just the right stage of decay, it can be ripped open for wood-feeding species; if the larvae are present, they can be taken home in an attempt to breed them.

(**g**) *Timber piles and cut stumps*: These should be examined, especially in the early morning, when the sap is still wet and fragrant. Even timber-stacks on dock sides yield good results at times, though it must be admitted that the beetles are, as a rule, not British.

(**h**) *Old manure heaps, compost heaps and hot-beds*: When old manure heaps and hot-beds have lost their smell, they can be pulled to bits for species of *Monotoma, Myrmechixenus, Atomaria, Cryptophagus, Anthicus*, etc.; these may be sieved out. *Philonthus jurgans* Tott. is most likely to be found at the very heart of a manure or compost heap. Some beetles may be found in the drops of water on the underside of the glass over hot-beds.

(**i**) *Dung*: All kinds of dung can be examined, either by turning it over, pulling it to bits over a sheet, or throwing it into water, when after a time, the beetles struggle to the surface and can be skimmed off. Many species are local and occur only in certain types of locality. The dung beetles are contained in the sub families Scarabaeinae, Geotrupinae and Aphodiinae of the family Scarabaeidae and some members of the Hydrophilidae, Staphylinidae, etc., also occur in dung. Dung, like carrion, is best worked when one is young, though perhaps the comments of the unenlightened are likely to be most irritating then. It is, perhaps, just as well to wash one's hands after working dung and before eating one's lunch in the field. One fastidious collector whom I used to know was wont to take a tablet of soap for this purpose to the scorn of some of his younger associates.

(**j**) *Cut grass and reed refuse*: This often yields excellent results. We wait until the top of the cut swathe is quite dry but the underside is still damp; tossed or dry hay is practically useless. The grass is then carefully lifted with as much of the damp materials as possible, put on a sheet and the greater part of the grass is shaken over it and then thrown back on the field. This is best done with herbage, and gives better results in damp seasons than in very dry ones. This is a good way for getting *Barynotus* spp., *Alophus, Apion* spp.

(**k**) *Shore refuse*: This is best when it has been lying for some time out of the reach of the waves. It is examined in the usual way over the tray, special watch being kept for the very small beetles. *Nebria livida* (L.) can be found in this way.

Heaps of *Zostera* tossed up by a gale in the Tees estuary have yielded *Aëpus marinus* (Ström, H.), etc. A dead porpoise, dead gull or other large animal tossed up on the shore may give us coast carrion beetles.

(**l**) *Seaweed*: Decaying seaweed thrown up by storms well above high-water marks should be rough-sifted and then sieved over a sheet.

The best results are obtained when the weed is still damp; dry weed seems to yield little of interest. Many species are to be found, some adventitious visitors, but many in their normal habitat, among them being *Broscus cephalotes* (L.), *Aëpus marinus* (Ström) (under heaps of decaying *Zostera*), *Cercyon litoralis* (Gyll.), *C. depressus* (Steph.), *Corylophus cassidoides* (Marsh.), *Ptenidium punctatum* (Gyll.), *Omalium rugulipenne* (Rye), *O. laeviusculum* (Gyll.), *O. riparium* (Thoms), *Actocharis readingii* (Sharp), *Anotylus* spp. *Philonthus fumarius* (Grav.) *Cafius* spp., *Heterothops binotatus* (Grav.), *Diglotta mersa* (Hal.), *Phytosus* spp. *Arena tabida* (Kies.), *Halobrecta flavipes* (Thoms., C.G.), *H. algae* (Hardy), *A. trinotata* (Kr.), *A. triangulum* (Kr.), *A. vestita* (Grav.), *Aleochara grisea* (Kr.), *A. phycophila* (Allen), *A. algarum* (Fanv.), *A. obscurella* (Grav.), *Cantharis rufa* L., *Corticaria punctulata* (Marsh.), *C. crenulata* (Gyll.), *C. impressa* (Ol.), *Corticarina fulvipes* (Com.) *Anthicus antherinus* (L.), *A. angustatus* (Curt.), *Phaleria cadaverina* (Fab.), *Cassida flaveola* (Thunb.).

(m) *Seaside timber*: At least two species of beetles are associated with timber that is washed by the sea in seaports and by the coast. *Nacerdes melanura* (Linn.) can be caught flying in the sunshine round docksides and groynes, while *Pselactus spadix* (Herbst) can be found crawling on posts by the seashore.

(n) *Clay cliffs*: These can be examined in three ways.
(1) On the part standing rather back from the waves, especially where there is a runnel of water, there may be loose patches which can be pulled off. Behind them we may find *Nebria livida* (L.), *Bembidion stephensi* (Crotch), *Chlaenius vestitus* (Payk.), etc.
(2) We may search for these same species with a torch at the foot of the cliffs at night.
(3) Careful search along the cliff-face may show beetle burrows, perhaps near a runnel or in a sandy part of the cliff. The clay may be dug out with a trowel and pulled to pieces over a sheet. Thus we may obtain species of *Dyschirius* and *Bledius*.

(o) *Moors*: By walking among sphagnum in moist (not too wet) spots, one may stir up *Carabus nitens* L.

Possibly the best way to collect on moors at this time of the year is to find a bare patch covered with damp felted algae. This is turned back and we find numerous moorland Carabids, e.g. *Bembidion nigricorne* (Gyll.), *B. unicolor* (Chaud.); under peat blocks we may find *Miscodera artica* (Payk.), *Nebria salina* (Fairm.) *Pterostichus lepidus* (Leske), *P. adstrictus* (Esch.).

(p) *River beds*: In the beds of swift flowing rivers with occasional falls, several methods of collecting may be used.
(1) The moss from waterfalls is pulled to bits over the sheet, and we shall probably find *Dianous* and desirable species of *Quedius* and *Stenus*.
(2) Shingle-beds are often rich hunting grounds; stones can be turned over or the fine shingle can be passed through a sieve. A sharp lookout should be kept for minute species of *Bembidion, Lesteva, Zorochrus*.

(3) Water may be splashed on the shingle-beds or on the steeper earth banks to dislodge ground beetles and *Bledius* spp., *Heterocerus*, smaller Staphylinids, etc.
(4) Stones at the shallow edge of the water may be turned over for *Platambus*.
(5) Stones should also be turned over in the almost dried-up beds; many specimens collect here when there is still a certain amount of moisture.
(6) The water net can be used as usual in the backwaters for *Oreodytes rivalis* (Gyll.), *O. septentrionalis* (Gyll.).
(7) The following is an excellent way of collecting small aquatic clavicorns and palpicorns. The net is placed where there is a very swift current of water, or a loosely woven cloth may be stretched across and through the stream. One then stands a little way up stream and turns over the bigger stones, rubs their undersides where the beetles collect, and stirs up the shingle with a stick. These small beetles are washed down into the net and cling to its meshes.
(8) Species of *Dryops* and *Heterocerus* are most numerous in the moss or among the roots of other plants that grow in the water; these should be pulled up and examined over the sheet. By ripping off the bark of partly submerged timber, especially alder stumps, we may find *Cyanostolus aeneus* (Richt.).

Caution should be taken while working in river beds in hilly country during stormy weather for freshets may come down unexpectedly and in great volume. Only by providential good fortune did the writer, G. B. Walsh, once escape death in the R. Swale when a sudden flood covered with 3 feet of swirling water the place in the river bed where he had been standing only 10 seconds before. Three years afterwards two boys were drowned by a freshet in the same river.

(q) *Slow flowing streams and ponds*: The banks of slow flowing streams and of ponds and mud flats, especially in salt marshes, will repay examination. Water should be splashed on the sides and we may tap the mud or walk over it and so start beetles walking. In this way *Panagaeus* has been taken at Wicken Fen, and it is an excellent way of obtaining numerous small Carabidae – (*Elaphrus, Dyschirius, Clivina, Bembidion, Tachys*) and Staphylinidae – *Tachyusa, Philonthus, Carpelimus, Stenus*); and from their galleries we may wash out species of *Dyschirius, Bledius* and *Heterocerus*.

(r) *Netting in flight*: Williams, Omer-Cooper and Tottenham have tried with great success the almost indiscriminate use of a butterfly net made of voile. Williams carried it whilst riding his bicycle along country roads and the edges of woods, mainly in the evening; the others swept it through the air when beetles were in flight.

Many common species are taken, but there are also many choice things to be so obtained.

Freude, Harde and Lohse (1965, *Die Käfer Mitteleuropas*, **1**: p. 109) describe the "autocatcher" – a large tapered net fitted above the roof of a car. This is effectively a mobile flight interception trap and is capable of capturing a range of beetles only rarely encountered by more conventional collecting methods. In Britain, Alex Williams had a deal of success with an autocatcher fitted to his vehicle, the car is driven slowly around country lanes and the catch, which collected at the apex of the net, examined regularly.

(s) *Dry bark*: Dry bark may be ripped off, and larvae or adults of *Ctesias serra* (Fabr.) found which are feeding on spider-webs; the larvae can be reared on this pabulum.

(t) *Mountains*: These are well worth working, especially if we can get above the 2000 ft. level. There we may find under stones *Leistus montanus* Steph., upland forms of *Calathus melanocephalus* (L.), and *Nebria gyllenhali* (Schön.); in dung, *Aphodius lapponum* (Gyll.), *A. constans* (Duft.); and in moss unusual Staphs., such as *Anthophagus alpinus* (Payk.), *Oxypoda soror* (Thoms.), *Geodromicus longipes* (Mann.). At the top of the highest mountains in Scotland one may find *Nebria nivalis* Thoms.

(u) *Fungi*: If we can find myxomycetes on forest trees, we may get *Enicmus testaceus* (Steph.), *Symbiotes latus* Redt., *Anisotoma humeralis* (Fabr.), and *Agathidium* spp.

Damp wood with fungoid growth on it may yield *Biphyllus lunatus* Fabr., *Anommatus 12-striatus* (Muell., P.W.J.), *Sphindus dubius* (Gyll.), etc.

(v) *Burnt areas*: A restricted but interesting beetle fauna is associated with burnt areas and burnt timber. *Pterostichus angustatus* (Duft) and *Agonum quadripunctatum* (Dg.) are associated with burnt woodland areas, especially conifers. *Dromius angustus* Brullé can be found under bark on burnt trees. The buprestid *Melanophila acuminata* (Dg.), in Britain restricted to the Surrey and Berkshire heaths, is attracted to burnt areas by the infrared radiation caused by the forest/scrub fire, not by the scent of pine resin/oils given off during burning (see Bílý, 1982 *Fauna Entomologica Scandinavica*, **10**: p. 44).

Ernobius mollis (L.), *pini* (Sturm) and *gigas* (Muls & Rey) have all been recorded by beating burnt pines, as have *Salpingus ater* (Gyll.) and in Finland *Cartodere constricta* (Gyll.). At old fire-sites the small fungus *Pyronema confluens* is the host of the diminutive histerid *Acritus homoeopathicus* Woll. and under pieces of charred twigs, etc. we might find *Micropeplus tesserula* Curt.

(w) *Potato fields*: When the potatoes are dug up, one may examine the inside of the shells of seed potatoes for *Anommatus 12-striatus* (Muell., P.W.J.). Rev. Theodore Wood (1896, *E.M.M.*, **32**, 258–9) gives the following species as occurring in decaying "seed" of early potatoes, lifted at Broadstairs, Kent, in the early summer – *Parabathyscia wollastoni* (Jans., E.W.), *Anommatus 12-striatus* (Muell., P.W.J.) (prefers dry and almost powdery potatoes), *Anotylus insecatus* Curt., *Langelandia anophthalma* Aubé (lives in the drier "seed", is extremely sluggish and very easily overlooked).

(x) *Withies*: During June and July, *Leptideella brevipennis* (Muls.) and *Gracilia minuta* (Fabr.) may be found on withies, either growing or cut.

(y) *Sawpits*: Many beetles associated with timber can be found in these areas and the surrounding stacked timber and nearby flowers. Cerambycids will be in evidence and a variety of other families can be found at the cut ends of logs exuding sap and under the bark. The accumulated sawdust is worth searching, especially in the Scottish Highlands for the imagines and larvae of Lycids.

(z) *Moss in woods, etc.*: Moss in woods and along hedgerows, near violet plants, may yield *Orobitis* in August and September; it looks much like a seed.

(aa) *Lepidopterists' sugar*: Many beetles come to this e.g. *Rhagium mordax* (Dg.), *Helops coeruleus* (L.), *Sermylassa halensis* (L.) but it is scarcely worth while sugaring for beetles as they can usually be captured more easily in other ways.

(bb) *Cossus burrows*: The unpleasant-smelling sap round the wounds in trees in which the larvae of *Cossus* are feeding is very attractive to some species of Coleoptera, e.g. *Silusa rubiginosa* Er., *Philonthus subuliformis* (Grav.), *Phloeonmus planus* (Payk.), *Epuraea guttata* (Ol.), *E. fuscicollis* (Steph.), and *E. thoracica* Tourn., *Soronia punctatissima* (Ill.) and *S. grisea* (L.), *Thalycra fervida* (Ol.), *Rhizophagus ferrugineus* (Payk.) and *R. parallelocollis* Gyll.

(cc) *Salt marshes*: A number of species are restricted to coastal marshes which have soils of varying degrees of saltiness. Some species are dependent upon certain plants that grow in these areas (see the beetle/host plant list, p. 183); some aquatic beetles inhabit brackish waters (see p. 73). The species listed here are more closely related to marsh litter, soil and related habitats. They may be collected by such methods as sifting strand-line refuse and turning drift wood. Bare sand and mud showing signs of burrows will repay examination; often walking over the area will, after a few minutes, bring species to the surface. I had an amusing experience of this whilst collecting *Bledius* and *Dyschirius* species at Berrow, Somerset – a beach surreptitiously frequented by nudists. On a hot sunny day, with jacket, collecting bag, garden sieve, 8" sheath knife and wellingtons, I crawled along the bare wet sand with pooter in mouth. Approaching slowly towards a reed bed, two nudists stood up, gave me a filthy look, dressed and left. Finding three species of *Bledius*, *Bembidion pallidipenne* (Ill.) and *Dyschirius* spp., my mind was closed to all else.

An alternative method is to scrape the surface to about 2cm depth and see what comes to the surface after a few minutes.

Dyschirius nitidus (Dj.)
Dyschirius politus (Dj.)
Dyschirius extensus Putz.
Dyschirius salinus Schaum
Bembidion ephippium (Marsh.)
Bembidion normannum Df.
Bembidion minimum (Fabr.)
Bembidion aeneum Germ.
Trechus fulvus Dej.
Pogonus luridipennis (Germ.)

Carpalimus halophilus Kies.
Bledius furcatus (Ol.)
Bledius tricornis (Herbst)
Bledius spectabilis Kr.
Bledius unicornis (Germ.)
Bledius bicornis (Germ.)
Bledius fuscipes (Rye)
Bledius fergussoni Joy
Staphylinus ater (Grav.)
Hypocaccus metallicus (Herbst)

Pogonus litoralis (Dufts.)
Pogonus chalceus (Marsh.)
Anisodactylus poeciloides (Steph.)
Dicheirotrichus gustavii Crotch
Dicheirotrichus obsoletus (Dj.)
Polistichus connexus (Geoffr. in Fourc.)
Carpalimus foveolatus (Sahlb., C.R.)
Heterocerus flexuosus Steph.
Heterocerus obsoletus Curt.
Heterocerus maritimus G–M.
Phylan gibbus (F.)
Melanimon tibialis (F.)
Opatrum sabulosum (L.)
Phaleria cadaverina (F.)
Crypticus quisquilius (L.)

(dd) *Keepers' lines*: The dead animals to be seen on a gamekeeper's pole or line are well worth tapping over a tray from the time when they are first hung up till they become mere dry skins. In addition to the ordinary *Necrophori* and *Cholevae* which we expected to get in corpses, we may get *Necrobia* spp., *Korynetes*, *Osmosita* spp., *Dermestes* spp., and *Trox* spp. Some of these we may also find if we can examine the bones in a bone mill.

In furs and dried skins may be found *Dermestes* spp., *Attagenus* spp., *Megatoma undata* (L.) and *Anthrenus* spp.

(ee) *Submerged logs*: Partially floating timber and submerged branches and roots in rivers should be carefully examined for *Macronychus quadrituberculatus* Muell., and *Cyanostolus aeneus* (Richt.). Both species are extremely local; but the latter is fairly widely distributed. With them may be found *Helichus subtriatus* Mull., and other related beetles.

(ff) *Spider-webs*: Spider-webs would not seem to be a very nourishing pabulum, but at least two species of Coleoptera are associated with them – *Ctesias serra* (F.) and *Trinodes hirtus* (F.). Their larvae feed on webs under dry bark, probably to a large extent on webs of the spider *Segestria senoculata*.

(gg) *Light*: In the tropics this seems to be a very good way of attracting many species of Coleoptera, but those recorded as coming to light in Britain are common species usually, but some quite scarce beetles have been captured regularly at light – e.g. *Odonteus armiger* (Scop.) at M.V. light, and *Platydema violaceum* (F.) at domestic light.

(hh) *Manure heaps*: These provide rich hunting grounds for many species of beetles, the best time being from early July to late September when the heap is dry and gives off little smell. The following beetles have been recorded; but there are many others to be found: *Cercyon* spp., *Euconnus fimetarius* (Chaud.), *Smicrus filicornis* (Fairm.), *Carpalimus* spp., and other *Oxytelus*, spp., *Lithocharis* spp., *Philonthus* spp., *Euplectus signatus* Reich., *Monotoma* spp., *Atomaria* spp., *Anthicus* spp. *Philonthus jurgans* Tott, should be sought in the middle of the heap.

5. Autumn

(a) *Sweeping*: Sweeping can be continued in favourable seasons right into November, especially in open woodland and along the glades and borders. We may hope to find *Apion pallipes* Kirby on *Mercurialis* – Dog's Mercury, and species of *Colon*. Sweeping at night may give us delectable species of *Leiodes*.

(b) *Fungi*: As Autumn advances fungi become more and more abundant. Every kind should be pulled to bits over the sheet, although new specimens may not yet be occupied. Puffballs give good results, e.g. *Cryptophagus lycoperdi* (Scop.), *Lycoperdina bovistae* (F.) *L. succincta* (L.) in the East Anglian Brecklands and *Pocadius ferrugineus* (F.).

Fungi may be gathered in woods and put into little heaps in the hollows at the bases of trees; these should be examined at frequent intervals.

(c) *Nests of bees, wasps and hornets*: These nests can be dug out and examined, though the task is an unpleasant one. Wasps' nests give results even after the wasps have all died. If the wasps are still present in small numbers, the mouth of the nest can be covered with turf and the returning wasps be drived off by beating them with twigs. If ammonia solution be poured over the nest, it will kill the inhabitants, both wasps and Coleoptera. We may thus take *Metoecus paradoxus* (L.) and *Leptinus testaceus* Müll. In subterranean nests of *Vespula vulgaris* (L.), *Cryptophagus pubescens* Sturm and *C. populi* Payk. may occur. In nests of *Vespula germanica* (F.) in trees we may take *Cryptophagus micaceus* Rey after the wasps have died.

The following species are associated with the nests of bees: *Cryptophagus pubescens* Sturm (*Bombus terrestris* (L.)); *Cryptophagus populi* Payk. (nests of *Colletes daviesana* Smith, F.); *Antherophagus* spp. (nests of *Bombus*); *Ptinus sexpunctatus* Panz. (bred from nests of *Osmia rufa* (L.)); *Meloe* spp. (hairy bees, e.g. *Andrena* and *Anthophora*).

Donisthorpe (1906, *Ent. Rec.*, **18**, 186) used a trap consisting of an ordinary jam-pot buried up to its neck in the ground at the foot of a tree in which was a hornets' nest. The jam-pot was charged with a small quantity of ordinary sugaring mixture which was frequently removed. The beetles, *Velleius dilatatus* (Fabr.), were attracted by this, fell into the jar, and were of course unable to get out.

In the south of England and especially in the Oxford district, we may possibly find in late summer one of the rarest of our beetles, *Apalus muralis* (Forst.), on the outside of a wall in which the nests of a mason bee (*Anthophora*) have been made.

W.H. Tuck gave the following advice for getting *Metoecus paradoxus* (L.), which is parasitic in the nests of *Vespula vulgaris*.

"Anyone attempting this work must have plenty of leisure and be prepared to undergo much labour. He should know of every nest within (say) a mile radius of his house. He should be able to tell at a glance, by the flight and manner of the wasps, whether they are of the right kind, viz. *V. vulgaris* (L.), and to form an opinion, by the number of workers going in and out, of the proper time to take the nest and also of course, to take it properly and bring it home intact for examination after digging it out.

"The nests in banks with rough herbage or inside and by the edge of woods are the most productive, and certain banks facing south often contain the host and parasite year after year. The nests in open fields rarely yield anything. A nest of *V. germanica* (F.), although close by, never contains the parasite, which I have taken in tree-stumps and once in a hot-bed in a walled garden (August 1–September 7).

"The parasite is reared either in the male or worker cells, generally either near the edge or quite in the centre, and it is very often impossible to tell, without tearing away the lid, whether the cell contains a wasp or a parasite, which, when liberated, rushes quickly out and often takes to flight".

(d) *Coastal sand-dunes*: Good results are obtained on the East coast hills when a S.W. wind blows towards the middle or end of October. With good fortune *Hydnobius* and *Leiodes* species especially *L. furva* (Er.) are found on the leeward sides of the dunes, often struggling up the slopes.

(e) *Fruits and seeds*: In the south of England, especially in the Isle of Wight and Dorset, larvae of the weevil *Mononychus punctum-album* (Herbst) may be found in the capsules of *Iris foetidissima* L.

(f) *Dung*: Certain species of onthophagous beetles are most common in the autumn, e.g. Beare mentions *Aphodius tessulatus* (Payk.) and *A. conspurcatus* (L.) in sheep dung in early November.

(g) *Cellars*: The insects to be found in cellars depend, of course, on the goods stored there, but there may be found *Sphodrus leucophthalmus* (L.), *Pristonychus terricola* (Herbst.), *Laemostenus complanatus* Dj. and *Blaps* spp.

STARTING AND BUILDING UP A COLLECTION

By J. Cooter

Unless you are lucky enough to have inherited a collection, the chances are that you will start from humble beginnings. My own grew from a motley mass of largely misidentified, gum-spattered specimens on curly cards housed in home-made or adapted cardboard boxes. It gave me great pleasure and certainly made me strive for better things.

As the number of specimens accumulated increases, a time will come when it is necessary to lay out the collection in proper taxonomic order following the latest published Check List. One has a choice of using store-boxes or a cabinet. Whichever is chosen, it should be of sound construction from good quality wood (or top grade interior ply wood), have almost air-tight lids and be of standard size.

CABINETS: Apart from a microscope, an insect cabinet will be the most expensive single item. However, good cabinets are always in demand, and they can, depending upon the make and condition of the cabinet, appreciate in value with the passage of time. The "Rolls Royce" of insect cabinets is a Brady or Gurney; both types are scarce and command high prices. The modern Hill Units made by Grange and Griffiths and the Watkins and Doncaster cabinets are very good quality. The "Hill" 20-drawer Oxford Unit is perhaps the most versatile with its removable drawer slats. Alas, the high degree of craftmanship and high quality materials are reflected in their high price. The 10-drawer Hill Unit is a very useful cabinet (beware of *Hill Type Units* which in the author's personal professional experience are to be avoided at all costs). The drawers and doors are interchangable making the system easy to expand at a future date.

There are generally a number of second-hand cabinets on the market and our own A.E.S. "Wants & Exchanges" list generally sees a small number on offer each year.

When purchasing a cabinet, even a brand new one, inspect it thoroughly. Check the carcase is sound, remove *all* the drawers, look all over inside, outside and underneath. Drawers must slide in and out freely and not jar, they must have wood-framed glazed lids, camphor cell (ideally running the whole front or back or both). A good cabinet is a pleasing piece of furniture and should be maintained as such. Rub candle wax along the drawer runners and polish the outside with a good quality wax (*not silicone*) polish. Although doors and locking side pillars are not necessary, they are an obvious advantage.

STORE-BOXES: These have several advantages, but more disadvantages than a cabinet. They have a lower individual cost (as a rough guide, one store-box can be regarded as equivalent to 1½–2 cabinet drawers of average size). A store-box housed collection can easily be expanded, extra boxes added where needed – for this reason it is best to purchase boxes supplied in a range of standard sizes. Points to look for, apart from sound construction, are provision of a camphor cell, tight fitting lid and a depth adequate to take double the length of pin used – the "extra deep" boxes being ideal; they cost only a fraction more than the lepidopterist's shallow box.

The store-boxes produced by Watkins and Doncaster are renowned for their uniform high quality and because of this have virtually cornered the market. One would be hard pressed to make these to the same high standard for a cheaper price. They come in standard sizes and two depths.

A universally accepted use of store-boxes is for the safe housing of unidentified material, a large back-log of which will soon be amassed.

LININGS: Traditionally cork or compressed cork covered with an opaque, thin, white, non-gloss paper was the standard cabinet drawer or store-box lining. Recently the use of cross-lined polethylene, marketed under the trade names "Plastazote" and "Freelite", has become more widely accepted. It has great advantages over cork/paper and does away with the need for paper and holds the pin better than cork. Both "Plastazote" and "Freelite" are available in a variety of densities and thicknesses and can be purchased in large sheets from a plastics wholesaler – see Yellow Pages for your nearest supplier. The two differ in the type of surface – "Freelite" having a smooth surface which shows pin holes and "Plastazote" a flat surface not smoothed which with the passage of time holds dust. Both materials are washable.

On no account use polystyrene as this holds a static charge which attracts dust, it is attacked by solvents, paradichlorobenzene and naphthalene and will not hold a pin.

Fumigants

A chemical to deter attack by insect pests should be in the camphor cell. Of the three major chemical used paradichlorobenzene is known to be the most toxic to humans and its use should be avoided. Naphthalene is also toxic to humans, but not so bad as paradichlorobenzene; nonetheless, its use is not recommended. Less toxic than naphthalene is camphor, its toxicity appears to be within health and safety limits and is the chemical fumigant used by the New Zealand National Insect Collection.

These chemicals are there to deter pest attack, they will not stop it. If an attack is noticed, a simple cure is to open the cabinet drawer or store-box and pour in a small quantity (say 10ml) of ethyl acetate, close the lid and allow the solvent to evaporate.

Further deterrents can be used. The inside of the cabinet carcase can be sprayed with "Nuvan Staykill" or similar preparatory product. These generally have an active life of anything up to one year.

Disposing of a collection

We can leave instructions as to the fate of our prized collections after death, a Will is the obvious place for this. However, many people die intestate and it is thus left to others to decide what to do with the insect collection – often these people will have no interest in entomology.

So, it is obviously good practice to attend to this matter in good time. A collection represents a huge investment of time, effort and money and, if properly organised with the *data with the mounts*, will be a valuable scientific record.

To my mind it is very pleasing to be able to pass a collection on to an interested contemporary, ideally one's own child or close friend; generally such persons are as familiar with the collection as its owner. Apart from continued private ownership, the local museum is an obvious choice, but be careful and take time to check out the organisation and staffing of any museum. Points to bear in mind are – is there a permanent full time natural historian on the staff, and is this post likely to continue in the foreseeable future? Is the storage adequate and suitable? Will the collection be maintained (including being incorporated with others), or will it be split up with specimens going on display or school loan (where they will deteriorate rapidly)? An alternative is to sell the collection at auction, preferably in a sale of natural history or other entomological material. The price realised will benefit one's surviving relatives, and will reflect the content of as well as the actual cabinet (or store boxes).

Similarly, instructions as to the disposal of one's library will be helpful. Far better to get an entomological book dealer to quote; they know the specialist market. The local antiquarian might well buy the lot then resell to a specialist.

If leaving a collection to a museum, it is beneficial not to tie it up with any clauses such as "it must be maintained as a separate collection". Museums are short of space and, if properly organised (i.e. with data labels), one's own specimens are easily recognisable after its incorporation into a museum's main collection. It should be possible to dispose of the collection only and sell the empty cabinets for the benefit of one's estate.

INDOOR WORK
By J. Cooter

This section covers the procedures entailed in preparing the live beetle for the cabinet. I hope it is comprehensive; it has been compiled from many years' background as an amateur, combined with a wide professional experience in Museums. There is no such thing as the correct method; each will have their own variation on the same theme. There are, however, INCORRECT methods which should be avoided at all costs.

A minority of Coleopterists keep their material in alcohol or other liquid preservative. Details of this are given later; here we deal with the preparation of a "dry" collection.

Killing

The secret to good preparation lies in the killing method. When removed from the killing bottle, the beetle should be perfectly relaxed. The medium should be quick acting and present a minimum hazard to the user (it is good practice to regard any chemical used in entomology as potentially dangerous).

Cyanide. Old textbooks on entomology recommend the use of potassium cyanide crystals under plaster in a glass bottle; it is best to regard this as inspired lunacy as cyanide in minute quantities will kill healthy adult human beings. When exposed to moisture, even the moisture in air, it gives off an equally lethal colourless and, contrary to popular belief perpetuated by detective novels, odourless gas – hydrogen cyanide. Cyanide can only be rendered safe by chemical methods, burial or burning are quite ineffective and totally irresponsible. AVOID IT AT ALL TIMES.

Ethyl acetate or acetic ether is the most widely used killing agent today. It is cheap, freely available from dispensing chemists and reasonably safe to use, though it and its vapour are highly inflammable. It is such a good agent that others are not worth considering.

When using ethyl acetate it is worth remembering that it will dissolve plastics and renders dimethyl hydantoin formaldehyde (D.M.H.F.) soft, prolonged exposure or immersion will dissolve the D.M.H.F.

Boiling water is a useful standby and particularly suitable, for example, with Histerids. Not suitable for species with scales or pubescence and not very practical to use with beetles over, say, 5mm. The beetles to be killed are turned out into a cup or similar heat-resistant object and allowed to move about. A minimum quantity of boiling water is poured in and the dead beetles are removed at once and dried on blotting paper or tissue prior to mounting.

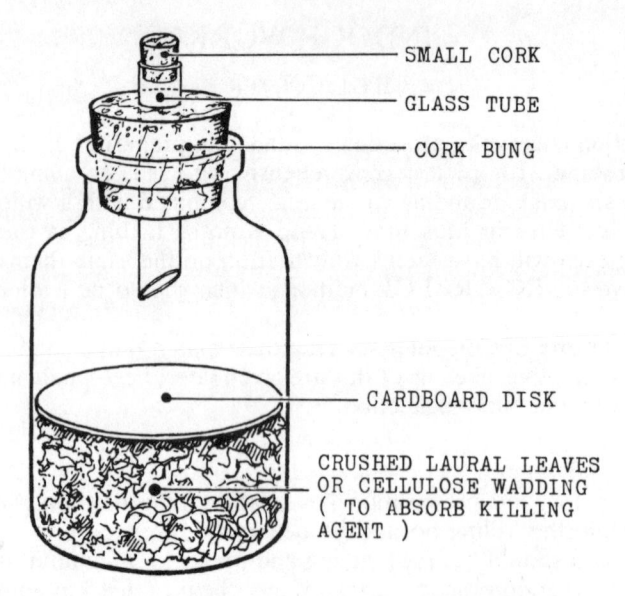

Figure labels: SMALL CORK, GLASS TUBE, CORK BUNG, CARDBOARD DISK, CRUSHED LAURAL LEAVES OR CELLULOSE WADDING (TO ABSORB KILLING AGENT)

KILLING BOTTLE WITH DOUBLE CORK FITTING TO ALLOW USE FOR SMALL OR LARGE BEETLES

Fig. 5 A Killing Bottle.

Killing bottle

The term "bottle" is here taken to be a jar or tube, any clear glass container with tight fitting lid will suffice but a larger tube of say 1"–2" diameter and between 3"–6" long is probably the ideal for most beetles. Larger species can be killed individually or in a large sized killing bottle; a ground glass stoppered spice or storage jar is preferable to a jam-jar or similar with a screw or metal lid. The rubber-like seals on these are destroyed by the vapour given off by the killing agent.

The bottle should be kept clean and dry, cotton wool should *never* be used as the beetles will get irretrievably entangled in it. White blotting paper, filter paper, kitchen roll, "soft" toilet tissue or cellulose wadding are all suitable and readily available, cheap and absorb the fluid any excess moisture forming within the bottle; these materials are readily replaced. An alternative is to pour an inch or so of plaster of Paris into the bottom of the bottle and leave it to set thoroughly before use. The ethyl acetate is absorbed into the set plaster and a ring of blotting paper inserted will absorb any moisture forming during use. (Fig. 5)

It is best to make up fresh bottles weekly by pushing an amount of the absorbent material tightly in the bottom of the tube, adding a few drops of ethyl acetate followed by another layer of tissue. Plaster bottles need only be recharged when the efficacy of the killing fluid requires it. A piece of crumpled tissue is then added to give the beetles a large surface to run about on, and to help absorb excess moisture. The beetles are added and the stopper put in place, the whole being kept away from direct sunlight.

The time taken for the beetles to die will depend upon the charge in the bottle, temperature and number and type of beetle. Histerids and other similarly hard-to-kill species should either be kept separate or put in first, the others added when these have stopped moving around. Experience is the only sure guide here; if in doubt, leave it another $\frac{1}{2}$ hour or overnight. Many collectors use the killing bottle to "knock out" their catch, after a few minutes they are removed and set, then the cards with set beetles returned to the killing bottle.

Points to remember are:
>Use a minimum of killing fluid.
>Keep the beetles away from direct contact with the killing fluid.
>Keep the killing bottle dry in use.
>Never over-fill the bottle with beetles.
>Keep large or voracious species separate (I still recall the agony of watching helplessly as my first specimen of *Hypebaeus flavipes* was devoured by a cantharid in the killing tube).

Laurel bottle

Like Joy's *"Practical Handbook of British Beetles"*, this is something coleopterists either swear by or swear at. Equally similarly, the laurel bottle, when used *correctly*, is a very handy aid that has stood the test of time. Although once recommended as a killing medium, ethyl acetate has superseded this use. For keeping freshly killed material relaxed it is very useful, though it should not be used for long periods as there is a tendency for its vapours to darken colours, and there can be a problem with excessive moisture.

Laurel mixture can be purchased from Watkins and Doncaster – this has a low moisture content. Alternatively the young leaves of the shrub cherry laurel (*Prunus laurocerasus*) are picked and cut then bruised and pressed down tightly in a large glass jar making a layer about one quarter the depth of the jar. Fresh laurel tends to be very moist so precautions must be taken. A layer of cellulose wadding is placed on top of the laurel layer and on top of this a blotting paper lining. The crushing of the leaves produces hydrogen cyanide and benzaldehyde – the latter, smelling strongly of almonds, keeps the insects relaxed indefinitely; indeed appendages may actually drop off on handling the specimens if they have been in direct contact with the laurel, or in the jar for excessively long periods.

The beetles should thus be kept in batches in screws of absorbent paper or in small air tight polythene self-seal bags. These should first be perforated with a pin to permit entry of vapour. A lining of kitchen roll is added and the beetles placed in one side of this and a data label inserted on the other. The bag is sealed and placed in the laurel jar (or in a plastic box containing crushed laurel).

Material that cannot be dealt with straight away

If the catch cannot be prepared directly there are several options available. If the delay is only a day, the beetles can be left in their tubes, or killed and left in an ethyl acetate killing bottle. Transfer to laurel will permit a longer period to elapse.

For long periods, the material can either be left to dry or put in 70% solution of alcohol (industrial methylated spirit (I.M.S.) or isopropyl alcohol). If alcohol is used it is advantageous to fill polypropylene stoppered tubes ⅔ to ½ full and add a piece of tissue. Live beetles can be dropped in directly, but often it will be found in their death struggle, their elytra will open and wings extend. Killing in ethyl acetate prior to transfer to alcohol will overcome this. A slip with the data must be added to each tube.

Dry material. It is of prime importance to ensure the material is dead before packing. This may seem obvious. On removal from the killing bottle, the beetles are laid neatly in rows on cellulose wadding in a box (stout cardboard is perhaps preferable as its walls do help by absorbing excess moisture to a degree; this is not so with plastics or metal). If the appendages are roughly arranged there might be no need to relax the specimens later on. The layers are built up, and data kept with the beetles. Never mix material from different localities on the same layer. Keep spare wadding on top of the last layer as this will ensure the contents are not disturbed and mixed up when the box is transported.

There is often a need to handle material in this way especially if collecting abroad where the aim is to amass as large a sample as possible in the time available. In the tropics the addition of silica gel or other moisture absorbing compound is necessary to minimise the risk of moulds. Some pesticide in the box is also advantageous to ward off insect attack.

Relaxing

The purpose of relaxing dry material is to render the musculature soft enough to permit manipulation of the appendages, and, if necessary, dissection of the genitalia.

This goal is simply achieved by putting a lining of tissue, cellulose wadding or other absorbent material (not cotton wool) in a plastic or metal box with tight fitting lid. Water is added soaking the lining, beetles added in rows, the lid replaced and the whole placed in an airing cupboard, on top of a radiator or similar warm place. After a few hours, depending upon the size of the beetles, but certainly overnight, the beetles will be ready for mounting.

A crystal of thymol or a few drops of phenol (carbolic acid) will help prevent mould, but it is good practice to relax at one time only sufficient beetles that can be mounted in one session. The relaxing box should be cleaned out between batches and a new lining used each time.

Boiling beetles in water will work, but is unnecessarily drastic.

Similarly, personal experience has shown the use of proprietory made up "relaxing fluids" are a waste of money.

Mounting

Whether the beetle has been removed from the killing bottle, alcohol tube or relaxing box, the subsequent preparation follows the same path.

Because our fauna is well documented and composed mostly of small to medium-sized beetles, British Coleopterists have traditionally carded all species. Some, however, prefer to pin larger species such as *Carabus, Necrophorus* or *Lucanus*.

Before describing the methods of mounting, we should first consider the equipment and materials needed.

Equipment

The basic necessary items are few. Always buy the best quality available; quality does indeed increase with price.

Brushes. Two, three or more artist's good quality sable hair brushes, size 0, 00, or 000 being the most useful. Personal preference will result from use.

Forceps. At least one pair of watchmakers fine pointed straight forceps size 5. Two pairs are advantageous, and a third pair of less fine, say size 2, will be found very useful. If possible, stainless (anticorrosive) and non-magnetic should be used. *Pinning forceps* are used for handling pins. Like all tools they vary greatly in quality and an inferior pair can easily lead to damage being caused when working the collection. They should close with gentle pressure, both ends meeting evenly over their length, a "peg" will prevent "shearing". The addition of ribbing on the finger grips will minimise slipping, and "knurling" on the ends will assist in gripping the pin, preventing slipping when pressing the pin into the cabinet. Pinning forceps with magnetic qualities are a confounded nuisance when using steel pins.

Dissecting needles of the standard biological type are too coarse for the bulk of entomological work, but a pair are at times useful. A very useful tool is a micro-lepidopterist's headless stainless steel pin. These are too short to be hand held, and easily slip when held in forceps. A watchmakers "pin vice" comes in very handy here, being about 6″ long and having good balance. The "business end" is a three-jawed chuck activated by a screw collar. Thus pins of various lengths and diameters can be changed easily.

Dissecting dish. The dissecting dish should not be too large, too small, too deep or too shallow. It must be heavy with a flat base. An excavated glass block is very useful; about 1½″ square with flat base. The spot-testing pallette used by biologists has advantages. They generally have a dozen or so shallow depressions and permit rapid transfer from one fluid to another, being of white porcelain the dissection and minute dissected parts are easily seen.

Scissors. Your own pair of top quality scissors should be kept well hidden from other members of the family. A good pair will make a sharp clean cut. Necessary for trimming labels and countless other tasks.

Instrument roll. The tie-up cloth dissecting instrument rolls are ideal for keeping the equipment safe when not in use or when in transit. Points of forceps and the heads of brushes are easily damaged if kept loose in boxes.

Knife. A scalpel with blade size 10, 10A or 11 is very handy, but any sharp craft knife will suffice.

Pins

Pins have three main uses:
> For manipulating the beetle during mounting, and for performing dissections.
> For holding the card-mounted beetle and its labels.
> For direct pinning of beetles.

The headless, stainless micro pins are available in sizes ranging from 0.0056″ diameter by 10mm long (A1) to 0.0179″ diameter by 30mm long (G3). Diameters from 0.01″ (D) are suitably stout for manipulating beetles and dissecting. These pins are sold in lots of 100 of each size by Watkins and Doncaster; a sample card is available upon request (send s.a.e.).

The size of pin used to hold a mounting card or to direct pin a beetle should be as long as possible. Alas, many old cabinets and store-boxes are too shallow to take Continental pins (38mm long) so English size 12 or stainless steel headless size G3 will have to be substituted. Brass and brass plated pins should NEVER be used, with time corrosion sets in at the point where the pin passes through the mounting card and into the cabinet/store-box bottom, eventually the pin will break at one of these points.

Continental pins are of two lengths, sizes 000 to 6 being 38mm long, and sizes 7 to 10 are 53mm long (intended for use with large tropical insects). Sizes 3 and 5 will be found adequate for British workers. They are made in stainless steel or black steel and the former is preferable.

Mounting cards

More and more British coleopterists are now using the die-stamped standard-sized cards sold by European dealers. They come in two types, with or without fine black lines at one end and sizes 4.5 × 11mm; 5 × 14mm; 6 × 17mm; 10 × 21mm; 12 × 27mm 13 × 30mm and 15 × 36mm; the first four sizes are also available in black card. They are sold in packets of 250 or 500 depending upon size. (Fig. 7.)

A collection mounted on these standard cards looks very neat and the printed lines are a useful guide for placing the pin and producing neat rows in the cabinet/store-box.

It is of course possible to cut one's own cards to whatever size is desired. A good quality Bristol Board (3 sheet for small cards and 4 sheet for larger) should be employed. A very sharp knife is needed for cutting as otherwise ragged edges or "burring" will result. A range of standard sizes should be employed and a stock of each cut prior to use.

Pinning stage

This simple device (Fig. 6) is a great help and the easiest way to ensure that mounting card, data and other labels are spaced evenly and standardly on the pin. Similarly, direct pinned insects will have their ventral surfaces at the same height.

The pin is pushed through the mounting card and then into the deepest hole in the stage, the data label in the next deepest and so on.

Having every specimen at roughly the same height – the cards will be at identical heights, but the height of the individual beetles varies – is a great help and saves a lot of refocusing the microscope when making identifications or comparisons.

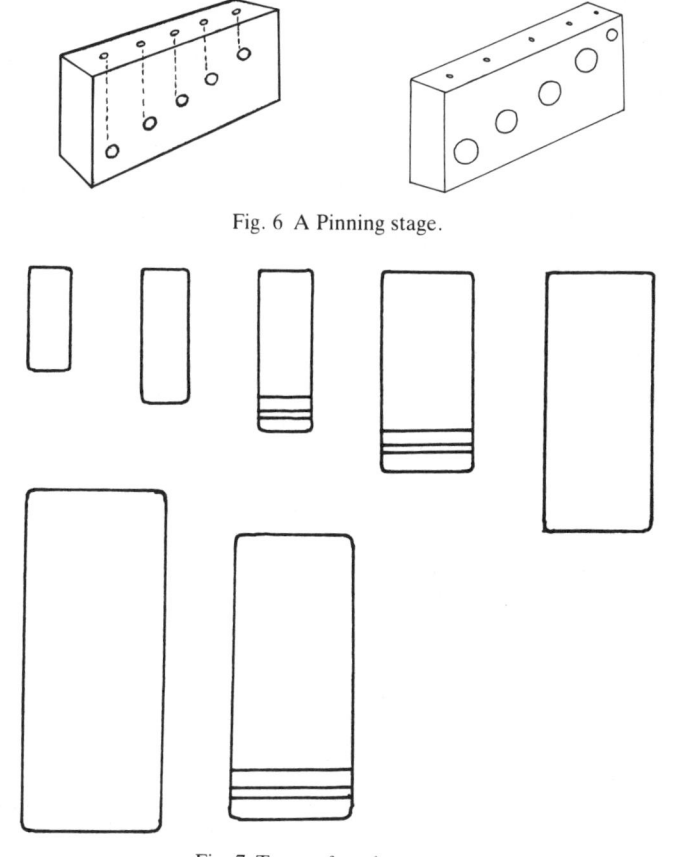

Fig. 6 A Pinning stage.

Fig. 7 Types of card mounts.

Labels

Labels are the most important device for keeping information relating to the specimens; diaries and journals very often get lost (often after the death of the collector), accidentally destroyed or fall to some other disaster. A collection without data is virtually useless. A collection with data in a journal is, once passed to a different owner, a very frustrating and tedious entity; a Museum Curator will not have the time to add such data individually to several thousand beetles. However, if done, piecemeal as the collection is built up, the task is simple. **THERE IS NO EXCUSE FOR NOT PUTTING DATA WITH SPECIMENS.** If you remember nothing else, please, please do remember this. Having spent my working life in Museums curating insect and other collections, I have learned this fact from sad experience. Colleagues too report seeing well set out collections with good series of uncommon species – but no data. Very often it transpires that after the death of the collector, the collection remains untended for a period and is then passed to a Museum. In the intervening time, personal effects are disposed and books sold off as job-lots. Thus the diary or journal has already gone. To the lay-person an insect collection is a collection of insects; to the entomologist it is a collection of insects with data – the difference is immense.

Data labels. These can be hand written or printed. Photocopies whilst cheap are not to be used as the ink is attacked by solvents. Likewise anything produced photographically is, with passing of time, liable to fade or cloud. Ball-point pen ink is also attacked by solvents and fumigants. **GOOD QUALITY INDIAN INK OR PENCIL** give the best proven results. Ideally the paper used should be "acid free" to avoid browning with age; such paper can be obtained from artist's supply shops.

The basic information to be recorded will include County, locality, date, captor's name and grid reference. At the same time, the label is best kept as small as possible and this will necessitate developing small writing with a fine pen (a 0.1mm mapping pen nib for example), use of both sides of the label, or more than one label for data. Recording of other detail such as food-plant or plant from which the beetle was taken – the two are not necessarily the same – or other microhabitat should be added if possible.

Determination labels. Likewise written in indian ink or pencil this label records the beetle's name, date determined and name of the person making the determination. If known, the sex of any particular specimen should be added to this label.

After the mounting card is pinned, add the data label(s) and then the determination label. Any information extra to this (as well as duplicated data) can be kept in a journal or diary.

Gum

Various coleopterists will swear by their own preferred gum and tell you theirs is superior to all others. In truth, there are many equally good gums and recipes available. There are, however, some glues that should never be used – these are non-water soluble types.

Leprieur's Gum is a widely used mixture –
>60 parts gum arabic dissolved in a minimum of water
>30 parts white sugar similarly dissolved, then mixed with the arabic. To this is added a mixture of 2 parts phenol (carbolic acid) dissolved in 8 parts alcohol.

The sugar renders the gum arabic less brittle and the phenol is a fungicide.

Gum arabic is readily available from a good dispensing chemist or herbalist. The powdered form is preferable and generally more pure than the lump form. It leaves a glossy residue on the mounting card.

Gum tragacanth is difficult to obtain in powder form but acquisition of a small (50g or 100g) bottle will supply one's life-time needs. A very small quantity should be placed in an egg cup and water added. It will take up a jelly-like consistency and with more water become creamy – the required consistency. A small quantity of powder will make a large quantity of gum. It is a weak gum and should be used very sparingly for sticking down the terminal segments of antennae, tarsi and other appendages. It leaves no visible residue on the card.

Solvite wall-paper paste is used by some coleopterists with success. It is readily available, cheap, strong and has a built-in fungicide.

It is well to note that some fungicides, with the passage of time, turn the mounting card brown.

Whatever gum is used a small quantity should be kept in a small "glue pot" – an egg-cup is of suitable size. When not used, a gum without fungicide can be left to dry out. When needed again, water is added.

Mounting methods

In order to determine a beetle, those characters upon which the identification keys are based must be displayed. This mostly involves the dorsal surface, legs, antennae and mouthparts, plus, where necessary, the genitalia and associated structures. There is no need to arrange the beetle symetrically in the upper centre of the card, but most people do, and a collection mounted in this way is not only pleasing to the eye, but affords better opportunity for comparing beetles with each other.

Beetles can be mounted individually directly on their pre-cut or die stamped cards, or "rough-set" in any order in small numbers on odd pieces of stiff white card (artist's mounting card is very suitable). Such

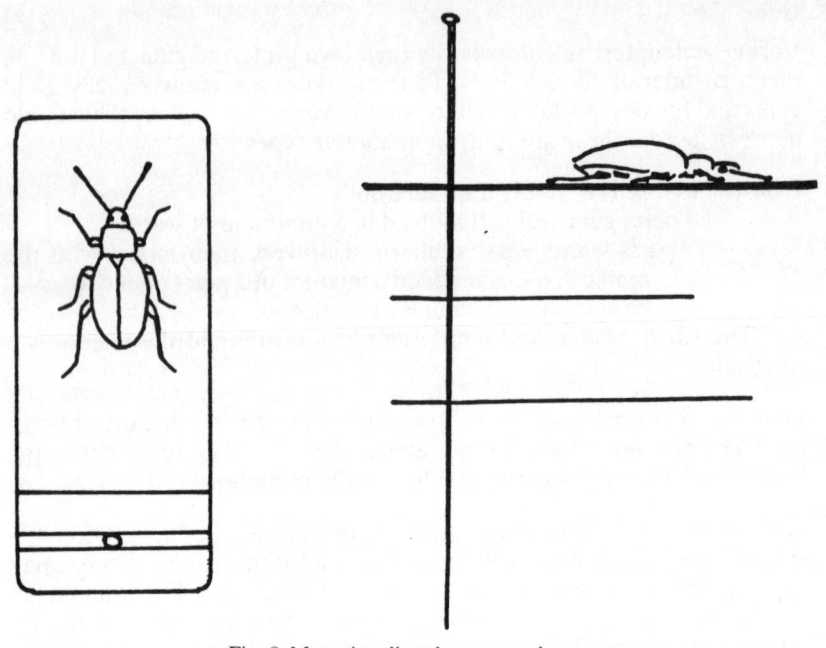

Fig. 8 Mounting directly onto card mount.

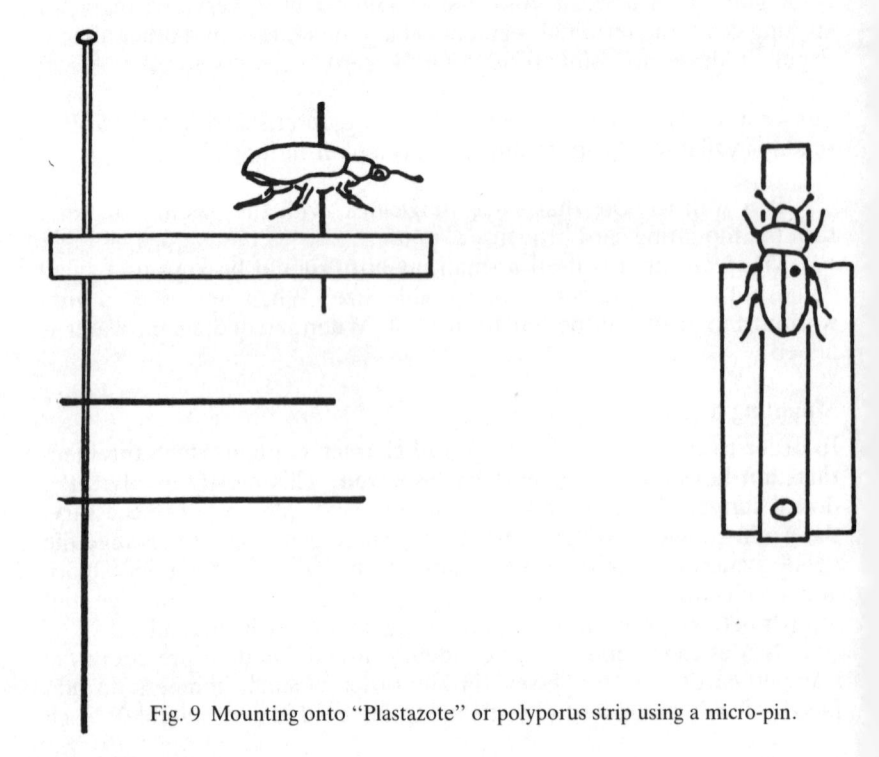

Fig. 9 Mounting onto "Plastazote" or polyporus strip using a micro-pin.

specimens are transferred, when dry, to their final cabinet mount. It has an advantage of speed, but time has to be spent later in the transferring process. However, the remounted beetle can be more easily located centrally on its card, and a minimum of gum can be used to anchor the specimen.

The fresh beetle, be it from the killing bottle, relaxing tin or alcohol tube, should be laid on a piece of clean blotting paper or thick tissue. Any dissection should be carried out at this point. The appendages are teased out and the mandibles opened – this can often be achieved by inserting the closed tips of fine forceps between them and relaxing pressure slightly, enough to open the points, and hence the mandibles. A blob of gum is placed on the card, and the beetle placed in this, dorsal side uppermost. Legs and antennae are positioned with the aid of a fine brush and/or setting needle, forceps, or all three in combination. A minimum of gum is used to anchor the appendages. Great effort should be made to keep gum off the upper surface of the beetle especially the tarsal segments. The antennae should receive gum on the underside. Often a very thin film of gum around segments serves to distort their relative proportions. A pin is inserted through the centre of the bottom edge of the card (Fig. 8) *Full data should be added.*

After "rough setting", data should be written on the card. When the beetles have dried, they are soaked off with cold water, dried on blotting or filter paper and mounted on their individual cards by using a small quantity of gum to attach the ventral surface of the abdomen pronotum and head. The appendages are lightly gummed. This is very important and an excess of gum can distort the relative proportions of segments and obscure fine detail, scales, pubescence etc. The card is then pinned *and data added.*

Beetles are pinned through the right elytron (Fig. 9) so that the pin emerges between the middle and posterior coxae. This is so for direct- or micro-pinned specimens. The beetle can be held or placed on a cork/"Plastazote" mat and the pin pushed through the elytron, checked to see if it is on the right course and, if necessary, adjusted before being pushed through the ventral exoskeleton. With direct pinned specimens, the pinning stage (see page 39) is employed to give the correct height. With micro-pinned beetles, enough pin should be left above and below to permit handling with forceps during study/determination and to permit secure fixing in the staging material – polyporus or cross-linked polyurethane strip. This generally works out as $\frac{1}{3}$ length above, inside, and below the beetle; the micro-pin should never protrude above the pin holding the stage (Fig. 9).

Degreasing

When the fat body contained within a mounted beetle breaks down, a thin film of oily "grease" is to be seen covering the beetle externally. Advanced cases show a tell-tale yellow-brown stain on the mounting card and the mounting gum's bond will be weakened or broken. An additional hazard results as fine dust and dirt, present in even the most air-tight cabinet drawers, will adhere to the beetle. Brass and plated brass pins decay more rapidly in the presence of "grease".

There is no way of knowing if a specimen will develop this condition, though some species are more prone than others. On the other hand one specimen of a contemporary series might "go greasy" whilst its neighbours will not. If the problem is not very advanced, perhaps noticed when the beetle was examined under the microscope, it will require only simple immersion in an organic solvent; more advanced cases will in addition need re-mounting on a clean card with a stainless steel pin.

Most organic solvents will tackle the problem well, but possibly carbon tetrachloride or toluene are preferable. Both are hazardous and the warnings on their bottles should be noted. The latter is inflammable, the former is not. A quantity of the solvent is poured into a glass jar with tight fitting lid. A disc of "Plastazote" or cork is cut to fit easily through the neck of the jar and into one side of this is pushed a long pin. On the other side pin the greasy beetles still on their mounting cards (this process is best done individually with heavily greased beetles as they might at this stage fall from their cards and become mixed). The disc is then placed in the jar, beetles first, and floated on the solvent. Thus the beetles are totally immersed. The lid is fitted and the jar left in a safe place for a few days or longer in severe cases.

Upon removal, allow the beetles to dry thoroughly in a well ventilated room. CAUTION make sure the solvent will not attack plastic of the lid, or if being treated, the acetate/celluloid strips containing genitalia or their mountant, or DMHF if used. The use of good quality Indian ink or pencil is once again stressed for label writing.

NOMENCLATURE

It is beyond the scope of this Handbook to describe the workings of the International Rules of Zoological Nomenclature; these are very complex. However, the notes given below will, it is hoped, explain the various conventions the student is likely to meet in publications, particularly Check Lists.

Typically a Scientific name consists of four parts which from left to right are generic name, trivial name, author and date (for example *Apion dispar* Germar, 1824).

Example 1

Dytiscus marginalis Linnaeus, 1758

This shows *Dytiscus marginalis* was described and named by Linnaeus in 1758 and has remained in that genus ever since.

Example 2.

Hydaticus transversalis (Pontoppidan, 1763)

Curved brackets around the author and date show that the species has been transferred from the genus in which it was originally placed by its author. In this case Pontoppidan placed *transversalis* in genus *Dytiscus* in 1763. Later, Leach (in 1817 to be precise) erected the genus *Hydaticus* and Pontoppidan's species was transferred to it from *Dytiscus*.

Example 3.

Dytiscus semisulcatus Muller, 1776
    = *punctulatus* Fabricius, 1777

This is a case of synonymy, Muller and Fabricius both published descriptions of the same species, both placing it in *Dytiscus*, but each giving the same beetle a different trivial name. As Muller was first to publish, Muller's name is used. This is an illustration of the Law of Priority. Alas, exceptions to this Law are numerous and various, one of the most common being that the first published description was not precise enough to define one species – several distinct species could as easily fit the description. The original description and type specimen may have become lost, so there is a need to redescribe it, if no previous description and name is available, or designate a lectotype/neotype so that the original author stands.

Example 4.

A beetle might have been described and given a name already in use in that genus:

Aphodius tenellus Say, 1823
    = *putridus* Herbst, 1789 *nec* Fourcroy, 1785
    = *foetidus* Fabricius, 1792, *nec* Herbst, 1783

Herbst described the beetle we now know as *tenellus* Say in 1789 as *putridus*. However, there was already an *Aphodius putridus*, that described by Fourcroy in 1785. The word *"nec"* means "not", that is to say Fourcroy's *putridus* is not conspecific with Herbst's *putridus*. To further complicate the issue, Fabricius also described, in 1792, what we now know as *tenellus* Say, but called it *foetidus* a name also already occupied by an *Aphodius*, that described by Herbst in 1783. Again the use of *"nec"* shows Herbst's and Fabricius's *foetidus* are not the same species. So, in 1824 Say redescribed the beetle that was previously known as *putridus* Herbst and *foetidus* Fabricius as *tenellus*. This illustrates the importance of always adding the authors name when referring to beetles (and indeed all insects).

Example 5.

Philonthus succicola Thomson, 1860
= *proximus* Fowler, 1888 *nec proximus* Kraatz, 1859

The interpretation is exactly the same as in example 4, the beetle we now know as *P.succicola* is synonymous with *proximus* Fowler, but not with *proximus* Kraatz.

Example 6.

Staphylinus melanarius Heer, 1839
= *globulifer* sensu auct. *nec* Fourcroy, 1785

The addition of "sensu auct" means "in the opinion of authors" and shows Fourcroy's description of *globulifer* published in 1785 was misinterpreted by other Coleopterists, in fact they supposed Heer's *melanarius* to be Fourcroy's species. ("auct.Brit" means "of British authors" and shows the beetle known to British Coleopterists is not the same as that referred to by workers in other countries.)

Example 7.

Carabus monilis Fabricius, 1792
v. *gracilis* Kuster, 1846
   = *consitus* auct. *nec* Panzer, 1792

The species *monilis* and, because it is set flush with the name above, the variety *gracilis* both occur in Britain. However, the variety was by various authors once called *consitus* in error, it is in reality not Panzer's species.

Example 8.

Carabus [*problematicus*] Herbst, 1796
ssp. *gallicus* Gehin, 1885
   = *catenulatus* Fabricius, 1801 *nec* Scopoli, 1763 .

The use of square brackets shows the nominal species *problematicus* does not occur in Britain, but is represented by sub-species *gallicus* Gehin. The beetle used to be known as *catenulatus* Fabricius, but Fabricius' species is not the true *catenulatus* of Scopoli.

Example 9.

Nebria gyllenhali (Schonherr, 1806)
v. *rufescens* Ström, 1768
v. *balbii* Bonelli, 1809

Thus Schonherr's *gyllenhali* was originally placed in a different genus, in Britain two varieties occur. However, to illustrate the unsettled nature of the nomenclature, the Scandinavians list the same beetle as:

Nebria rufescens Ström, 1768
   = *gyllenhali* (Schonherr, 1806)
   = *hyperborea* Gyllenhal, 1827

IDENTIFICATION

BY J. COOTER

It is an assumed objective that the student will want to know the names of the beetles collected. Many of the initial captures will be familiar from common knowledge (ladybirds, cock-chafer for example) and others from illustrations in text books (for example the majority of our Cerambycidae). Others will look familiar, but seem somehow subtly different from their congeners. With increasing experience an increasing number of beetles will be identified in the field or at least placed in their correct genus or thereabouts. The smaller species, especially those in large genera such as *Cryptophagus*, *Longitarsus* and *Apion*, will generally need careful examination under magnification, and very often will need dissection too.

By far the best method of identifying beetles is to use a reliably identified collection in conjunction with up-to-date identification keys and manuals. Identifying beetles in batches of the same (larger) genus is sound practice. The work will be quicker and the results more accurate; the key better understood. Looking through a well set out collection will increase one's knowledge and show clearly the subtle differences so difficult to express in a key or even in longer text. Thus after a short time, the collection itself becomes a very useful aid for comparative work.

It is always worthwhile enquiring at the local Museum to see if a collection is available for consultation; very often these will not be on public display but can be seen by appointment. Some identifications may be suspect, especially so in older collections and Museums that do not have an entomologist on their staff. Very often it is to the "private sector" that one most profitably looks for help – local coleopterists if approached will generally be found to be very helpful and full of encouragement. The Advisory Panel of our own Society was established as a means to assist Members with their identification and other entomological problems. Whenever sending material away for expert opinion it is good manners to INCLUDE SUFFICIENT STAMPS TO COVER RETURN POSTAGE, and good policy to write beforehand to enquire if it is convenient to send such material. Do not be surprised if the odd specimen is retained by the person identifying your material. You might at the time feel a little aggrieved, but in reality this should not be the case; the process is one of mutual benefit, you have got all your material reliably determined, the identifier has a beetle which he needs. Identifications, as every coleopterist knows, are often onerous, and if dissection needs to be done, can be very time consuming. Most members of the Advisory Panel are not professional entomologists and so give freely of their limited spare time, (indeed many professionals through necessity carry out such identification work during their own time). The gift of the odd specimen is a small price to pay, and more often than not, the identifier will make good your loss with a few duplicates of his own.

Most coleopterists will, with time, amass a very useful library; a selection of the most useful books and journals is listed below. Although most of the systematic papers published in the journals of the Royal Entomological Society of London tend to deal with foreign faunas, the Society possesses an unrivalled entomological library for the free use of Fellows. Details of Fellowship can be obtained from The Registrar, Royal Entomological Society of London, 41 Queen's Gate, South Kensington, London SW7 5HU.

In order to facilitate identification some form of magnifying device is required. Many coleopterists make do and indeed do good work with a simple ×10 and/or ×20 hand lens. Points in favour of a lens are its low cost and being small can be carried on one's person. Points against are that being small it is easily lost, use puts a strain on one eye and there is a greater risk of damaging specimens. Further, with very small beetles, or those where dissection is required, even ×20 magnification is insufficient.

The ideal is a stereoscopic wide-field binocular microscope with a range of magnifications of say, ×10 to ×80 or ×100 (occasionally limited instances will require an even higher magnification). A rotating objective is preferable to the type that has to be slid in and out in order to change magification. Zoom devices are also available and give a constant range of magnification between set limits. Although these instruments are not cheap, they are often no dearer than a video recorder, compact disc player or home computer. A well made model will last several life times, will retain its value and can be readily sold at a later date. The better models are sold by specialist dealers, either at their showrooms or via a regional representative that will make house calls. Microscope sales people know their business so explain fully your likely requirements and try out a range of models offered. It is a good idea to take to the showroom a box of mounted beetles covering a range of sizes – for example *Amara*, *Bembidion* and *Apion* plus a very flat and a very convex species. Run these through an identification key to see if the described characters are easily seen under magnification. Some dealers will permit home trial upon receipt of a suitable deposit. Most microscopes are supplied with a pair of ×10 eye-pieces and acquisition of a pair of ×20 will greatly enhance the versatility of the instrument. Eyepieces should always be purchased in pairs.

Other useful accessories include an eye-piece micrometer for taking absolute measurements and for comparing dimensions and ratio. Rubber eye-cups are useful for those not wearing spectacles. These and other accessories can be purchased with the microscope or at a later date. A good heavy base is a very handy feature.

With a microscope, a light source is needed. Some have a built in light, others have none. Invariably a separate light is more useful and can be moved freely without having to move the subject and re-focus. Direct unfiltered light is too intense, causes too much reflection from the subject and obscures detail. Placing a filter between the light and subject will greatly enhance the lighting capacity. The diffuse light will show up surface sculpture and other features. Moving the light source closer to or further away from the subject is also very beneficial, revealing subtle characters. Moving the light "around the clock" so that the light shines from different directions will pick out characters such as pubescence. With practice the use of a filtered light source will be better understood, it is one of the secrets of determination.

Coleoptera keys are somewhat unusual in typically offering comparative choices in each couplet. Other Orders, such as the Diptera and Hymenoptera more often give a clearcut "either/or" "with/without" choice. Subtle differences "less closely but more intensely" "more closely but less intensely" are difficult to appreciate in isolation, but if authentic material is available so that direct comparison can be made, the differences at once become very clear. Thus the importance of being able to refer to reliably identified material cannot be overstressed and the value of determining generic batches becomes obvious.

Some identifications will run smoothly from the start to an obviously correct conclusion; in many works there are additional diagnostic characters listed under species headings. At other times an obviously incorrect conclusion will be reached, or indeed a point in the key where none of the choices seems to fit. At other times one will run through the whole key without success. The best advice here is "go back to start", this time take it more slowly and make sure *all* the characters in *each* couplet are considered. However, there is little to be gained, other than complete frustration and lack of confidence, in persisting. If continued lack of success results, it is better to put the problem beetle to one side, perhaps with a brief note of the difficulties, and try again another day or with a different, often older, key. Alternatively, put all such problematic beetles in a box ready for your next visit to the Museum or a more experienced friend. By "keeping at it" and not getting too dispirited, these probables WILL be overcome, and others will arise!

Literature

As interest in the Coleoptera develops, there will be a need to obtain literature and build up a working library augmented with interesting, but out-of-date texts, back volumes of the entomological journals, foreign works and a variety of reprints.

The standard work on the British beetle fauna is "*The Coleoptera of the British Islands*" by Canon W.W. Fowler published in five volumes between 1887 and 1891 with a sixth, supplementary volume by Fowler and Donisthorpe appearing in 1913. The larger format illustrated volumes are very scarce and expensive, but the smaller format unillustrated edition, although again somewhat scarce, is generally reasonably priced. Although somewhat out of date, "Fowler" is still a necessary reference and its continuing usefulness is a glowing testament to the thoroughness of its author.

Joy's "*A Practical Handbook of British Beetles*" (two volumes published in 1932, reprinted 1976) is the most widely used identification manual although, with the passage of time, it too is somewhat out of date. Its keys are easy to follow, but great use is made of comparative differences and often these are only appreciated if the student has a series of reliably identified specimens to hand; it contains little information otherwise.

The Royal Entomological Society of London's *"Handbooks for the Identification of British Insects"* – now published by the British Museum (Natural History) are very good, if somewhat technical and concise (some say "too concise"). The first volumes to deal with beetles appeared in 1953 and, to date (1990), only twenty handbooks devoted to the Coleoptera have appeared. Very often the student is forced to turn to foreign publications. Reitter's *"Fauna Germanica"* (5 volumes, 1908–1916) is nowadays perhaps too out-of-date to be of much use to the British coleopterist, but Freude, Harde & Lohse's *"Die Käfer Mitteleuropas"* (11 volumes, 1965–1983, with catalogue and first supplement – vol. 12 published in 1989 and a second supplement in preparation) is widely used in Britain. So too are the six volumes so far published in the Fauna Entomologica Scandinavica Series which cover Coleoptera.

In recent years, popular beetle families have been the subject of excellent, readable well illustrated and reasonably prices books. These include Richmond Publishing Company's *"Common Ground Beetles"* by T. Forsythe and *"Ladybirds"* by Majerus and Kearns. The Field Studies Council's AIDGAP series has *"A Key to the Adults of British Water Beetles"*.

Other books which should be in everyone's library include *"British Water Beetles"* (3 volumes) by F. Balfour-Browne published by the Ray Society, 1940–1958 and, as a set, very scarce. The Nature Conservancy Council's *"British Red Data Books, 2: Insects"* and their long expected, soon to be published, *"A Review of British Coleoptera"* will be useful references.

The *Entomologist's Monthly Magazine, Entomologist's Gazette* and the *Entomologist's Record and Journal of Variation* all regularly carry papers and shorter articles devoted to the Coleoptera. The Balfour-Browne Club's *"Bulletin"* and the *"Coleopterist's Newsletter"* are two useful publications with helpful and interesting articles, notices of meetings and other news.

Useful addresses

E.W. Classey Ltd., P.O. Box 93, Faringdon Oxon, SN7 7DR – carries a very wide stock of entomoligical books, British and foreign, back volumes of journals, and publishes the *Entomologist's Gazette.*

Richmond Publishing Company Ltd., P.O. Box 963, Slough, SL2 3RS.

Entomologist's Monthly Magazine, Brightwood, Brightwell-cum-Sotwell, Wallingford, Oxon., OX10 0QD.

Entomologist's Record & Journal of Variation – 4 Steep Close, Orpington, Kent BR6 6DS.

Balfour-Browne Club, Dr. G.N. Foster, 3 Eglinton Terrace, Ayr KA7 1JJ.
Coleopterist's Newsletter, J. Cooter, 19 Mount Crescent, Hereford HR1 1NQ.

General Bibliography for Nomenclature and Identification
Handbooks for the Identification of British Insects. Coleoptera – **Vol. 4** (Pts. 1, 2, 6a, 8a, 9, 10) **Vol. 5** (Pts. 1(b), 2(c), 5(a), 5(b), 10, 11, 16). Royal. ent. Soc. London.
Crowson, R.A. 1955. The Natural Classification of the Families of the Coleoptera. Reprint 1967. E.W. Classey Ltd.
Fowler, W.W. 1887–91. Coleoptera of the British Islands. 5 Vols. Reeve & Co. London.
Fowler, W.W. & Donisthorpe, H.St.J.K. 1913. Supplement to the above. (**Vol. 6**.)
Joy, N.H. 1932. A Practical Handbook of British Beetles. 2 Vols. Reprint 1976. E.W. Classey Ltd.
Linssen, E.F. 1959. Beetles of the British Isles. 2 Vols. F. Warne Ltd., London.
Pope, R.D., Kloet & Hincks. 1977. A Checklist of British Insects (3) – Coleoptera and Strepsiptera. Handbooks for the Identification of British Insects **Vol. 5**. Pt. 11.

THE SYSTEMATIC IMPORTANCE OF THE MALE GENITALIA IN COLEOPTERA

By R.A. Crowson

In beetles, as in other insects, it has long been established that in adult males the most striking differences between closely related species are often to be found in the aedeagus. Many systematists, indeed, have taken the male genitalia by themselves to provide a sufficient basis for the distinction and classification of species in their own particular groups. Quite apart from its practical disadvantages when dealing with female specimens, this approach is unsound in principle. If we define a species as an assemblage of potentially inter-breeding populations, then there are many species known in which the aedeagus shows a great deal of variation, in respects which would elsewhere be relied on to distinguish species, and others where "good" species show little or no detectable difference in the male genitalia. A further practical difficulty arises from the fact that the male genitalia are 3-dimensional and somewhat flexible structures, whose appearance may be considerably influenced by the processes of preparation and by the precise angle from which the finished preparation is viewed. When all these qualifications are allowed, it remains true that the examination of the aedeagus is a very useful, if not indispensable, procedure for the coleopterist seeking reliable species determinations in "different" genera. The external genitalia of the females as a rule show far less species-variability than does the aedeagus, but a wholly internal structure, the spermatheca, not infrequently differs notably between related species and has been used for their discrimination.

The structure of the aedeagus may also provide important evidence for classificatory relations at levels much higher than the specific, though this aspect of it is rarely likely to be of much concern to the average beetle collector. In the suborder Adephaga, the absence of a true basal piece to the tegmen seems to be a basic feature; in the vast group Cucujiformia, including the Cleroidea, Lymexyloidea, Cucujoidea, Chrysomeloidea and Curculionoidea, a special "ring" or "Cucujoid" form of the tegmen is a basic feature – though the basic pattern is often so much modified as to be hardly recognisable. The division of the penis (median lobe) into separate dorsal and ventral sclerites characterises the superfamily Dascilloidea, and the incompleteness of the "ring" of the tegmen on the ventral side (morphologically) is the most reliable single feature by which to distinguish those Heteromera with 4-4-4 tarsi from the Clavicornia of the Cerylonid group. The orientation of the aedeagus, both in the retracted and the extruded position, may be of systematic importance. An aedeagus which is normally oriented when extruded very often lies on one side when retracted, for example in many Cerambycidae, Erotylidae, Chrysomelidae etc.; in a number of families, the entire structure is inverted, both in the retracted and the extruded condition, as for example in many Cucujidae, many Heteromera such as Pyrochroidae, certain Staphylinidae etc. In beetles with an inverted aedeagus, the normal copulatory position is for the male and female to be facing in opposite directions and only in contact with each other at the end of the abdomen, whereas with a normally oriented aedaegus, the male climbs onto the back of the female.

Systematically useful characters of the aedeagus may be found in the tegmen (basal piece plus parameres), the penis (median lobe) or the internal sac. The last named structure is apt to be overlooked, unless the aedeagus is cleared and mounted in canada balsam or a similar medium, permitting its study by transmitted light. The internal sac is a more or less membranous tubular structure, usually invaginated into the penis and not easily everted from it; it often has more or less complex and characteristic structures on its walls, and may be extended at its apex (in the everted condition) into a long thin tube known as a flagellum, for example in Cerambycidae and Cucujidae.

Besides the aedeagus proper, the last 2 abdominal segments (nos. 8 and 9) normally partly or wholly withdrawn inside segment 7, show more or less marked modifications in the sexes which are liable to be of systematic importance, particularly in forms like Staphylinidae and Cantharidae where these segments are more or less exposed. The last visible sternite (ventrite) in most beetles is that of segment 7, and this too often shows distinct sexual differences; in some beetles, notably in the Scraptiid *Anaspis*, various species of *Stenus*, and some Heteromera, there may be species-characteristic sexual modifications of some of the more anterior abdominal sternites.

EXAMINATION OF COLEOPTEROUS GENITALIA
By J. Cooter

Equipment

Most of the instruments needed for extracting and dissecting the required parts of the genitalia should already be possessed by the coleopterist.

1. *Fine brush* similar to the type used for mounting, ideally a good quality artist's sable hair, with fine point; size 000 or 00 are suitable.

2. *Needles* – undoubtedly the best are made from fine gauge tungsten wire. However, their manufacture may be beyond the capabilities of the average coleopterist as it requires special equipment. The wire is cut to length and held in a metal clamp fixed to a rotating eccentric. To shar-

pen them they must be dipped into molten sodium nitrite (not a solution) with an electric potential passing through the wires and electrolyte. The degree of point can be regulated by time and depth of immersion. Too long immersion will result in flat cut ends. I was lucky enough to obtain a small stock of these needles and used them for over 20 years; they do not blunt or bend in use, but can split or be damaged accidentally. Re-sharpening is as described above. Tungsten wire is very expensive. Alas, without the necessary equipment, my small stock became "life expired" and I turned to using microlepidopterist's stainless steel headless pins of 10mm/12.5mm length. These can be held in forceps, attached with a strong glue such as araldite to a match-stick or similar, or better still held by a pin vice with screw-activated chuck.

3. *Forceps*. One or preferably two pairs of very fine pointed stainless steel antimagnetic watchmaker's forceps (No. 5A or 5) with a pair of less fine-pointed (No. 2) and pinning forceps close to hand. The points of dissecting forceps should be protected when not in use.

4. *Microscope* or hand lens.

5. *Lighting*. The microscope light is very useful and the filter can be left out to give an intense beam. A desk light is useful, but generates too much heat which, apart from making the worker uncomfortable, dries up the dissection.

6. *Dissecting dish*. Some form of shallow vessel is needed. An excavated glass block placed over a white paper background is good; being heavy it will stay put and has a flat base. A better alternative is the white porcelain palette used by biologists for spot-testing. This is heavy, white and opaque with shallow depressions. Too much water in the dissecting vessel will cause everything to float about and possibly move out of the field of view; too little water leads to drying out and dissected parts being accidentally flicked out of the dish. The spot-testing pallette has another great advantage in having usually twelve depressions. As these are set out in rows of three, top to bottom, four dissections can be undertaken in one session. The top row may have the beetle plus water, the next, the genitalia plus alcohol and the third, if needed, may be used for xylene or another reagent. Care must of course be taken not to mix up the dissections.

7. *Boiling vessel*. This should not be of too small a diameter or too deep, thus a test-tube is quite unsuitable. A stainless steel egg-cup or a chemist's porcelain crucible are ideal, the latter more so as they are supplied with lids. Both are heat and chemical resistant. If boiling, two or three pin-heads can be added to prevent "bumping" – the formation of large bubbles which can, and do sometimes, eject the contents of the vessel.

8. *Heat.* For boiling, a spirit lamp is suitable. Boiling though is not strictly necessary; warming and keeping warm for a time will suffice. Indeed cold water is quite adequate, but a much longer time is needed. I have experienced good results by soaking the dissected genitalia in hot water drawn from a kettle. This is then simply left overnight, the initial heat being sufficient to get the softening process started. Alternatively the gentle heat of a radiator or airing cupboard overnight are very useful alternatives (if using the airing cupboard do make sure the pot is covered and secure against accidental damage).

9. *Hydroxide.* Potassium hydroxide solution is more potent than sodium hydroxide solution of comparable concentration. It acts to break down the musculature enabling the component parts to be dissected from the surrounding tissue. It must be remembered too that prolonged immersion will eventually dissolve everything – chitin included. Thus masceration in hydroxide should never be left unattended. This latter property is useful when the internal structure of the median lobe of the aedeagus is to be studied. The chitin of the lobe can be rendered thinner and thus more transparent by immersion in hydroxide.

Any dissection exposed to hydroxide solution must be washed in dilute acetic acid and then in several changes of water in order to expell any hydroxide residue. IMPORTANT – both sodium and potassium hydroxide in solution and as solids (which are deliquescent) cause burns to skin and will ruin clothing. Great care in the use of such solutions must be exercised.

10. *Cork mat* or postcard-sized piece of plastazote should always be to hand.

Dissecting the Genitalia

The student should become familiar with those species and genera that need to be dissected, the degree of dissection involved, and the orientation of the parts when finally mounted. These last two points vary greatly; there is not, indeed cannot be a hard and fast rule.

It may be possible to determine the sex of a beetle from external characters; again these vary from family to family, but where present, generally a reference to the differences is given in the literature ("Fowler" being very good in this respect). As an example, I take the genus *Choleva* – the males have the segments of the anterior tarsi dilated, females more linear. The male aedeagus shows useful specific differences, and with the females tergites of the last segments of the abdomen are characteristic, females of the majority of the Catopinae have a poorly chitinised spermatheca. If the sex of the beetle is known, care can be taken and the required part only dissected.

The component parts of the genitalia are very delicate, often tubular, structures so extreme care must be taken throughout dissection and mounting.

With dry material it is necessary to soften the tissues before attempting to remove the genitalia. If this is not done, damage is almost certain to occur. The tissues may be softened quite easily in water, the abdomen or part of it can be removed and immersed. The length of time required for the softening to occur will vary according to the nature and size of the abdomen but can be speeded up if the water is heated or boiled. The addition of hydroxide will also help, but if used, the parts must be thoroughly washed in several changes of clean water in order to remove the caustic residue which will continue to attack the chitin.

Any freshly killed beetles or those removed from fluid preservative will not need softening; it is always a more straightforward task to dissect fresh material and this can be done prior to mounting or soon afterwards. I have often carded small Staphylinidae and dissected them the next day while the tissues are still soft, but the beetle is firmly held by the mounting gum. If the beetle is not mounted it should be held gently but firmly with forceps or a needle. With very small beetles such as Ptiliidae, the whole beetle can be placed under water and the last two segments of the abdomen carefully removed with a fine pin.

If the abdomen has been removed, the genitalia can be extracted by pulling out the tissues and genitalia through the opening with a pair of fine forceps. In many cases all that is required is to tear the pleurites or tergites open and pull out the genitalia surrounded by a mass of tissue. The more heavily scleritised beetles such as Histerids and most weevils generally permit removal of the genitalia via the raised pygidium – this being held open with a pin or forceps. However, it must be borne in mind that very often the apical segments themselves have characteristically shaped sclerites (especially useful in some small Staphylinidae). In such cases a needle inserted under the posterior edge of tergite of abdominal segment 7 can be moved posteriorly so that the anterior edge of abdominal tergite 8 is pulled clear. In such a way segments 8 and 9 will be removed from the abdomen. These are transferred to the dissecting dish, the genitalia removed with pins and forceps and the sclerites carefully separated and great care taken not to confuse tergites and sternites. The occasional beetle will die with aedeagus partly extruded and with careful use of pin and forceps the whole armature can be removed.

Having removed the genitalia and their associated tissues, the next stage can begin. This involves the removal of the unwanted musculature and fatty tissue. If the genitalia are to be used solely for confirming an identification, there may be no need to progress further, but if the genitalia is small or weakly chitinised, it will be necessary to clean off the unwanted tissues that will otherwise obscure the "hard parts". The genitalia and tissue are placed in water which may be heated or have hydroxide added. The dish is then placed under the microscope and the scleritised genitalia teased out from the tissue, some of which might have to be carefully pulled away with the forceps.

With many genera, for example *Stenus* and *Catops,* the relative lengths and attitudes of the parameres and median lobe are of great importance and so should not be separated from each other. With others, for example *Olibrus, Atheta* s.l., *Philonthus* and *Quedius*, the dissection should be taken further as the diagnostic points are at this stage not revealed. This will generally consist of separating the median lobe and paramere(s) and any other genital sclerite and great care must be taken not to damage the individual structures.

The dissected parts can be removed from the water with a fine brush or needle and a useful tip is to soak up the bulk of the water with blotting paper leaving a thin film in the dish. With this minimal water, the brush will retain its point and the risk of minute genital sclerites being drawn into the hairs by capillary action is avoided.

Mounting

In a few cases it will be necessary to treat the dissected aedeagus further in order to reveal its internal structure as is necessary with xantholine staphylinids and most *Leiodes* species as well as some Carabidae. Details of this are given on p. 85. Alternatively it is sometimes necessary to view the dissected parts in more than one orientation a fact which effectively rules out mounting and in such cases the van Doesburg method should be employed, see p. 59.

With the bulk of species, the genitalia can be lightly glued to the mounting card behind the beetle from which they were dissected along with any sclerites dissected during the process, see figure 10. Tottenham (1954) recommended gluing dissected tergites on the left and sternites on the right; it is sound practice to adopt a systematic standard method such as this as it gives a degree of uniformity and avoids unsightly labelling.

Mounting in a medium

A growing number of Coleopterists use the synthetic water soluble transparent mountant *dimethyl hydantoin formaldehyde* (D.M.H.F.). It is crystal clear and permits observation of minute detail whilst setting rock-hard and thus affording maximum protection. Being water soluble, it permits removal of the genitalia for reorientation or recarding should the need arise. Alas, it sometimes does not attach too well to the mounting card and can break away with the included dissection; this habit should be overcome if the card is scratched minutely with a pin prior to adding the D.M.H.F. It is a difficult chemical to obtain in small quantities and whilst its use is advantageous, it is not a basic necessity.

Mounting in a medium is only necessary in order to study the genitalia in detail using transmitted light. Before describing the processes involved, some mention of the final mount should be made.

It is good practice to keep all dissected parts, data and other labels *with* the beetle they relate to. The practice of preparing genitalia for transmitted light as a microscope slide is not to be recommended. Other authors describe how to make a slide by punching a circular hole in mounting card and gluing acetate on one side – the "slide" is then pinned under the mounting card before addition of data labels; to view, the "slide" is moved to one side. This method is acceptable, but in light of experience, perhaps unnecessarily complicated. The recommended final mount is described below after the sequence of preparation.

Clove oil preparation. With a few genera, or certain species within certain genera, the internal sac of the aedeagus displays diagnostic characters. It is not necessary to evert the sac, but to observe the fine detail the chitin of the median lobe has to be rendered more transparent. The dissected aedeagus is immersed in cold 10% potassium hydroxide solution for several hours during which time it must be examined under the microscope regularly in order to assess the degree of detail being rendered visible. When the internal detail is sufficiently clearly visible, the median lobe in removed and washed in acetic acid to neutralise the caustic solution, after this it is washed in two or more changes of water and then dehydrated in 70% alcohol and, ideally, then absolute alcohol. Next it is transferred to xylene and finally to clove oil for clearing. After about one hour or when the internal detail is clearly visible, it can be transferred back to xylene and then to Canada Balsam or a modern synthetic mountant such as Euparal, alternatively it can be transferred directly from clove oil to Euparal.

For mounting, the recommended method is to spread a small quantity of mountant on an acetate strip cut a little longer and a little narrower than the mounting card. This mountant should be as shallow as possible to cover the aedeagus but as flat as possible to avoid optical distortion. Orientation is checked as the mountant dries and finally the acetate strip is pinned under the beetle mounting card, followed by the data and determination labels, see figure 10. Strips of the acetate (or celluloid) can be cut before hand and kept as a stock with one's mounting cards and setting equipment.

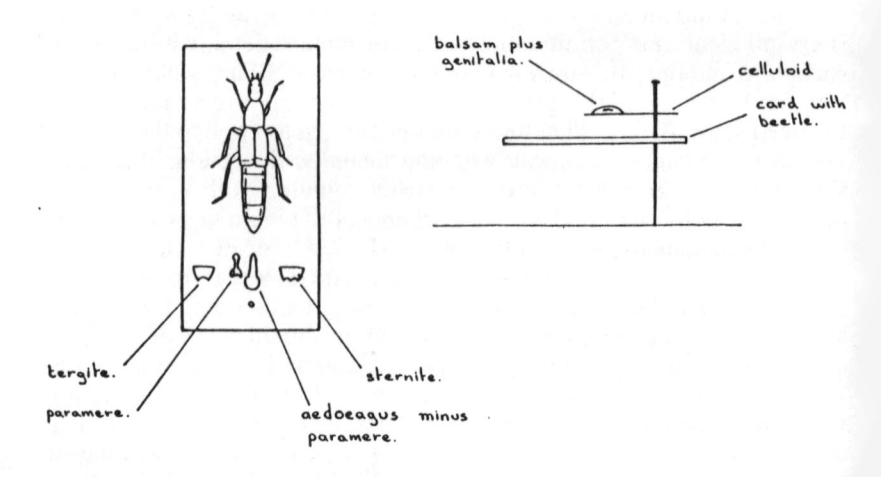

Fig. 10.

Presentation

The final mount should be kept on the same pin as the beetle, usually underneath it (Fig. 10). With small beetles, e.g. Ptiliinidae, the aedeagus or spermatheca can be placed in a patch of Balsam on a piece of celluloid sheet of suitable size, making sure that enough Balsam is added so that the armature is covered.

THE VAN DOESBURG METHOD

By Ashley Kirk-Spriggs

In the case of poorly sclerotised or complicated genital structures which require examination at different orientations, it is desirable to store genitalia in a wet state.

Several storage methods have been used in the past which ensure genitalia are always associated with the specimen from which it came, for example glass "Durham" tubes. These, however, prove expensive to buy, and the cork stoppers become impregnated with glycerol and eventually leak down the pin and stain the cabinet lining. The Dutch entomologist, P.H. van Doesburg, developed a method which is now used extensively by dipterists and hemipterists and increasingly by coleopterists in U.K. Institutions and abroad.

Procedure: Polythene tubing (available from Arnold-Horwell, see addresses of suppliers, below) of 3mm bore and 5mm outside diameter is heated gently over a spirit lamp until the polythene becomes soft and sticky, this should be done in a well-ventilated room and care taken not to heat the polythene too much or it will ignite. Having softened the tubing, place the heated end between the forceps a little way from their tips and squeeze the forceps together and hold for a few seconds to give the polythene chance to bond. With the tube now sealed at one end, cut off approximately 15mm.

After cooling, check the tube is properly sealed before going on to the next stage. Fill a pipette or syringe with glycerol; then insert its point in the tube as far as it will go and fill half the tube. Continue to hold the tube vertically to ensure the glycerol settles into the base of the tube. The polythene caps for these tubes are made from polythene rod (from Portex Ltd., see list of suppliers, below). A 4mm length of rod is cut with a scalpel or razor blade (scissors will burr the rod). The cut rod is then inserted into the open end of the tube to seal. After a dissection, the genitalia are placed in the tube with glycerol and the rod cap replaced. The tube is pinned through the flattened heat-bonded end under the dissected beetle's mounting card.

If genitalia are mounted on microscope slides or otherwise kept separate from their parent beetle, the two must be cross-referenced and both should carry full data. Slides can be stored in proprietory microscope-slide boxes or cabinets.

Materials

3mm diameter polythene rod from: (order code 800/100/605/100)	Portex Ltd., Hythe, Kent CT21 6JL
3mm bore/5mm outside diameter tubing: (order reference 800/010225)	Arnold-Horwell 73 Maygrave Road, West Hamstead, London, NW6 2BP
Transparent tubing is best suited for entomological use.	

As these items are available in bulk, it is hoped, as the van Doesburg method gains popularity with entomologists, a supply will be available to meet small orders. This will most likely be announced in the entomological press or the Bulletins issued by the various Recording Schemes.

References

Tottenham, C.E., 1954. Coleoptera Staphylinidae (a) Piestinae to Euaesthetinae. *Handbk. Ident. Br Insects*, 4(8(a)):5, Royal Ent. Soc. London.

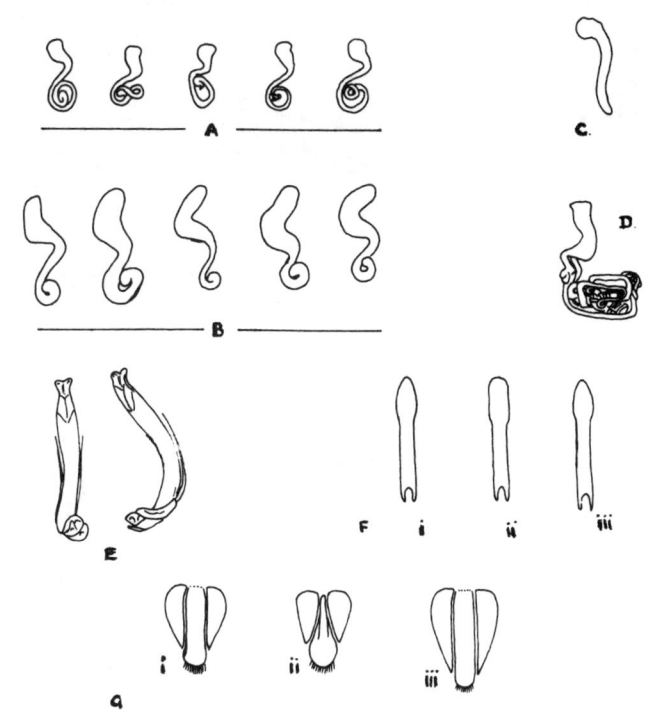

Fig. 11.

A. Spermathecae of *Atheta fungi* (Grav.).
B. Spermathecae of *Boreophilia islandica* (Kraatz)
   = *eremita* (Rye).
   Both species have variable, but presumably, constant form of spermatheca.
C. The spermatheca of *Liogluta nitidula* (Kraatz).
D. Spermatheca of *Atheta (Stethusa) incognita* (Sharp).
E. The aedeagus of *Catops fumatus* Spence. It shows a strong asymmetry, and is the only British species to do so.
F. Aedeagi of *Bagous tempestivus* (Herbst) (*i*) and (*ii*); *Bagous czwalinai* Seidlitz (= *heasleri* Newbery) (*iii*). Note the variation in the two forms of *tempestivus* and the similarity of (*i*) with that of (*iii*) *czwaliniai*.
G. With the absence of a sclerotised spermatheca, other sexual features can be used in many cases to separate the females. In the genus *Choleva* Latr. tergites of the apical abdominal segments are specifically distinct. (*i*) *C. jeanneli* Britten; (*ii*) *C. oblonga* Latr.; (*iii*) *C. fagniezi* Jeannel.

A, B, C, and D, after Strand, A., and Vik, A., 1964. Die Genitalorgane der Arten der Gattung *Atheta* Thoms. (Col., Staphylinidae) Norsk. Ent. Tidsskr., Bind XII (5–8).
E. After Kevan, D.K., 1945 The Aedeagi of the British Species of the Genus *Catops* Pk. (Col., Cholevidae). Ent. mon. Mag., **81**; 69–72.
F. After Dieckmann, L., 1964. Die mitteleuropaischen Arten aus der Gattung *Bagous* Germ. Ent. Bl. Biol. Syst. Kafer, **60**: 88–111.
G. After Kevan, D.K., 1946. The Sexual Characters of the British Species of the Genus *Choleva* Latr., including *C. cisteloides* Frohl. New to the British List (Col., Cholevidae). Ent. mon. Mag., **82**: 122–130.

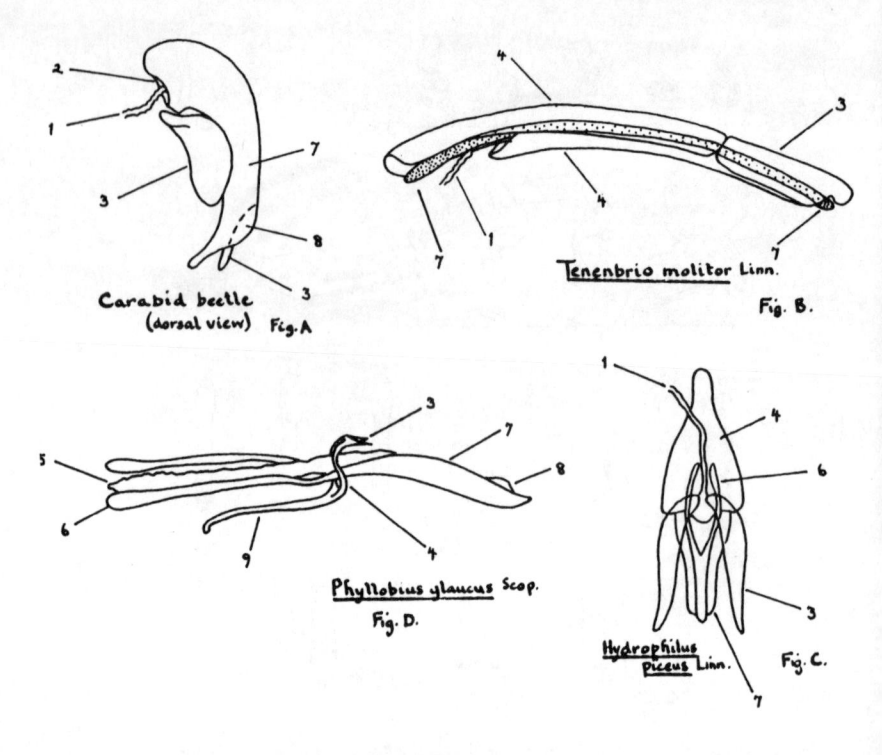

Fig. 12.

Explanation of figures A, B, C, and D.
(1) ductus ejaculatoris; (2) basal orifice; (3) paramere; (4) basal piece; (5) internal sac; (6) apophysis; (7) median lobe; (8) ostium; (9) manubrium.

AEDEAGAL TYPES

Figure A. That of a Carabid beetle, dorsal view. *The Articulate type*, Usually asymmetrical with basal piece often unsclerotised and then seemingly absent. Parameres articulating to the median lobe by a true condyle. Found in most adephaga and Staphylinoidea.

Figure B. Aedeagus of *Tenebrio molitor* L. lateral view *The Vaginate type*. Basal piece and parameres forming a pipe or a (dorsal or ventral) channel through which the median lobe moves. Parameres unable to move much. Sometimes the median lobe is so reduced that the tegmen assumes a large part of its copulatory function. Found in certain Heteromera etc.

Figure C. Aedeagus of *Hydrophilus piceus* L., dorsal view. Illustrating the *Trilobate type*, in which the structure is symmetrical, with basal piece well sclerotised, and the parameres articulate to the basal piece. The first connecting membrane allows very little independent movement of the median lobe and basal piece. Sharp and Muir refer to this type as *the Byrrhoid type* and is found in many families; regarded as the most primitive form within the Coleoptera.

Figure D. The aedeagus of *Phyllobius glaucus* Scop., lateral view, an example of the *Annulate type*. Basal piece forming a complete ring around the median lobe, as the two are only loosely connected, considerable movement of the median lobe is possible. Parameres are usually reduced, in some instances to a pair of processes firmly fixed to the basal piece. Phytophaga and some other families.

Transitional types also occur; aberrant types of aedeagus are found in the families Oedemeridae and Helodidae.

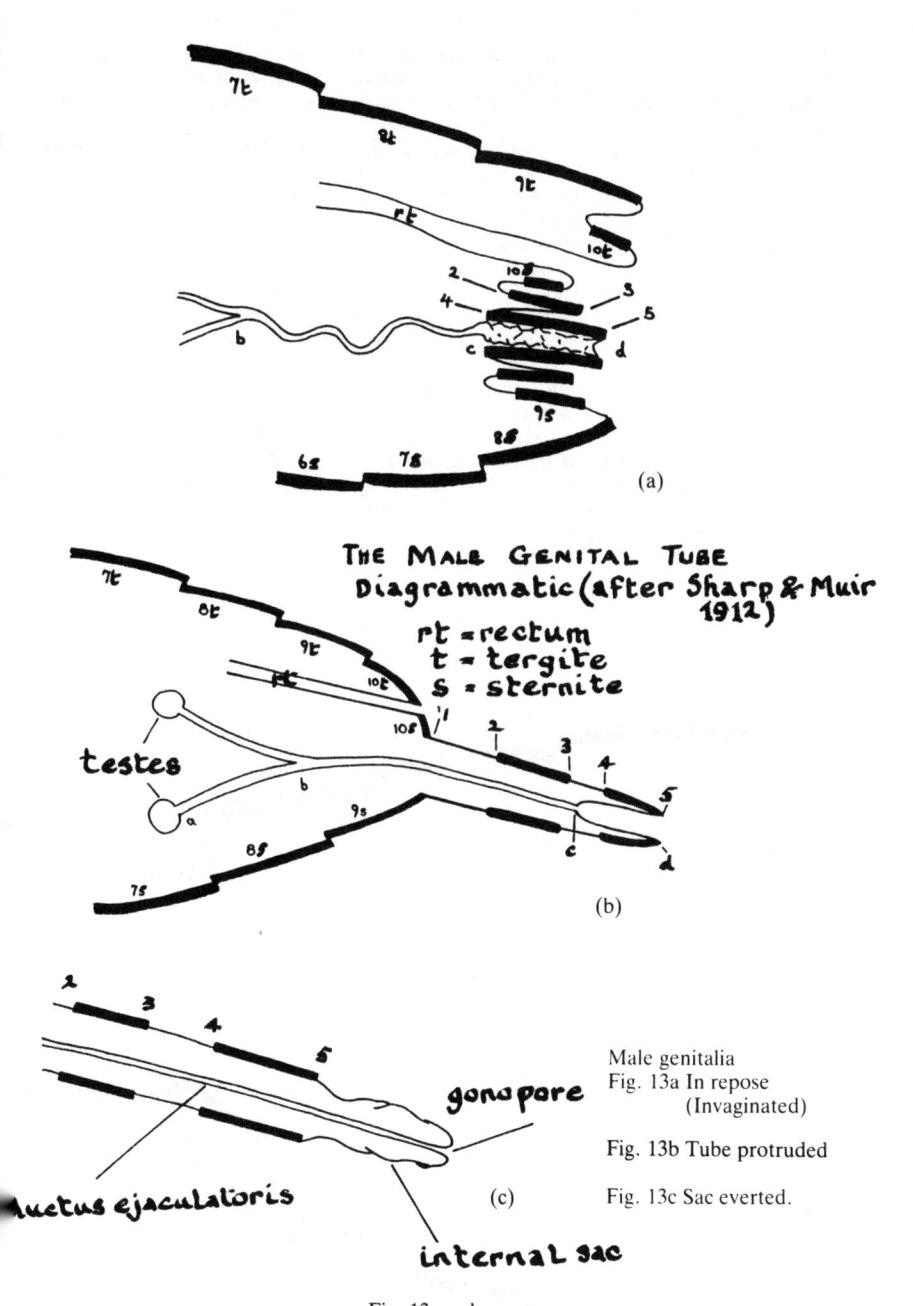

Fig. 13. a, b, c.

Figures 13 a, b, c. Outline of the Morphology of the Male Genital Tube.

The features described below relate to a general case, the accompanying illustrations are diagrammatic and do not represent a paraticular beetle. The Society is most grateful to the Royal Entomological Society of London for permission to use these diagrams, taken from Sharp and Muir (1912).

A pair of SEMINAL DUCTS lead from the TESTES and form what is known as the ZYGOTIC PORTION, (a–b). These join to give a long, single, highly irregular tube that is joined to the body wall – the AZYGOTIC PORTION, (b–d). The azygotic portion is divided into a long slender STENAZYGOTIC PORTION, (b–c) beyond which the tube enlarges to form the EURAZYGOTIC PORTION (c–d and 5–1). That part of the eurazygos that is not external (c–d) is the INTERNAL SAC. In all cases observed by Sharp and Muir the internal sac is evaginated during copulation, and forms a continuation of the external parts of the genital tube. In many cases there is no demarcation between the stenazygotic and eurazygotic portions, in such a case the internal sac is said to be undifferentiated. The portion of the tube that is external is termed "phallic".

The sclerites situated on the "phallic" portion fall into two groups.

i. Those on the distal portion, are known as the MEDIAN LOBE (5–4).

ii. Those nearest to the base are known as the TEGMEN, (3–2). The membranes between these two groups of sclerites are the FIRST CONNECTING MEMBRANE (4–3) and the SECOND CONNECTING MEMBRANE (2–1). The median lobe plus the tegmen forms the aedeagus. The MEDIAN ORIFICE (4–d), and the MEDIAN FORAMEN (4–a) corresponding point below).

FEMALE GENITALIA

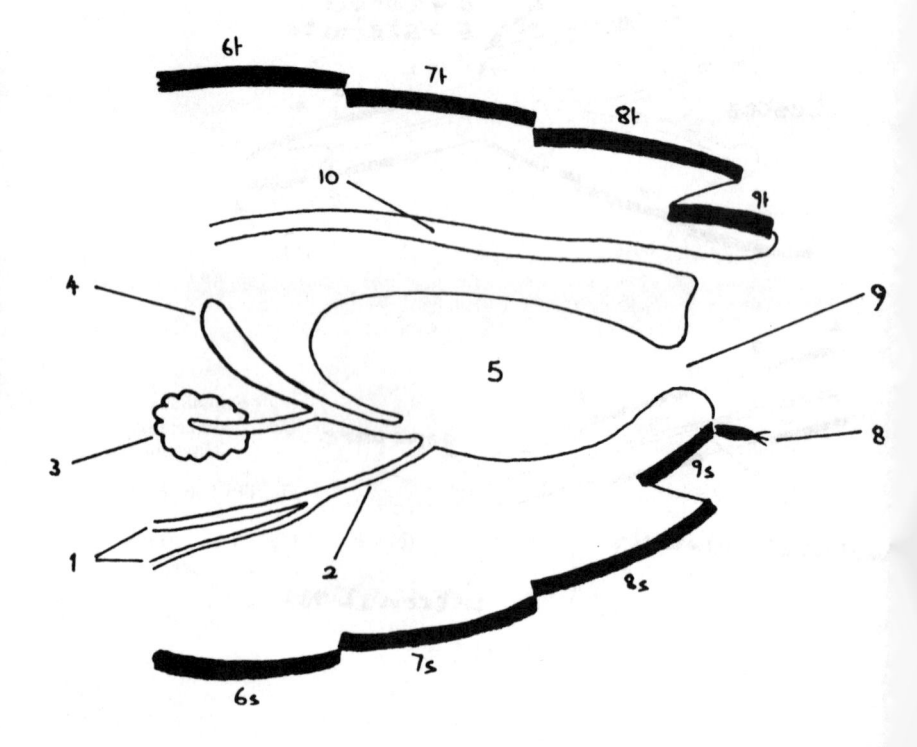

Fig. 14.

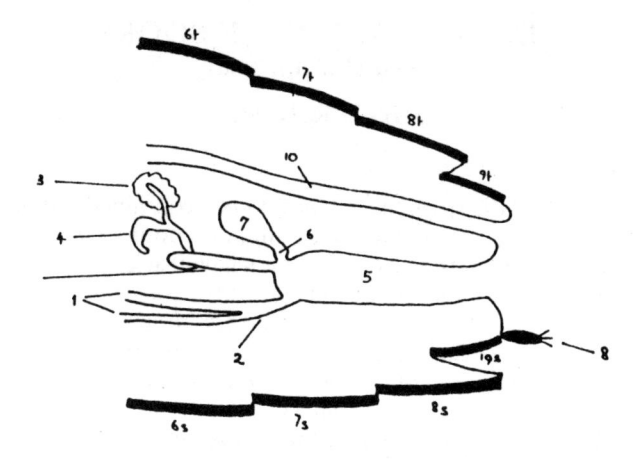

Fig. 15.

FEMALE GENITALIA.

Figure 9 *Saccular type*, without separate bursa copulatrix.
Figure 10 *Tubular type*, with a separate bursa copulatrix, generally combined with the development of an ovipositor. (Sketches; after Heberdey, 1931).

(1) lateral oviducts; (2) common oviduct (oviductus communis); (3) spermathecal gland; (4) spermatheca; (5) vagina; (6) ductus bursae; (7) bursa copulatrix; (8) genital style; (9) vulva; (10) rectum.

An aberrant type occurs in the genus *Cyphon* Pk., the female plays the active part during coitus and possesses special copulatory apparatus known as the *prehensor* which develops from the ventral wall of the vagina.

Different names have applied to the same parts of the genitalia by various authors. Wattanapongsiri (1966) gives two very useful tables of synonyms used by many authors. [Table 1, covering the male, and table 2, the female genitalia.]
  Tuxen (1970 – 2nd ed.) contains a glossary with 5,400 terms giving synonyms used by other authors.

Reference:
Tuxen, S. L. (Editor) Taxonomist's Glossary of Genitalia in Insects. Copenhagen 1956
  (Coleoptera section by C.H. Lindroth & E. Palmer). (2nd ed. 1970).

DIARIES AND NOTE BOOKS
— a Reappraisal
By D.R. Nash

The first edition (1954) of this "Handbook" contained extensive sections on keeping a journal and a determination book as well as suggesting the need for a field note-book in one's haversack.

The journal or "Coleopterist's Day Book" was considered (*op. cit.* p. 47) *"an adjunct of special importance for those prepared to make detailed studies of the Coleoptera of the areas they investigate, for within are recorded not only lists of species and numbers of examples collected on specific dates and at described places, but also those valuable pieces of extended data which comprise the observations made in the field, hastily jotted down in the field note-book"*

It was recommended that the journal should contain dates of determination against each species recorded or captured which, together with a number on the specimen label, would enable cross-reference of specimen and journal entry to a second fast-bound book – the determination book. This second book would not only duplicate all of the information in the journal (except habitat and field data relating to an individual specimen) but would supply exact dates upon which specimens were named, provide details of the current location of specimens e.g. determiner's/collector's collection, donation to National/Provincial Museum, donation to a private collector, as well as giving a reference to the transfer of the record to a card index, county/local list or a publication.

This summary should serve to indicate the importance placed upon these recording books by the authors of the first edition. As a tyro, the writer followed their advice and maintained an impressive-looking journal/determination book for some fifteen years. During that time I gradually came to question more and more the value of recording in this very time-consuming and cumbersome way. I also discovered that only one coleopterist known to me kept a journal and separate determination book as recommended in the "Handbook".

Reflecting upon the reasons for the great emphasis placed upon such books by the highly competent coleopterists who produced the "Handbook" over thirty years ago, one wonders if they deliberately recommended that an excess of written, cross-referenced information be associated with a collection to ensure that the collections of novices following their advice were more scientifically valuable than many of those in the past. These writers, moreover, were unaware of the great upsurge of national recording of precise data on all living organisms which was to occur in this country over the next few decades and which was to result in a more immediate sharing of basic information among naturalists. At the present time, much data (particularly that relating to less-notable species) which in the past would probably have remained buried in a collector's journal, is being sent almost as soon as it is gathered, to the Biological Records Centre local recording scheme organisers or used to compile site lists for the Nature Conservancy Council, county conservation trust, etc.

Although only brief mention is made of a card-index in the original "Handbook", the writer considers that this is a practical alternative to journals and determination books and that it is probably the most efficient recording system for those without a computer database and the requisite skills and knowledge to use one as a data base.

A separate card (at least 6" × 4") should be headed up for each species on the British List which the coleopterist intends to record. Whilst these cards can be arranged exactly as in the latest "Check List" to provide a system which facilitates the compilation of locality lists in systematic order, it is far quicker to extract or insert data if the families have their genera arranged alphabetically.

The front of the card can be used for recording all personal captures and field data relating to the species. As fine a pen as possible should be used, with the writing as small and clear as can be conveniently managed, so that as much information as possible can be compressed onto each card. It is suggested that a single line be used for each entry, as shown below, except where additional field or rearing data needs to be included:

DATE	FIELD DATA	LOCALITY AND GRID REFERENCE
12.v.1987	2ex off *Inonotus hispidus* on elm	Lawford, Essex TM098310

The reverse of each card can be used to list such things as all known county records, data from other sources (*e.g.* museum collections), information from other coleopterists, published references to the species, and so on.

When the front or back of a card is filled, additional cards can be headed with the species name and affixed with a rustless paper-clip.

Part 2 – The Beetle Families

INTRODUCTION

This section of the "Handbook" gives specific information about the members of the bulk of the beetle families represented in Britain. It is meant to supplement the earlier notes on general collecting and preparation methods. In several instances acknowledged authorities have very kindly contributed information about their particular speciality; the rest, I have tried to cover as best as I can. Alas, some families receive no mention here. Generally species belonging to these omitted families will turn up in plenty by adopting a range of the more general collecting methods (see Part I p. 12) and will need, more often than not, the standard preparation.

Our knowledge of beetles is increasing daily and the student should always glean what information is available by attending meetings, visiting specialist libraries and in general conversation with colleagues. New keys to species groups, genera and, less often, whole families are published from time to time in a wide variety of places. Foreign journals and entomological literature are often very useful and sometimes our only source of up-to-date information on a particular group of beetles.

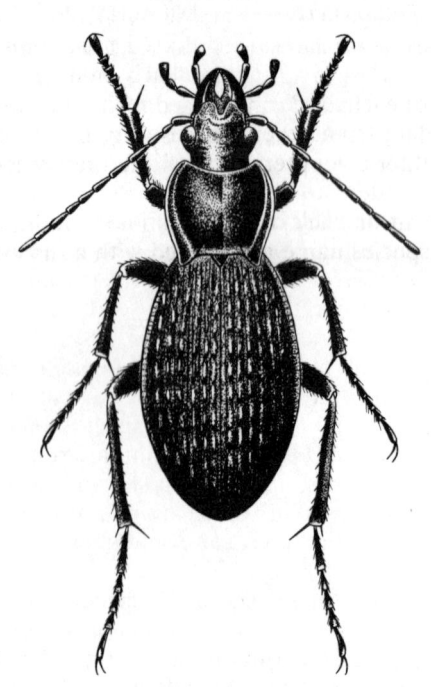

CARABIDAE
Carabus   problematicus
Herbst.

CARABIDAE

BY M.L. LUFF

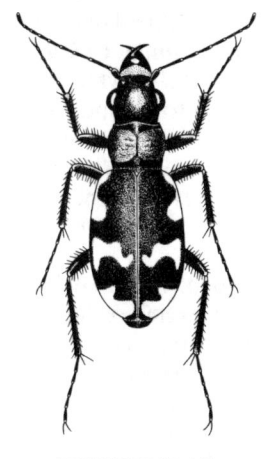

CICINDELIDAE
Cicindela hybrida.

The Carabidae or "Ground Beetles", with about 350 British species, are one of the most popular beetle families with collectors: they are readily recognisable as a family, and many species are common and easily identifiable while some, especially of *Carabus* itself, are large, conspicuously coloured and sculptured. As now recognised, the family includes the Cicindelinae or Tiger Beetles (5 British species), which are treated as a separate family in earlier works. There are two other small subfamilies, the Bombardier Beetles or Brachininae (1 established species) and the Omophroninae (1 species, doubtfully established): all the remaining ground beetles are in the large subfamily Carabinae.

The Cicindelinae are diurnal predators, active in bright, sunny conditions in dry, open situations where they run and (except for *Cicindela germanica*) fly readily. They are commonest in spring and early summer, spending the rest of the year as larvae in vertical burrows in the soil. The adults may be caught by hand or net, or even by pitfall trapping (q.v.). Our single Bombardier beetle occurs locally in chalky and other dry districts of southern Britain under stones, often in aggregated groups. Little is known of its biology, but the larvae are probably ectoparasitic on pupae of other beetles, including Hydrophilidae and Staphylinidae. Our only Omophronine, *Omophron limbatum*, is a semi-aquatic species which burrows in sand at the margins of flooded sand-pits, etc. in the extreme south of England.

The majority of the remaining ground beetles, the Carabinae, are ground living predators or scavengers, many being mainly nocturnal. Exceptions are some metallic coloured species such as *Notiophilus*, *Amara* and some *Bembidion's* which commonly run in bright sunshine. Nocturnal species are best sought in their shelters, such as under stones, logs, etc. Isolated stones tend to be more productive than heaps of boulders, but even small pebbles can sometimes cover a specimen hiding beneath.

By rivers or lakes many species, especially of the large genus *Bembidion*, occur in shingle, or at the roots of waterside vegetation on sandy, sloping margins. They run rapidly, so the collector must be ready with tube or pooter as the area is searched. In marshy habitats, sieving or searching among litter on the soil surface will yield species of *Agonum*, *Acupalpus* and *Stenolophus*, some of the smaller *Pterostichus*, further *Bembidion's* or even *Chlaenius* species, *Oodes helopioides* or *Badister unipustulatus*. Searching in gardens or rough grass will produce the commoner large species of *Carabus, Nebria, Pterostichus, Calathus* and some *Harpalus*. Drier habitats, especially if exposed to the sun, may yield the better *Amara* and *Harpalus* species as well as *Notiophilus, Badister bipustulatus* and possibly *Licinus* species. In woodland one may find *Abax parallelopipedus, Leistus* species, *Agonum assimile, Calathus piceus* or the characteristic snail-feeder, *Cychrus caraboides*, which can stridulate loudly if handled. Many species are confined to upland habitats and stone turning in moorland can be very rewarding, as well as searching at the roots of heather which will yield *Trichocellus, Bradycellus, Metabletus* and *Notiophilus* species. Some widespread upland species, such as *Nebria gyllenhalii*, also occur at lower altitudes, but in more restricted habitats, in this case on stony river banks.

Coastal habitats are particularly rich in Carabidae. Some such as *Dicheirotrichus, Pogonus* and *Dyschirius* species are exclusive inhabitants of salt marshes; the latter genus burrows in sandy or clayey banks where they are predatory, usually on the staphylinid genus *Bledius*. Other carabids, such as *Broscus cephalotes* and *Nebria complanata* are found under litter at the top of sandy beaches, while several *Bembidion's* occur in salt marsh litter. Sand dunes support *Calathus mollis* as well as local *Amara* and *Harpalus* species. The two small species of *Aepus* actually live between the tides, either under stones on sand, or in sand-filled crevices in rocks which are covered by the tides. They can be collected, together with the staphylinid *Micralymma marinum*, by splitting open such crevices with a hammer and cold chisel. In addition to these specialist coastal species, many carabids which are widely distributed in the south become partly or wholly coastal in their distribution as one moves north in Britain.

Although stone-turning or hand-searching will discover all these beetles in their resting places, they can be caught during their normal activity by pitfall trapping. The simplest traps suitable for carabids consist of plastic beakers or cups set into the soil with their rim flush (or slightly below) the soil surface. If visited frequently the traps can be left dry, but with some risk of the smaller specimens being eaten by the larger ones. A bait can be used, but is not essential. For longer periods of unattended trapping, the most readily available preservative is commercial antifreeze (preferably the blue variety, not the red/orange). About 1cm depth in the bottom of the trap will maintain specimens in identifiable condition for two weeks or more, even if the trap is subsequently filled with rain water.

Most ground beetles are annual, breeding either in spring and early summer (larvae in summer, overwintering as adults) or autumn (larval overwintering). A few large, mainly autumn-breeding, species of *Carabus, Pterostichus* and *Harpalus* live for more than one breeding season, so that both larvae and adults overwinter. Overwintering adult carabids may be found, sometimes in large numbers, in habitats such as grass tussocks, which can be dug up and shaken out over a sheet. As well as common *Pterostichus, Trechus, Notiophilus, Amara* and *Bradycellus* one may come across *Lebia chlorocephala*, whose larva is parasitic on Chrysomelidae. In marshy areas, large numbers of *Carabus, Pterostichus* and *Agonum* can be collected in winter sheltering above water level under bark of old logs and tree stumps as well as in moss on trees and walls. Flood refuse is also very productive at this time.

A few British carabids are found more commonly on plants than on the ground. These include *Amara aulica*, which can be swept from rough herbage where it feeds on seeds of Compositae, and the rare *Zabrus tenebroides* which feeds on cereals and is actually a crop pest in central Europe. Also on plants one finds *Demetrias* and the smaller *Dromius* species, which feed on other insects on the foliage. Other *Dromius* are arboreal, found under bark of trees in the day and on the foliage, when they can be collected by beating, at night. Also above ground, but mainly on walls, occurs *Bembidion quinquestriatum*, as well as rare or local species of *Tachys* and *Trechus* in crumbling sand and brickwork. In old cellars or outbuildings one may be lucky enough to find the very large and nowadays rare *Sphodrus leucophthalmus*, which is believed to prey on the cellar beetle *Blaps* (Tenebrionidae). Also associated with buildings are *Laemostenus* species, one of which (*L. terricola*) also occurs in and near small mammal burrows and runs.

Larvae of most Carabidae occur in the soil or other substrate where the adults are also found, although some species (some *Amara, Chlaenius*) migrate by flight between breeding and overwintering sites. The only larvae commonly found active on the surface are those of *Nebria, Notiophilus*, some *Carabus* and *Pterostichus* (particularly those of *P. madidus*). The larval period of spring-breeding species is quite short, but autumn breeders may spend almost a year as larvae. The pupal stage is short lived in all species. Most larvae are predatory, but those of some *Harpalus* are known to be seed feeders.

Following killing by ethyl acetate vapour, most adult Carabidae are robust enough to be relaxed by floating in water if this is needed. Although some collectors prefer to pin the larger species (especially *Carabus*), the slender antennae and legs of such specimens are easily broken: there is in fact no reason why all British Carabidae should not be card mounted. Underside characters (which are seldom needed) can usually be seen if the specimen is viewed from the side, due to the convex shape of the underside of most species. The body should be well glued down, especially if larger specimens are to be sent by post. Antennae and legs can also be well glued, but only on the side nearest to the card; it is difficult or impossible to see essential pubescence or setae on the segments if the appendages are completely covered in glue. The maxillary palpi should if possible be displayed when mounting the specimen. Most species can be sexed by the basal segments of the front tarsi, which are expanded in the male. Some *Pterostichus* have a tooth or ridge on the apical ventral segments of the abdomen, also in the male. The aedeagus is a useful confirmatory feature in many species, especially of *Bembidion* and *Harpalus*, subgenus *Ophonus*, but is seldom an essential feature for ground beetle identification. It can readily be extracted from freshly-killed and relaxed specimens by pressure on the abdomen to extend the pygidium, together with prising open the apical abdominal segments, which can be pushed back under the elytra when the specimen is mounted. The most useful orientation of the aedeagus varies according to the species, but it is best mounted dorso-laterally, glued onto the card in a drop of DMHF. The parameres are markedly asymmetrical and should be pulled away from the median lobe so as not to obscure it. The female spermatheca and external genitalia may sometimes show useful identification features, but these are seldom, if ever, used in the British species.

For identification, a magnification of up to ×50 is needed to see features such as the microsculpture on small species of *Bembidion, Acupalapus* and *Agonum*. Bright but diffuse light is often essential in order to see small pores or setae on the elytra or pronotum. The keys in the R.E.S. handbook (Lindroth, 1974, *Handbk. Ident. Brit. Ins.*, 4(1)) are the standard means of identification of adult carabids. The main key to genera in this work is, however, difficult to use by inexperienced collectors as it makes no use of subfamilies or tribes. There is also a serious error in the key between couplets 45 and 55 (see *Antenna*, 1977, 1(1): 25, for a corrected version). The key to genera in Joy gives a much better idea of the taxonomic "structure" of the family, although the ordering of tribes and position of some genera is not as generally accepted nowadays. The keys to species in Lindroth are generally reliable, although some further notes on particular species have appeared in the *Coleopterist's Newsletter* (May 1981, *Carabus violaceus/problematicus, Nebria brevicollis/salina*: November 1981, *Bembidion lampros/ properans, Notiophilus biguttatus/quadripunctatus*).

Comprehensive keys to carabid larvae do not exist, but there are keys to genera of the world (van Emden, 1942, *Trans. R. Ent. Soc. Lond.*, 92: 1–99) and Europe (Hurka, 1978 in Klausnitzer, *Ordnung Coleoptera, Larvern*, 51–69, Junk, The Hague). Keys to species of larvae of particular tribes are being produced by the author in *Entomologist* and *Entomologist's Gazette*: at the time of writing all tribes up to and including the Trechini have been published.

WATER BEETLES

By G.N. Foster

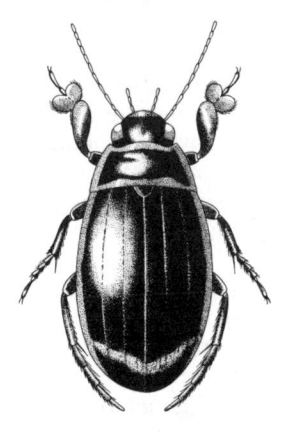

DYTISCIDAE
Dytiscus marginalis L.

About 300 species may be conveniently classed as "water beetles". These may be split into the aquatic Adephaga (or Hydradephaga), five families of truly aquatic, swimming beetles, and a roughly equal number of species from several families of the Polyphaga. With the Polyphaga the Hydrophilidae once included the Hydraenidae and these two families can still be collectively referred to as "palpicorns". A more natural combination occurs in the Dryopoidea, a superfamily embracing the Elmidae, Dryopidae and three very small families, each with only one genus in Britain.

Some species of ground beetle, such as *Oodes helopioides* and *Odacantha melanura*, and many other beetles, particularly among the Staphylinidae, are exclusively associated with water but this account concerns only the families dominated by aquatic members.

Water beetles have more than their fair share of identification problems. This partly owes to their adaptation to life under water, the streamlined body form not lending itself to accurate description. Another reason is that coleopterists regularly divide into those who are terrestrial and those who are not. Water beetling has become a specialist activity and those who dabble occasionally experience difficulty in familiarising themselves with the field characters of the commoner species.

In many genera identification can be made certain only by dissection of the male genitalia (see Part I) and these are signified in the following account.

Haliplid beetles are all small, between 3 and 5mm long, with a yellowish ground colour and large flat hind coxal plates. In America they are known as the crawling water beetles, a reference to their method of swimming with the legs operating alternately, but they have not attracted a common name in Britain. Most species are associated with slow-running water or with lakes though at least one species, *Haliplus heydeni*, is mainly found in dense vegetation in small water bodies. The larvae feed on algae and duckweeds. At least two species, *H. confinis* and *H. obliquus*, are dependent upon stoneworts found only in hard water. The adults are less restricted in their diet and are partly carnivorous, *H. lineolatus*, for example, feeding on Hydrozoa, the freshwater relatives of jellyfish. Haliplids are largely confined to lowland habitats and one species, *H. apicalis*, is rarely found away from brackish water. Members of the *Haliplus ruficollis* group (subgenus

Haliplinus) are best identified by examination of the male genitalia, and the male fore and mid tarsi.

The Hygrobiidae are represented by a single species, the squeak or screech beetle (*Hygrobia hermanni*), a distinctively convex, large (9mm), yellow and black insect. This species stridulates loudly when disturbed. Both the larva and adult are predatory, grubbing around in soft mud in pools and ditches.

The Noteridae are generally considered to constitute a valid family, with distinctive adults and larvae and a most unusual feature in that pupation takes place in cocoons attached to the roots of aquatic plants. All other beetles with aquatic larvae and adults must pupate above the water. The two species of *Noterus* are found in stagnant water, usually in association with dense vegetation including floating rafts of species such as bogbean (*Menyanthes trifoliata*).

The diving beetles (Dytiscidae) are the most attractive species of water beetle to the young coleopterist. Both adults and larvae are carnivorous, though it is certain that many adults are vultures rather than birds of prey. They divide into four tribes. The Laccophilinae, with three species of *Laccophilus*, are easily recognised in the field by their jumping ability. The larvae, and to some extent the adults, have a greenish coloration that disappears with death. The other three tribes, Hydroporinae, Colymbetinae and Dytiscinae, roughly conform to small, medium-sized and large diving beetles. They tend to be coloured black, yellow and reddish brown, with black stripes on a yellow background being a feature of species associated with sand or gravel. The small diving beetles include the notorious genus *Hydroporus*, with 28 British species found mainly in shallow bog and fen habitats. This genus poses the greatest identification difficulty to those without access to a wide range of material; dissection of the male genitalia is desirable and is best achieved by extrusion from the freshly killed insect. Many of the other members of the tribe are found over gravel, sand or mud in running water or in ponds. *Coelambus* species, with the exception of the northern lake species *novemlineatus*, are pioneers, being the first colonists of new ponds.

The medium-sized diving beetles first demand attention from would-be coleopterists. The commonest species in Britain is *Agabus bipustulatus*, remarkable for its ability to colonise all forms of aquatic habitat. If anything it most closely resembles in form *Ilybius* species, in which careful attention to size and the extent of metallic reflection is essential to avoid identification errors. Other bronze *Agabus* also pose problems, in particular in the species pair of *chalconatus* and *melanocornis*, which can only be distinguished with safety by examination of the tips of the parameres, an exercise not necessarily involving full dissection. Most species are found in the edges of pools or among vegetation at the edges of rivers. *A. guttatus* is a common species of small streams, *biguttatus* being associated with subterranean water including intermittent chalk streams. *A. brunneus* is a rare species of gravel beds in temporary streams in the New Forest and Cornwall.

The subterranean habitat, by virtue of its mystery, has a special appeal to coleopterists. Most species might better be described as "interstitial", moving down amongst gravel and sand as the water level falls. Only one species, *Hydroporus ferrugineus*, appears to be fully subterranean throughout its range whereas others, such as *H. obsoletus* and *A. brunneus*, occur away from springs south of Britain. Nevertheless all such species are best located in Britain by working the net in backwaters or in the outflows of springs after heavy rain; some occur in flood refuse.

The Dytiscinae, the larger diving beetles, are characterised by a strong difference between the sexes, the males having the fore tarsi modified into circular sucker pads and the females often having "sulcate" or "fluted" elytra, being provided with hairy grooves. The voracious larvae of this tribe are largely of the free-swimming type (except *Hydaticus*, whose larvae feed on the bottom like most Colymbetinae). They are the most obvious larvae to be found in the net and can easily be reared through to adults by supplying live food and a pupation site, such as a piece of bark pressed to the side of the aquarium.

There is no shame in experiencing difficulty in identifying the larger water beetles, errors being all too common with the great diving beetles (*Dytiscus*). The rarest species belong to the genus *Graphoderus* and it was only in 1974 that it was recognised that three species, not one, were represented in British collections. The male mid-tarsi provide important characters in this genus and should on no account be covered with glue.

Abroad it is common practice to operate underwater traps to capture the larger dytiscines. The basic principle is that of the minnow trap, with an inverted funnel or funnels leading into a larger chamber baited with rotting meat. Low water temperatures in Britain demand that these traps are operated for weeks rather than days to have any success, one reason for their lack of popularity. If provided with floats, such as blocks of polystyrene foam, they attract undesirable attention but this is the only way to keep specimens alive. This is not only necessary for conservation of unwanted invididuals but also because drowned specimens rapidly disintegrate under water.

The whirligig beetles (Gyrinidae) are of characteristic form and habit. They are carnivores, feeding on items falling on the water surface. The larvae are also predatory and are characterised by breathing through filamentous gills rather than by rising to the water's surface to renew their oxygen supply, a characteristic of the larger Dytiscidae. This feature presumably explains why whirligigs are largely confined to deep water bodies. The "common" species live in flotillas in full view and much time can be wasted in attempting to capture whirligigs staying just beyond the reach of the net. In fact the rarer species, if rarer they really are, live among dense emergent vegetation. The hairy whirligig, *Orectochilus villosus*, lives mainly in running water or in lakes; it gyrates on the water's surface at night and spends the day hiding at the water's edge, often clinging under rocks or bridge culverts. Most of the *Gyrinus* are found on still water. Individual species can be difficult to identify with several species-pairs – *aeratus* and *marinus*, *bicolor* (*payKulli*) and *caspius*, *substriatus* and *suffriani* – necessitating examination of the male genitalia. A species that appears to have become extinct in Britain, though it is still present in Ireland, is *natator* and it can be most easily be distinguished from the commonest British species, *substriatus* by examination of, most unusually, the *female* genitalia.

The Hydrophilidae, as recognised in Britain, include beetles of a wide range of forms. Most are able to trap a large bubble of air on their undersides and then perform the strange feat of being able to walk, upside down, immediately below the water's surface. They can also float the right way up and in this position they use their clubbed antennae to penetrate the water's surface to renew their air supply, in distinct contrast to the Adephaga, all of which use the bubble trapped under the elytra, the air supply of which is renewed by protruding the tip of the abdomen through to the atmosphere.

Some hydrophilids scattered through several genera have swimming hairs on the mid and hind tibiae and/or tarsi and these species can dive below the water's surface, e.g. the *Berosus* species, performing as agilely as the dytiscids. "Floaters" would be a useful collective name for the group were it not for such species and it would emphasise the ease with which most hydrophilids can be caught by disturbing the mud or vegetation at the extreme edge of the water and then skimming off the beetles.

Hydrophilids are most common in shallow, undisturbed water in pools and ditches, with *Helophorus* species characteristic of impermanent grassy pools and, at the other extreme, *Enochrus* characteristic of permanent water with mosses. All spin silken cases for their eggs, often with ribbon-like masts. *Helochares* species carry their egg cases on the underside of the body. Hydrophilid larvae are predatory but most adults appear to be either plant-feeders or scavengers.

One tribe, the Sphaeridiinae, cannot be classed water beetles, with all *Sphaeridium* and most *Cercyon* being associated with dung. It is, however, misleading to exclude this tribe from consideration as *Coelostoma orbiculare* and several of the more local *Cercyon* are exclusively aquatic. Some *Cercyon* are associated with specific habitats such as rotting seaweed (*C. littoralis*, *C. depressus*) or are perhaps nidicolous (*C. laminatus*). At the other extreme, a few species of *Helophorus* are associated with dry ground. A particularly interesting species is *H. tuberculatus*, associated with burnt heather, the fragments of which it resembles.

When identifying hydrophilids one must constantly be on guard for discrepancies between the published descriptions and the beetles under examination. Species additional to those in the last review (Balfour-Browne, F. 1960. *British Water Beetles* 3. Ray Society) have been or are in process of being added to nine of the British genera. When one realises that it is possible to misidentify the great silver water beetles, represented in Britain only by *Hydrophilus piceus*, then the importance of maintaining a critical stance is appreciated. Dissection of the male genitalia is highly desirable for *Hydrochus, Helophorus, Sphaeridium, Cercyon, Laccobius, Helochares* and the smaller *Enochrus*. It is also important to mount *Anacaena* species upside down so as to view the posterior femora.

The British arrangement of Hydrophilidae into six tribes is about to change. The Spercheinae, represented by *Spercheus emarginatus*, now possibly extinct in Britain, and Georissinae, represented by *Georissus crenulatus*, a local species of drying mud and silt by rivers, are to be upgraded within the Hydrophiloidea.

The British checklist of Hydraenidae has proved much more resilient with only one species, *Limnebius crinifer*, being added recently. Hydraenids are all small, between 1 and 3mm long, the largest being the male of *L. truncatellus*. They are frequently overlooked because of their size and sluggishness. *Limnebius* and *Ochthebius* are mainly associated with mud, most of the latter in brackish water. *O. auriculatus* is most often encountered in the drier parts of saltmarshes by sieving grass roots. *O. poweri* occurs on wet sandstone cliff-faces by the sea and *O. lejolisi* can be abundant in rockpools just above the high tide mark. *O. exsculptus* is unusual in that it lives in fast-running water, a habitat that it shares with several *Hydraena* species. Most of the more interesting *Hydraena* occur in the *edges* of streams, rather than, as is generally perceived, among stones in midstream. Thus "kick samples", such as are employed by biologists and inexperienced coleopterists, when the net is held downstream of the collector's shuffling boots, rarely catch anything other than *H. gracilis* and the elmids. It is better to search for clean backwaters and then to wash down the silt and wait for the hydraenids to float to the surface. It is often claimed that bags of debris and moss will yield good species if taken home and submerged completley so that the beetles are forced to the surface; heat extraction of the dried material is more effective and even then it is rare to find species that could not have been detected in the field.

Assuming that the hydraenid list remains unchanged it is feasible to identify all species without examination of the genitalia, but this is probably *why* the list has remained static! Apart from the realistic possibility of new species, the extraordinary structure of *Limnebius* and *Hydraena* male genitalia must be seen to be believed, if not understood.

Elmids have been well named "riffle beetles" as most are found amongst gravel and stones in the swifter parts of rivers and streams, or on the wave-washed shores of lakes. Two species that had been overlooked, *O. rivularis*, once thought extinct, and *O. major*, recently added to the British list, are mainly associated with slower stretches of rivers and with stone-lined fenland drains. *Stenelmis canaliculata* occurs in deep water in rivers and *Macronychus quadrituberculatus* appears to be most easily found on submerged logs. The two British species of *Riolus* are found in highly calcareous water and other elmids are commonest in harder waters.

Elmid adults are partly covered with a hydrofugic (water-repellent) pile that functions as a plastron, enabling the beetle to take oxygen from

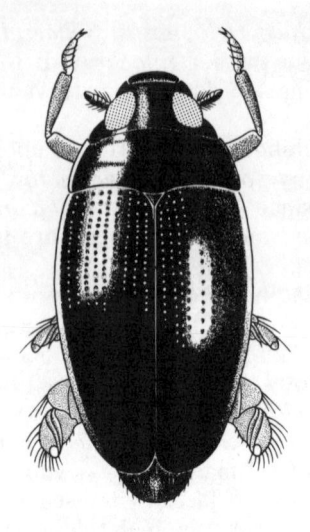

GYRINIDAE
Gyrinus substriatus Stephens
Whirligig

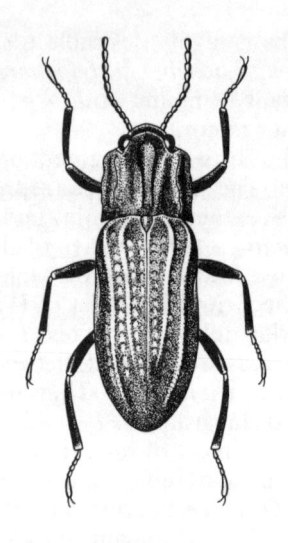

ELMIDAE
Stenelmis canaliculata
(Gyllenhal)

the water without the need to rise to the surface. They will nevertheless float to the surface if disturbed because they are buoyant. As with the Dryopidae, care should be taken when mounting elmids to ensure that they are thoroughly dry before using glue, otherwise it can spread over and disfigure the entire insect.

The wireworm-like larvae of *Dryops* live in wet soil or sodden wood but the adults are so hydrofugic that some coleopterists will have difficulty in accepting that they are water beetles at all. The commonest British species, *D. luridus*, occurs on wet mud by open water whereas the next commonest, *D. ernesti*, may often be found in muddy fields well away from truly aquatic habitats. However the rarer species are mainly associated with fenland. Only *Pomatinus* (*Helichus*) *substriatus* lives below water, usually in deep streams.

Heterocerus species are similar to *Dryops* in general appearance but with large mandibles, often with bright elytral markings and with a tendency to escape by flight. The larvae feed under algal mats on mud and most species are restricted to saltmarshes.

Limnichids, represented in Britain only by *Limnichus pygmaeus*, are not aquatic but, living on rocks in splash zones by running water, are most likely to be encountered by water beetlers. Our only psephenid, *Eubria palustris*, lives in the same habitat but is present so fleetingly as an adult that it is much more likely to be detected as its larva, known as a "water-penny", with the distinctive form of a flattened woodlouse.

The Scirtidae (also known as Helodidae) also have aquatic larvae and soft-bodied, ephemeral adults. The latter are mostly taken by sweeping vegetation, often well away from water. Their fragile structure and uncertainties about the British checklist ensure that they remain in comparative obscurity.

Several leaf beetles (Chrysomelidae) are associated with aquatic plants but only one tribe, the Donaciinae, is highly adapted to the aquatic life with larvae that breathe air from spaces in the roots of aquatic plants, pupae being attached in the same way. The highly attractive adults can be elusive and prove difficult to identify, particularly the commoner *Plateumaris*. Adults rest on floating and emergent leaves of aquatic plants and appear to feed mainly on pollen. Two species of *Macroplea* occur in Britain, living underwater both as adults and larvae, one species on milfoil (*Myriophyllum*) and *Potamogeton pectinatus* in lakes and canals, the other on eelgrass (*Zostera*) in coastal pools; despite this substantial difference in ecology it is advisable to dissect males.

Many weevils feed on aquatic plants and some have plastrons to facilitate respiration whilst crawling on plants below the surface. *Bagous* are perhaps the most elusive of all water beetles, being mainly associated with plants in very slow moving water. Extreme patience is required to detect them as they take so long to move when vegetation is dumped on a plastic sheet to dry. Several species are associated with milfoil in lakes. Many weevils are nocturnal but it is difficult to see how this could be turned to advantage in speeding up their capture. Plant associations are given in Part III.

HISTEROIDEA

By Michael Darby

The superfamily Histeroidea includes three families of which two occur in Britain: the SPHAERITIDAE, which is represented by a single species *Sphaerites glabratus* (Fabricius), found rarely in the north of England and Scotland in fungus, and the HISTERIDAE. The latter includes forty-nine species of small to medium-sized beetles easily distinguished by their heavily chitinised exoskeletons, their ability when disturbed to withdraw their heads and retract their legs as if feigning death and their geniculate antennae with solid club.

The Histeroidea are the subject of a "Handbook" by D.G.H. Halstead, (1963). Since its publication there has been the addition of *Plegaderus vulneratus* Panzer, which is found under bark of pine, mainly in the south.

The Histeridae are divided into six subfamilies as set out below but numerous changes at subfamily and generic level need to be made to the British List resulting from recent work by European authorities. Here Pope (1977) is followed.

TERETIINAE: Includes only one species *Teretrius fabricii* Mazur, which has not been taken in this country since 1907 and may be extinct. It should be looked for in association with *Lyctus* spp. on which it is a predator.

ABRAEINAE. This subfamily includes eight species in five genera, all of which are very small, the largest measuring only 1.5mm. Several occur in interesting habitats: *Halacritus punctum* Aube is found excluvively on sand and under seaweed just above high water level; *Aeletes atomarius* (Aube) in the burrows of the lesser stag beetle *Dorcus parallelopipedus* (L); and *Acritus homoeopathicus* Wollaston on burnt ground associated with the fungus *Pyronema confluens*. The commonest members of the subfamily are probably *Abraeus globosus* (Hoffmann) and *Acritus nigricornis* (Hoffmann) found in rotten wood and rotting vegetation respectively.

SAPRININAE. This subfamily also includes five genera. Two, *Baeckmanniolus* and *Myrmetes*, are represented by single species, the former found on the coast in dung, and the latter with the ant *Formica rufa* L. Most of the species of *Hypocaccus* and the closely related genus *Saprinus* are also found in dung and carrion, the former being confined to coastal sandhills.

DENDROPHILINAE. This subfamily includes only seven species divided between four genera. *Kissiter* and *Carcinops*, which are closely related but may be separated by the number of dorsal striae on the elytra, include only one species each, the former being found rarely at the roots of grass in sandy places. *Dendrophilus pygmaeus* (L) and *D. punctatus* are found in nests of the ant *Formica rufa* L., and species of *Paromalus* (formerly *Microlomalus*) under bark, *P. flavicornis* being fairly common in the south.

TRIBALINAE. Represented by the single genus *Onthophilus* which may be quickly separated from all other British histerids by the presence of longitudinal keels on the elytra. *O. striatus* (Forster) is the commoner of our two species being found in dung, rotting vegetation, etc.

HISTERINAE. The species of the five genera represented in this subfamily are of similar appearance and have at one time or another all been placed in *Hister* Linnaeus. Most are found in dung, carrion, rotting vegetation, etc. *Hister quadrinotatus* Scriba and *H. illigeri* Duftschmid are both thought to be extinct, and *H. quadrimaculatus* L., although not uncommon on the Continent, is now very rare in this country, being known from one or two localities in the south only. The single representative of *Grammostethus, marginatus* (Erichson) is the only British histerid which is confined exclusively to mole's nests.

HETAERIINAE. Represented by the single species *Hetaerius ferrugineus* (Olivier), a rare, small, reddish-coloured beetle found in the nests of the ants *Formica rufa* L. and *F. sanguinea* Lat. mainly in the south.

Reference
Halstead, D.G.H. 1963. Coleoptera, Histeroidea. *Handbk. Ident. Br. Insects*, **4**(10) 16 pp. Royal Ent. Soc. London.

PTILIIDAE

BY MICHAEL DARBY

A large family which includes the smallest known insects. In Britain it is represented by seventy-one species, none of which measures much more than 1mm. Adults and larvae feed on fungal hyphae and spores and are consequently found in a wide range of habitats including rotting vegetation, under bark, in rotten wood, dung, leaf litter, and fungi.

Because of their minute size special techniques of collecting Ptiliidae are needed. The author's methods, involving a kitchen sieve and micro-pooter, are described in detail in *Coleopterists Newsletter* No. 13, (1983). The use of traps, particularly grass piles and baited pitfall traps, has proved very effective in catching many species. Although not much used in Britain, flight intercept traps have been used in America with impressive results.

The identification of many species, particularly those belonging to the genus *Acrotrichis*, often requires study of the genitalia under high magnification. Slides are best prepared using a mountant such as Euparol on small pieces of perspex pinned with the specimen, as described on p. 57.

Because of lack of knowledge about the world fauna, work on Ptiliid phylogeny is still in its early stages. Consequently, no attempt is made either in Pope (1977), or in Freude Harde and Lohse (1971 & 1989) (which includes the most up-to-date keys to the identification of the British species) to assign genera to particular tribes. The notes which follow, therefore, are based on genera.

NOSSIDIUM. Represented in this country by one species *pilosellum* (Marsham) found in rotting elm stumps. This is the only British Ptiliid in which the male aedeagus possesses parameres.

PTENIDIUM. Includes eleven species which may be quickly recognised on external characters allied to knowledge of habitat. *P. pusillum* (Gyllenhal) and the much smaller *P. nitidum* (Heer) are the two commonest species, being found, often in large numbers, in grass piles, etc.

ACTIDIUM and OLIGELLA. These two genera include five species, three of which have not been taken in this country for many years. *O. foveolata* (Allibert) is found in compost heaps and old dungheaps.

PTILIUM and MICRIDIUM. *P. caesum* has been added to the list given in Pope (1977) although there are no recent records of this species in Britain. All the species of these genera may be quickly distinguished by the presence of three impressed lines on the thorax. Males of one of the commonest species, *P. horioni* Rosskothen, may be quickly separated from the similar but rarer *P. exaratum* (Allibert) by the presence of a fringe of long hairs between the hind coxae in the former and a long process, swollen apically, in the latter. The very rare *Micridium halidaii* (Matthews) is found only in old oaks in relict woodland. *P. myrmecophilum* (Allibert) is usually found, as its name suggests, in ants' nests.

EURYPTILIUM. The single species of this genus, *saxonicum* (Gillmeister), may be quickly separated from other British Ptiliids by the overall form and presence of elytral humeri. It is found more commonly in the north than the south, in grass heaps particularly.

PTILIOLA. This is the name now given to the genus listed in Pope (1977) as *Nanoptilium*. *P. kunzei* is the commoner of the two British species, being found in dryish dung, grass heaps, etc.

PTILIOLUM. Represented in this country by six species, the females of which may be readily separated on spermathecal characters. The rarest species are *P. caledonicum* (Sharp) and *P. sahlbergi* Flach, which are known from a few species only, and the commonest is *P. fuscum* (Erichson) which is found in a wide range of habitats.

MICROPTILIUM. *M. palustre* Kuntzen should be added to the list given by Pope (1977). Both it and *M. pulchellum* are very rare being known from one or two localities only.

PTINELLA and PTERYX. These genera now include *Ptiliolum subvariolosum* (Britten) which has recently been shown to be *Ptinella simsoni* (Matthews). Most of the species are yellowish in colour and live under bark and in rotten wood. Populations are frequently polymorphic, specimens occurring with or without eyes and wings. Some species, e.g. *P. errabunda* Johnson which is abundant everywhere, are parthenogenetic, and appear to be immigrants from Australasia.

BAEOCRARA. Immediately recognisable by the presence of the large punctures on the thorax. The single British species *variolosa* (Mulsant and Rey) is found in mouldy horse dung in woods.

SMICRUS and NEPHANES. These two closely related genera are each represented by one species. *S. filicornis* (Fairmaire and Laboulbene) is found uncommonly in sedge litter at the edge of ponds, and the much smaller *N. titan* (Newman) is common in old dung heaps.

ACROTRICHIS. This is the largest genus with twenty five species. They are immediately distinguishable from other Ptiliid genera on external characters, but many require study of the genitalia for individual determination. *A. sanctaehelena* Johnson, has recently been added to the list given in Pope (1977) and may be quickly distinguished from all other species except *A. grandicollis* (Mannerheim), by the presence of long protruding setae, particularly noticeable on the thorax and elytra. The student should refer to the keys and illustrations given in Freude, Harde and Lohse, for information about individual species until the publication of the "Handbook" which Colin Johnson is writing.

LEIODIDAE

BY J. COOTER

The higher classification of the Leiodidae and their taxonomic position has for a long time, and indeed still is, unsettled. Whilst a complete revision is needed, most authors today regard the Leiodidae, Catopidae and Colonidae as quite separate families. Purely for convenience, the scheme followed here is that adopted by Pope (1977) where each is afforded the rank of subfamily within the Leiodidae.

The uniform dull black or brown coloration, similar shape and small size, difficulties with identification and in obtaining sufficient material have all helped to make the Leiodidae unpopular with the amateur. They are, however, an interesting group with much for the student to discover about their biology and ecology. In recent years they have been the subject of much taxonomic research, as a result of which their identification is more straightforward, though in general they are regarded as a "tricky" family.

LEIODINAE: Conveniently divided into two tribes, the Leiodini and Agathidiini, in Britain comprise about 52 species divided unequally between eleven genera; many species are excessively scarce. These beetles are mostly mycophagous and the biology of most is quite unknown.

Leiodini: *Triarthron maerkeli* is reputed to live on subterranean fungi, but there are no reliable records. *Trichohydnobius* and *Hydnobius* species, on the other hand, are certainly associated with hypogeal fungi; *Hydnobius punctatus* has been found on an unidentified subterranean fungus. For a long time it has been known that some species of *Leiodes* are associated with truffles, *Leiodes cinnamomea* having been found, sometimes in numbers, adults and larvae together, in at least four species of truffle (*Tuber* spp.). *Leiodes ciliaris, furva, picea* and *dubia* have all been recorded from hypogeal fungi (not truffles), and it is more than likely that *Leiodes oblonga, rufipennis* and *polita* are similarly associated, but this needs confirmation. Fleischer used truffle as a bait, but attracted only *Leiodes cinnamomea* despite several other *Leiodes* species being on the wing in the vicinity at the time; he also noted as many as ten species of *Leiodes*, adults and larvae, on or near moulds growing on grass roots in a meadow.

Some *Leiodes* and *Cyrtusa* (sensu auct.) have been found on epigeal fungi, but none is known to breed there, though perhaps this might prove likely with *Cyrtusa* (sensu auct.). *Colenis* and *Agaricophagus* have been found on unidentified (probably *Tuber* sp.) truffles but are also associated with epigeal fungi. *Colenis immunda* has occasionally been noted on Agaricales and *Polyporus squamosus*. *Aglyptinus agathidioides* is an enigma. The Type material, two specimens, was found in a moorhen nest in Hertfordshire; no further examples have been found and the genus is otherwise confined to Central America; they are in all probability associated with fungi.

Leiodes furva and *Hydnobius punctatus* are typically found on coastal dunes; *Leiodes polita* is the most widespread and frequently encountered member of the genus. Some species are excessively scarce, for example, *Leiodes pallens* is know in Britain from three specimens captured together at Deal during 1873; *Leiodes silesiaca* is confined to Scotland and has been taken on two occasions in the Highlands. *Leiodes flavescens* is known in Britain from eight specimens only. *Leiodes picea* is another northern species, but its range extends as far south as Lancashire and there is an old (possibly unverified) record for Kent.

Agathidiini: All species of *Anisotoma* are associated with myxomycete sporophores (not plasmodia), this being so for all recorded instances of breeding, though adults have been found in Basidiomycetes. *Agathidium* species might be primarily associated with myxomycetes, but the evidence is less clear. Adults have been found in sporophores of myxomycetes and as often on sporophores of diverse Basidiomycetes especially Polyporaceae. *Amphicyllis* has been found several times on myxomycetes.

Agathidium arcticum is almost confined to Scotland where it can be found in Caledonian pine forest remnants under fungoid bark of fallen pine trees, especially those attacked by myxomycetes. *Agathidium seminulum, rotundatum* and *nigripenne* likewise occur under bark, but are more widespread and frequent deciduous trees as well. *Agathidium atrum, convexum* and *laevigatum* are often captured by pit-fall trapping or by sifting moss, tussocks and litter. *Anisotoma glabra* and *castanea* are northern species often found together, sometimes with *Agathidium* species under fungoid bark. *Anisotoma humeralis* is widespread and often found in numbers in and about myxomycete sporophores.

Occasional specimens of Leiodinae will be turned up by sifting grass-tussocks and flood refuse, or even simply by turning stones or examining sand-dunes, but better results will be had by evening sweeping. Pit-fall trapping is often very productive (see Cooter, 1989) but by far the best method of collecting is flight interception trapping (see Part I, p. 9). Agathidiini can be found, sometimes in numbers, under fungoid bark, especially that attacked by myxomycetes as well as in myxomycete sporophores.

The mature larva and pupa of *Agathidium varians* have been described by Angelini and DeMarzo (1984).

In the Leiodinae, the genitalia of the males exhibit reliable and often marked characters useful in the separation of species and it is highly advisable to extract the aedeagus from every male *Leiodes* specimen as a matter of course prior to mounting. This organ should be prepared as a clove-oil mount (see Part I, p. 57). Whereas all female Agathidiini have a sclerotised spermatheca of diagnostic value, in the Leiodini only in the genera *Liocyrtusa, Colenis* and *Agaricophagus* is one to be found. This must be dissected very carefully as it is bulb-like and often possesses a long flagellum; it is prepared for examination by transmitted light as a clove-oil mount. The aedeagi of male Agathidiini also exhibit specific characters, but it is not necessary to prepare these via clove-oil, they can be glued to the card next to the beetle or kept in glycerol in a van Doesburg tube attached to the insect pin; this enables dorsal, ventral and lateral aspects of the aedeagus to be examined (Part I p. 59).

Male Agathidiini have dilated anterior tarsi and/or 5–5–4 tarsal formula; females have rather linear anterior tarsi and 5–4–4 tarsal formula. Sexes of the Leiodini are often difficult to assess externally, especially with smaller examples of any particular species; if in doubt dissect. Typically males have broader anterior tarsi than the females and the posterior femora furnished dorsally and ventrally with a plate-like extension which may be rounded, angled or developed into one or more tooth-like projections. Generally the ventral plate is more strongly developed than the dorsal.

The antennae often exhibit important diagnostic characters and great care must be taken when setting these. An absolute minimum of gum should be added to the lower surface of the terminal segment only. All segments, especially those forming the club, must be set flat on the card and checked to ensure they are in the same plane. Being oval in cross-section, a slight twisting will give a marked distortion in relative widths. With the terminal and penultimate segments this is critically important. It ought to go without saying but glue must not reach the dorsal surface.

Some species of *Agathidium* have the ability to roll into a ball with head pressed against the venter. They can be "unrolled" with little difficulty if properly relaxed using fine-pointed forceps and a pin. The ventral surface of all Leiodinae must be set flat on the card, care being taken to ensure the head in particular is flat. A slight departure can obscure detail and distort other features.

CATOPINAE: These distinctive beetles inhabit various types of organic matter. 32 species occur in Britain.

Species of *Ptomophagus, Nargus* and *Sciodrepoides* occur in leaf-litter; *Catops* are associated with carrion and decaying fungi. *Nemadus colonoides* frequents the rotholes in trees that have been used by birds for nesting, both the nest material and associated debris should be searched for this diminutive species. Most species of *Choleva* inhabit mammal nests, especially those of the mole (see Part I, p. 16) and *Catopidius depressus* frequents rabbit burrows, from which it has been taken in large numbers by the use of baited pit-fall traps (see Welch, 1964).

Catopids can be taken occasionally during the course of general collecting, especially evening sweeping, sifting tussocks and litter. They occur in numbers in baited pit-fall traps (see Part I, p. 8). If a bait is used, the trap must be inspected frequently and a mesh screen or perforated lid placed securely over the trap will prevent admission of large and potentially destructive beetles such as *Necrophorus* spp. Material thus collected is often quite dirty and will need careful cleaning with water or ethyl acetate (to disperse fatty deposits). Once again, flight interception trapping is another very productive method of collecting.

Male Catopinae are easily recognised by their strongly dilated anterior tarsi. Their aedeagi are very distinct and provide reliable characters to assist identification. The females have rather linear anterior tarsi and in only a few genera is found a sclerotised spermatheca of diagnostic value; even these are not widely used as an identification aid. In *Choleva*, the males have characteristically-shaped posterior trochanters and it is often worth-while removing one posterior leg whilst mounting the beetle in order to display this feature. Female *Choleva* species possess characteristic well chitinised genital sclerites, which in repose are withdrawn into the abdomen and need to be dissected out and glued to the mounting card along-side the beetle or kept in a van Doesburg tube attached to the pin carrying the mounting card (see Part I, p. 59). Once again it is of prime importance to keep the antennal segments flat and in the same plane and to use an absolute minimum of glue on the terminal segment only. Catopinae are quite pubescent beetles and care must be taken to keep this clean, especially free from glue. Matting of the pubescence obscures the underlying surface ornament. For degreasing see Part I, p. 43.

COLONINAE: Represented in Britain by one genus, *Colon* Illiger with nine British species. They are very poorly known biologically and as yet their larvae have not been found. It is likely and widely assumed that they are associated with hypogeal fungi or moulds, but confirmation is needed.

Odd specimens can be found by general sifting of flood refuse, leaf-litter, moss at the base of trees and grass-tussocks; they occasionally turn up in pit-fall traps. Evening sweeping is more productive and good results have been obtained by the use of flight interception traps (see Part I, p. 9). *Colon* species have been found in rather barren looking areas, a point made by Fowler (1889 page 66) long ago and apparently overlooked subsequently. Joy (1933) describes a novel method of observing *Colon* species; the trick is to find the correct habitat favoured by the beetles!

Colon brunneum is without doubt the most frequently encountered species of the genus, with perhaps *serripes* next. Many collections made by competent coleopterists contain only these two, or even no *Colon* species, such is their scarcity.

In the sub-genus *Eurycolon*, the anterior tarsi of the males are dilated; males of many species have the posterior femora armed with a spine or tooth, but the size and shape of this does vary within species limits. Fortunately *Colon* species are easily sexed externally as the males exhibit five or six abdominal sternites, the females only four. The aedeagi are specifically characteristic and often afford the most reliable characters to facilitate determination. They should be prepared by the van Doesburg method (Part I, p. 59) or mounted in Euparal (after passing through alcohol and xylene only). Preserved by either method the setae and other features are not distorted by drying or obscured by glue. The antennae should be very carefully mounted with all segments in the same plane flat on the card and a minimum of glue on the terminal segment only (ideally no glue if the antennae can be set straight and flat). The relative widths of the terminal and penultimate segments and the shape of the club are important. Gum must be kept off the finely pubescent dorsal surface otherwise valuable diagnostic features will be concealed. When mounting *Colon* species it is good practice to pull the head forward clear of the pronotum exposing the "neck", thus the dividing line between the punctured area of the head and the unpunctured area of the "neck" is displayed; the head should be glued flat to the card.

References and Further Reading

Leiodinae

Allen, A.A., 1966. Annotated Corrections to the List of British species of *Leiodes* Latr. (Col., Leiodidae). *Entomologist's mon. Mag.*, **101** (1965): pp. 178–184.

Allen, A.A., 1968. *Leiodes clavicornis* Rye (Col., Leiodidae) new to England; with diagnostic notes. *Entomologist's mon. Mag.*, **103** (1967): pp. 262–263.

Angelini, F., & DeMarzo, L., 1984. Morfologia della larva matura e della pupa in *Agathidium varians* Beck (Col., Leiodidae, Anisotomini), *Entomologica* Bari **19**: pp. 51–60.

Cooter, J., 1978. The British species of *Agathidium* Panzer (Col., Leiodidae). *Entomologist's mon. Mag.*, **113** (1977): pp. 125–135.

Cooter, J., 1989. Some notes on the British *Leiodes* Latreille (Col., Leiodidae). *Entomologist's Gazette* **40**: pp. 329–335.

Daffner, H., Revision der paläarktischen Arten der Tribus Leiodini Leach (Col., Leiodidae). *Folia Ent. Hung.*, 44(2): pp. 000–000.

Fowler, W.W., 1889. *Coleoptera of the British Islands* **3**: 66 L. Reeve & Co., London.

Joy, N.H., 1911. A Revision of the British Species of *Leiodes* Latreille (*Anisotoma* Brit. Cat.). *Entomologist's mon. Mag.*, **47**: pp. 166–179.

Joy, N.H., 1933. *British Beetles their homes and habits*. pp. 71–72. F. Warne & Co.

Catopinae

Kevan, D.K., 1945. The aedeagi of the British species of the genus *Catops* Pk., (Col. Cholevidae) *Entomologist's mon. Mag.*, **81**: pp. 69–72.

Kevan, D.K., 1945. The aedeagi of the British species of the genera *Ptomophagus* Ill., *Nemadus* Th., *Nargus* Th., and *Bathyscia* Sch. (Col., Cholevidae) *ibid.*, **81**: pp. 121–125.

Kevan, D.K., 1946. The sexual characters of the British species of the genus *Choleva* Latr., including *C. cisteloides* (Frölich) new to the British List (Col., Cholevidae), *ibid.*, **82**: pp. 122–130.

Kevan, D.K., 1946. *Catops nigriclavis* Gerh. (Col., Cholevidae) new to the British List. *ibid.*, **82**: pp. 155–157.

Kevan, D.K., 1946. The aedeagus of *Catopidius depressus* (Murray) (Col., Cholevidae). *ibid.*, **82**: pp. 308–309.

Kevan, D.K., 1964. The spermatheca of the British species of *Ptomophagus* Ill., and *Parabathyscia wollastoni* (Jansen) (Col., Catopidae). *ibid.*, **99**: p. 216.

Welch, R.C., 1964. A simple method of collecting insects from rabbit burrows. *Entomologist's mon. Mag.*, 100: pp. 99–100.

Welch, R.C., 1964. *Catopidius depressus* (Murray) (Col., Anisotomidae) in large numbers in a rabbit warren in Berkshire. *Entomologist's mon. Mag.*, **100**: pp. 101–104.

Coloninae

Kevan, D.K., 1947. A revision of the British species of the genus *Colon* Hbst. (Col., Cholevidae). *Entomologist's mon. Mag.*, 83: pp. 249–267.

Szymczakowski, W., 1969. Die mitteleuropäischen Arten der Gattung *Colon* Herbst (Col., Colonidae). *Entomologische Abhandlungen* 36(8): pp. 303–339.

SILPHIDAE

BY J. COOTER

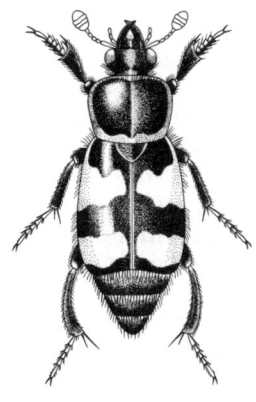

SILPHIDAE
Necrophorus vespillo (L.)

Members of this distinctive family are commonly known as the "true carrion beetles"; in Britain it is represented by just over 20 species.

The genus *Necrophorus* is familiar to the general naturalist as the Burying or Sexton Beetles. The largest species, *Necrophorus humator* is, except for its bright orange antennal club, wholly black; other members of the genus have distinctive orange bands on their elytra (*N. germanicus* is another wholly black species, including its antennal club, but is in all probability extinct in Britain).

Necrodes littoralis bears a superficial resemblance to *Necrophorus humator* but differs most markedly in having a much less abrupt antennal club, and in the male, thickened posterior femora. It is more frequently noted on the coast, but many inland instances have been recorded.

Thanatophilus species are associated with carrion, as is *Oeceoptoma thoracium* which is also to be found in the fruiting bodies of the Stinkhorn fungus. Both species of *Aclypea* are vegetarian and have been reported as pests of beet and turnip crops; *Dendroxena quadrimaculata* is a nocturnal predator of certain lepidopterous larvae in trees and bushes; during the day it can be found at the roots of grass and in litter at the base of trees. Some species of *Silpha* are predatory on snails, but frequent carrion as well. *Silpha atrata* can be found, sometimes in small numbers, under the bark of trees, particularly *Salix* species during the winter months.

Necrophorus species are large enough to be direct pinned (see Part I, p. 43), but most people opt to card them as this gives added protection from accidental damage. Identification poses little problem and with experience the majority will be recognised in the field or simply run down with the aid of a hand lens and key.

SCYDMAENIDAE

By Michael Darby

The Scydmaenidae is a moderate-sized family of small Coleoptera represented in this country by thirty species divided between four tribes. It has often been linked with the Pselaphidae, but although there is no dispute that both are correctly placed in the Staphylinoidea, the similarity of the adults is now thought to be the result of parallel adaptation, and the Scydmaenidae are considered to be more primitive than the Pselaphidae. Henry Denny's *Monographia Pselaphidarum et Scydmaenidarum Britanniae*, which was published in Norwich in 1825, still remains the only monographic treatment although the British fauna has been included in two major publications on the European species, by Croissandeau (1891–1900), and more recently by Franz and Besuchet in Freude, Harde and Lohse, (1971 & 1989). The latter provides the basis for the listing in Pope (1977), and up-dates the keys and other information given by Joy (1932), and by Allen (Notes on some British Scydmaenidae with corrections to the list in *Entomologist's Rec. J. Var.* **81**, 1969, pp. 239–246). Serious students will also find Caryl Brown and R.A. Crowson's "Observations on Scydmaenid larvae with a tentative key to the main British genera" in *Entomologist's mon. Mag.*, **115**, 1979, pp. 49–59 useful.

Brown and Crowson's research confirmed that the Scydmaenidae are probably placed correctly as near relatives of the Leiodidae, and suggested that two genera *Eutheia* and *Cephennium* deserve subfamily ranking. These groups are treated as tribes by Pope along with the Stechnini, with by far the largest number of species, and the Scydmaenini. More information about individual genera and species is set out below. Scydmaenidae may be quickly separated from Pselaphidae by the form of the elytra which completely cover the abdomen.

EUTHEINI. Includes the single genus *Eutheia* represented by five species. These may be quickly distinguished from all other British Scydmaenids by their flattish appearance and truncated elytra. All the species are rare, with the exception of *scymaenoides* Stephens, which is found in cut grass, manure, compost and other rotting vegetation.

CEPHENNINI. Some dozen or so species of the genus *Cephennium*, which is the only representative of this tribe, are found on the continent, but only one in Britain, *gallicum* Ganglbauer, one of our commonest Scydmaenids.

STENICHNINI. This tribe includes twenty five species in five genera most of which are rare. In Joy (1932) they are divided between *Neuraphes*, *Euconnus* and *Stenichnus*, but more recently *Microscydmus* has been shown to be the correct genus for the minute *nanus* (Schaum).

This genus is further represented by *minimus* (Chaudoir), which is also found locally in rotten wood; *helvolus* (Schaum) and the larger and lighter *sparshalli* (Denny), formerly assigned to *Neuraphes*, are now placed in *Scydmoraphes*. Both are recorded from a number of southern counties.

SCYDMAENINI. *Scydmaenus rufus* Müller and Kunze, and *S. tarsatus* Müller and Kunze are the only species in this tribe. Of the two, the larger *tarsatus* is the more common, being found in vegetable refuse. The two species may be separated by the presence of fovea at the base of the thorax in *tarsatus* which are absent in *rufus*.

SCAPHIDIIDAE

BY J. COOTER

A small family represented by only five species in Britain, one of which, *Scaphium immaculatum*, has not been taken for many years. The strikingly marked and unmistakable *Scaphidium quadrimaculatum* is not uncommon under bark. The species of *Scaphisoma* should be looked for in moss, rotten wood and vegetable litter, especially in rotten leaves under logs. The three British species resemble each other quite closely and should be set to show the antennae clearly, as the relative lengths of the antennal segments are important, as too are the characters displayed at the base of the elytra. The male genitalia are characteristic and should be dissected, ideally, before mounting. These are figured by Freude (1971) and by Lobl (1970) who, in addition, figures the antennae of each species. *Scaphisoma assimile* is in Britain a very scarce beetle, with only a single specimen being captured this century; *S. agaricinum* is the most frequently met with and widespread, *S. boleti* being more scarce and localised.

REFERENCES

Freude, H., 1971. Scaphidiidae in Freude, H., Harde, K.W., and Lohse, G.A., *Die Käfer Mitteleuropas*, **3**: pp. 343–347.

Löbl, I., 1970. *Klucze do Oznaczania Owadow Polski* (Keys for the Identification of Polish Insects) Coleoptera, zeszyt **23** Scaphidiidae: 16pp.

STAPHYLINIDAE

By S.A. Williams

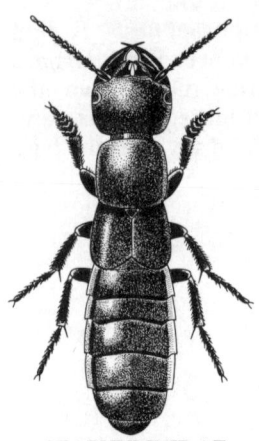

STAPHYLINIDAE
Staphylinus caesareus
Cederhjelm

Until quite recent times the Staphylinidae were often regarded by coleopterists in the same vein as the microlepidoptera so often are by the lepidopterists. The writer well remembers a visit to a distinguished coleopterist and was shown a fine collection of everything except Staphylinidae . . . these were in two mostly empty store-boxes. No doubt this was due in some part to a reluctance on behalf of many coleopterists to dissect out the genitalia which often afford excellent characters and, combined with the availability of many well illustrated revisional papers and texts, the majority of the Staphylinidae are relatively no more difficult than any other large family of Coleoptera. Staphylinidae, or "staphs" as they are informally known, are not on the whole particularly attractive to look at and the majority are quite small; this too has not helped their popularity but there are some fine exceptions, notably amongst the genus *Staphylinus* and its close relatives.

Many genera, in addition to having characteristic genitalia, possess terminal abdominal tergites of specific diagnostic value, for example *Gyrophaena* and *Placusa*. Often the female spermatheca is remarkably formed in certain *Atheta* spp.

As by far the majority of "staphs" are predaceous on other invertebrates, one does not require a good botanical knowledge to aid their study as is necessary with, for example, the weevils. A small number, as one might expect in such a large and diverse family, (in the region of 900 species in Britain), are excessively scarce and are known in Britain from one or two, or a small handful of old specimens. While "staphs" as a family are quite ubiquitous, certain genera and species have very precise and specialised habitat preferences and must be sought with some effort and persistence. Indeed most particular habitats, except open water and in general flowering plants, will have certain staphs associated with them. Some, for example *Hydrosmecta* spp., occur typically deep under river gravel; wet moss under water-falls is the haunt of *Stenus guynemeri, Dianous coerulescens* and *Quedius auricomus*. Sea-weed at the tide mark can be sifted for a wide variety of "staphs", including *Omalium laeviusculum, Anotylus marinus, Cafius* spp. *Halobrecta* spp. and *Aleochara algarum*, whilst crevices in rocks between the tide levels harbour *Micralymma marina*; drift-wood on damp sand near the high tide mark often hides *Phytosus* spp. The nests of many species of ant have their own particular "staph" fauna (see Part III, p. 210) as do mammal nests, especially those of the mole (*Quedius puncticollis, nirgocaeruleus* and *Pycnota paradoxa* for example). Bird nests in hollow trees are home for *Quedius brevicornis* and artificial bird-nest traps have produced a wide range of scarce species (*Quedius aetolicus, Philonthus addendus, Atheta taxiceroides, Aleochara sparsa* for example). *Velleius dilatatus* is a prize worthy of searching debris near hornet nests. Rotten wood, especially red-rotten oak (for *Quedius scitus* etc.), under bark (for various small omaliines, *Phloeopora* spp.) and sap runs on damaged trees (*Thamiaraea* spp. etc.) will all produced a variety of characteristically specialised "staphs".

Staphylinidae are mostly ground-living beetles, so the majority are collected by sieving rotting vegetation, in decaying carcases, moss on the ground or on old wood, under rubbish such as that thrown into a hedge. The sides of streams, rivers and ponds are excellent places for "staphs", some species (*Bledius, Manda* for example) burrow into the wet mud/sand near water. If the collector is able to get up mountains he may find several very scarce "staphs" that occur nowhere else. With the exception of water-netting, all forms of collecting will produce some "staphs". When sifting grass- and compost-heaps remember that different parts of the heap are more popular with certain species than others; this includes the very bottom of the heap where the refuse is often almost soil-like and is a likely place for *Oligota picipes*. The same is true for the piles of farm manure one sees dumped in fields prior to spreading, any pockets of hay in these piles are well worth investigating and are generally richer in species than the straw/manure material. Grass tussocks, especially isolated ones will produce a wide variety of "staphs" throughout the year, but are less productive at times of hard frost when the inhabitants will go deeper than the saw will reach (see Part I, p. 14). The "autocatcher" may produce "staphs" that are otherwise seldom encountered, including uncommon *Atheta* and *Trichophya pilicornis*. It can be regarded as a mobile flight interception trap; flight interception traps (see Part I, p. 9) are another very productive method.

Whilst it is a great experience to visit the classic collecting grounds: New Forest, Windsor Forest and the Scottish Highlands for example, much can be achieved locally in the garden, local woods and around edges of ponds. Often a good knowledge of these places will produce more and better material than a day at one of the well-known localities. The long lists of rarities and very local species from such places are the results of collecting over many years by numerous coleopterists. It is possible to build up a fair knowledge of the Staphylinidae of one's local area by putting down pit-fall traps (see Part I, p. 9) in a variety of places, baited and unbaited traps will produce different species.

Staphylinids are killed and prepared in the normal way (see Part I, p. 34). As usual, always use a minimum of glue which must not reach the upper surface of the body as this will make identification difficult. Care should be taken to set the beetle as naturally as possible to avoid stretching out the abdomen and antennae, easily done with "staphs". The genitalia should be removed wherever possible and mounted alongside the beetle, taking care to include the apex of the abdomen if this was removed during dissection, and to display any tergites that were separated during the process. Experience will tell if the median lobe and parameres of the aedeagus should be separated during dissection; there is no hard and fast rule. In some genera, such as *Xantholinus*, the male genitalia is sac-like and it is necessary to examine the internal structure. If kept dry the aedeagus is useless for diagnostic purposes and it is usual in such cases to make a clove-oil preparation or to keep in a small vial of glycerol attached to the mounting pin (see Part I, p. 59).

Great care must be taken with the smaller species as they can be very delicate (for example *Myllaena* spp.) and will easily fall to pieces if they are either roughly handled or kept in relaxing boxes too long. Ideally they should be mounted directly after killing. Another situation to avoid is the keeping of small species in the same collecting tube as, say, a large *Philonthus*; at the end of the day the tube will contain only a well fed *Philonthus*.

When identifying "staphs" it is a sensible idea to collect good samples of all the types encountered and then sort them out with the naked eye or lens before using a microscope. Then if we have perhaps a storebox full, pick out the larger ones which will probably be mostly *Philonthus* or *Quedius* and work them out using Joy (1932) and the figures in Freude, Harde & Lohse (1964, 1974 and 1989) if you are lucky enough to possess or have access to it. It is often easier to study and thus name several different specimens of the same genus than the odd one, so it is perhaps a good idea to wait until you get several of any particular genus; the characters employed in the identification keys are thus better appreciated. *Stenus* are an easy genus to recognise and can be named without too much difficulty with Tottenham (1954). It will take you some time to recognise the main groupings but once this is achieved everything becomes a lot easier. Within its limits (which are expressed by the author in his introduction) Joy (1932) is a most useful book and the basis for most British coleopterists.

REFERENCES

Freude, H., Harde, K.W., & Lohse, G.A., *Die Käfer Mitteleuropas* Goeke & Evers.
    1964, volume 4 Staphylinidae 1, Micropeplinae to Tachyporinae.
    1974, volume 5 Staphylinidae 2, Hypocyphtinae and Aleocharinae, Pselaphidae. 381pp.
    1989, volume 12 (1st supplement by Lohse, G.A. and Lucht, W.H.) pp. 121–240.
Tottenham, C.E., 1954 Handbooks for the Identification of British Insects **4 (8a).**

(Useful works for identification purposes:
Palm, T., Svensk Insektfauna. Family Staphylinidae. Issued in seven parts from 1948 to 1972. Well illustrated with numerous figures of male and female genitalia.
Szujecki, A., Keys for the Identification of Polish Insects (Klucze do Oznaczania Owadow Polski). 1961 vol. 24b Steninae; 1965 vol. 24c Euaesthetinae and Paederinae; 1976 vol. 24d Xantholininae; 1980 vol. 24e Staphylininae. The aedeagi of most species are illustrated, plus many other useful diagnostic figures.)

PSELAPHIDAE

BY MICHAEL DARBY

The Pselaphidae, like the Scydmaenidae, are small beetles (0.7–2mm) which feed on mites. Unlike the Scydmaenidae, the Pselaphidae have long been popular with entomologists. On the continent Reichenbach published an important monograph on the family as early as 1816, which was quickly followed by another by de Leach in the following year. Henry Denny's *Monographia Pselaphidarum et Scydmaenidarum Britanniae*, published in Norwich in 1825, has already been mentioned. The males and females of many Pselaphidae exhibit sexual dimorphism which caused early authors to attribute them to different species. In spite of this, six of the new species described by Denny still survive in the British list, which now numbers fifty one species and one subspecies.

Students of the British fauna will find the Reverend Pearce's R.E.S. *Handbook* (1957) an invaluable guide to species and their identification. Many useful tips are provided about collecting and habitats, but it is important that the *Handbook* be read in conjunction with the same author's "A revised annotated list of the British Pselaphidae" published in the *Entomologist's mon. Mag.*, **110**, 1974, pp. 13–26. The latter brings the *Handbook* up to date in the light of research by Dr. Claude Besuchet, the leading continental authority. Besuchet's chapter on the Pselaphidae in Freude, Harde and Lohse (1974 & 1989), which includes all the British species, is essential reading for the serious student not least because it includes drawings of many aedeagi not illustrated by Pearce, which are essential for the accurate determination of some species.

Pearce divided the Pselaphidae into two sub families, the Clavigerinae and the Pselaphinae and the latter into two tribes. Besuchet, however, favours a division of the family into seven tribes, the Euplectini, Batrisini, Bythinini, Tychini, Brachyglutini, Pselaphini and Clavigerini, and this is the classification used by Pope (1977).

The following more specific notes will aid in the capture and identification of specimens.

EUPLECTINI. This is the largest tribe with seven genera including twenty six species and one sub-species. Morphologically it is distinct, all representatives having the hind coxae adjacent or contiguous, and the hind body more linear and parallel sided (with the exception of the rare *Trichonyx sulcicollis* (Reichenbach)). Like *Trichonyx* two of the other genera are represented by single species and are rare. Since the publication of Pope (1977) a second species of *Plectophloeus, erichsoni* (Aube), has been added to our list. Representatives of the two largest genera *Euplectus*, with fourteen species, and *Bibloplectus*, with six, are found commonly in a wide range of habitats, but the different species are notoriously difficult to separate and dissection of the aedeagi is essential for most males. Some females, not associated with males, cannot be identified with certainty.

BATRISINI. Contains the single genus *Batrisodes* with three species normally found associated with ants, although *B. venustus* is sometimes found in other habitats, particularly under bark and in rotten wood of oaks.

BYTHININI. Includes three genera of which *Tychobythinus* is also myrmecophilous, being found most commonly with the ant *Ponera coarctata* (Latrielle) in the south and south east. The species of *Bythinus* and *Bryaxis* exhibit sexual dimorphism, the males having the first two segments of the antennae enlarged. This is particularly noticeable in *Bryaxis bulbifer* (Reichenbach), which is one of the commonest British Pselaphids.

TYCHINI. Includes two species in the genus *Tychus*, of which *T. niger* (Paykull) is the commoner. The males of this species may be quickly distinguished from *T. striola* Guillebeau by the presence of the enlarged fifth segment of the antennae. *Tychus* species generally may be distinguished by the second segment of the maxillary palpus which is almost as long as the third, apparently last, segment.

BRACHYGLUTINI. Includes four genera of which two, *Reichenbachia* and *Trissemus*, are represented by one species each. Both are found in association with water. This is also the habitat favoured by the species of the genera *Rybaxis* and *Brachygluta*, *B. helferi* (Scmidt-Goebel) and *B. simplex*, (Waterhouse) being confined to salt marshes where they may be found crawling on the mud amongst vegetation.

PSELAPHINI. Includes two species formerly ascribed to the genus *Pselaphus*, but one of which, *dresdensis* Herbst, is now placed in *Pselaphaulax*. Both are found in moss and tussocks, especially in bogs.

CLAVIGERINI. The two species in this tribe, both in the genus *Claviger*, may be quickly distinguished from all other British Pselaphidae by the absence of eyes and the truncated antennae. They are both myrmecophilous, being found in the nests of ants of the genus *Lasius*.

SCARABAEOIDEA

BY D.B. SHIRT

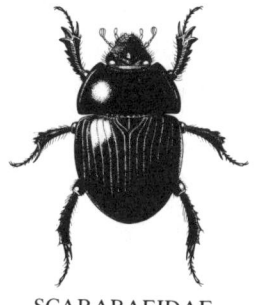

SCARABAEIDAE
Geotrupes vernalis (L.)

The scarabs constitute a clearly-defined natural grouping of families, which include the Lucanidae (stag beetles), Trogidae, Geotrupidae (dor beetles) and Scarabaeidae (dung beetles and chafers). The Lucanidae are sometimes split off into a separate superfamily, the Lucanoidea. The scarab's most obvious distinguishing feature is the plate-like (lamellate) form of the antennal club, which gives rise to the former name of "Lamellicornia". The large size and conspicuous ornamentation of many species, particularly in the tropics, has ensured man's interest over a very long period. The sacred scarab of the Ancient Egyptians, *Scarabaeus sacer*, was seen as representing the sun god Khepri, a word perhaps related to the Greek *kopros* ("dung"). The word "scarab" could be used for any beetle in the past; it still is in the Latin countries (e.g. French *scarabée*), as is "chafer" in the Germanic countries (e.g. German *käfer*). They have often been popular with amateur coleopterists in the past, and a recovery in interest is perhaps overdue.

The scarabs are a largely tropical group, with numbers increasing rapidly as one travels south through Europe. The full British list amounts to some 100 species, but it must be stressed that a number of these have not occurred in Britain for 60 years or more, and some of these may never have been true natives. Several more are now very rare, if not already extinct.

The three families Lucanidae, Trogidae and Geotrupidae together contain only fifteen species. The remainder are all included in the Scarabaeidae, though its major divisions are often themselves elevated to family status on the Continent. The Aphodiinae and Scarabaeinae are mainly dung beetles and number about 68 species, two-thirds of which belong to one of the largest of beetle genera, *Aphodius*. Chafers form the remaining four subfamilies, numbering about seventeen species. The pest status of some of these has ensured the financing of much research on their biology and control in recent years.

Most adult scarabs are heavily-built, their flattened anterior tibiae and powerful spiny posterior tibiae being adaptations for burrowing. They tend to walk slowly and clumsily, but most are capable of strong flight. Sexual dimorphism is often strongly marked, with the males having greatly enlarged mandibles (stag beetles), or horn-like projections on the head and pronotum (e.g. some Geotrupidae and Scarabaeinae, and many tropical groups). The whitish, fleshy larvae have a characteristically C-shaped or "scarabaeiform" appearance. The larval life may be unusually long in some species – $3\frac{1}{2}$ years or more in the stag beetle – in contrast to the few weeks of active life in the adult. However, the adults of more species remain inert in the pupal cell from late summer until the following spring or summer before emerging. The habitats in which the larvae and adults live are diverse: the following accounts are arranged according to the larval food source.

1. Dead Wood

The Lucanidae and most Cetoniinae breed in decaying trees, and excavation in suitable wood will often reveal both adults and larvae of the commoner species such as *Dorcus parallelipipedus* (the lesser stag) and *Sinodendron cylindricum*, and sometimes *Cetonia aurata* (rose chafer). Others are more restricted, such as *Lucanus cervus* (stag beetle) in the south-east, *Trichius fasciatus* (bee beetle) in birch in Wales and Scotland, and the very local *Gnorimus nobilis* in old fruit trees in the fruit-growing counties. *G. variabilis* is confined in Britain to ancient oaks in Windsor Forest and Great Park. The small aphodiine, *Saprosites mendax* is an Australian species established in Arundel Park and the London area, and has been found in the borings of *Dorcus* and *Sinodendron*. If larvae of any of these species are found they may be brought home with a quantity of the substrate and reared through to adult with a minimum of attention (and a maximum of patience!). Adult lucanids may be taken in flight at dusk in June and July, whereas most of the cetoniine chafers (which are very active fliers) may be taken on flowers in sunshine.

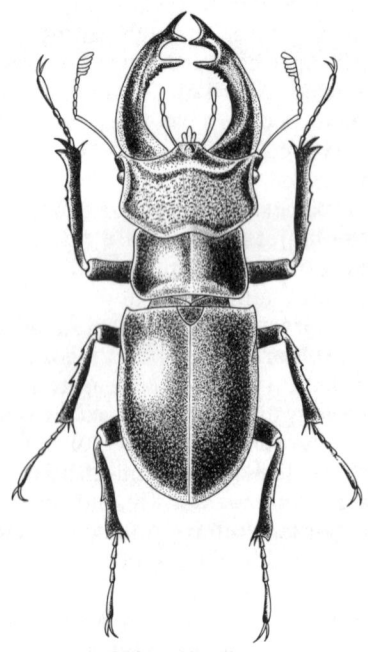

LUCANIDAE
Lucanus cervus (L.)
Stag beetle

2. Decaying Vegetation

Most of the Aegialiini and Psammobiini feed on decaying plant debris in sandy soils, a habit sometimes regarded as ancestral to feeding on dung or on live roots. The members of these groups usually occur among coastal sand dunes, an exception being *Aegialia sabuleti* which is found under stones on damp sandy soils by river banks. The rare *Diastictus vulneratus*, found only on the sandy heaths of the Breckland, may also belong to this category: its suggested association with rabbit dung is unconfirmed. *Aphodius niger* feeds on decaying matter in damp soil, and in Britain is apparently confined to the muddy edges of a single New Forest pond used by horses and cattle. *Oxyomus sylvestris* and *A. granarius* have catholic tastes and are often found among decaying vegetable matter, as are occasional individuals of other *Aphodius* species more usually associated with dung. The chafer *Cetonia cuprea* also belongs in this section. Its larvae should be sought in the nest mounds of wood ants, particularly those of *Formica aquilonia* and *F. lugubris* in the old Caledonian pine forest of Scotland, and are quite easy to rear. Its relative *C. aurata* has sometimes been found in similar situations in the south.

3. Plant roots

These constitute the larval food of chafers of the subfamilies Hopliinae, Melolonthinae and Rutelinae, most of which appear to be less common than formerly. In some years, however, *Melolontha melolontha* (cockchafer or maybug), *Amphimallon solstitialis* (summer chafer) and *Phyllopertha horticola* (field chafer) can be locally abundant, and may become serious pests of turf or cereal crops. The adults feed on leaves. Two species are usually found on the coast: *Anomala dubia* among sand dunes, and *Amphimallon ochraceus* on grassy clifftops at a few sites on the western and southern coasts. Many of the chafers may be found flying about actively by day, whereas the dusk fliers (*Serica brunnea, M. melolontha* and *A. solstitialis*) may be attracted to light in large numbers.

4. Fungi

One small species of the Geotrupidae, *Odontaeus armiger* is said to breed in underground fungi such as truffles; apparently the most primitive habit seen in the scarabs. It has perhaps always been rare, with most records coming from the chalky and sandy areas of the south and southeast. It flies readily to light at dusk and many records are from this source. *Aphodius plagiatus* does not feed on dung, unlike its relatives, but on small fungi growing in damp hollows among sand dunes, etc.

5. Animal Remains

Members of the Trogidae specialise in decaying animal material, and are usually found among the dry skins of old carcases and the nest debris and pellets of hole-nesting birds such as owls. *Trox sabulosus* is a very local day-flier confined to sandy areas, whereas the commoner *T. scaber* flies at night and may be taken at light. Some of the smaller dung beetles occasionally appear in carcases.

6. Dung

The remaining geotrupids (*Typhaeus* and *Geotrupes*), together with the Scarabaeinae (*Copris* and *Onthophagus*), excavate burrows beneath the source of dung. The burrows are 1–1.5m deep in *Typhaeus* (the minotaur), 40–60cm in *Geotrupes*, 10–20cm in *Copris*, and 5–20cm in *Onthophagus*. Pairs cooperate in excavating the burrows, which have one or more brood chambers, and in furnishing them with supplies of dung for the larvae. Parental care is taken still further in the very rare *Copris lunaris* (horned dung beetle), whose female remains in the brood chamber until the young adults emerge. The larvae of such burrowers may be reared, with a greater chance of success where a brood chamber has been removed intact. Sometimes the burrows of *Geotrupes* are found to contain smaller scarabs which appear to behave as "cuckoo parasites" (kleptoparasites); others may simply use the burrows as overwintering sites. *Geotrupes* species fly at dusk and may come to light, which has also been known to attract *Copris*.

Most of the Aphodiini, in contrast, remain in the surface dung throughout their larval life and pupate nearby. They hibernate as egg, third instar larva, prepupa or adult according to species. The numerous species lend themselves to rewarding studies in succession, competition, habitat and seasonal distribution. There has been some success in rearing their larvae, but surface dung is very liable to become mouldy (see the second edition of this Handbook, p. 70). Only three or four species fly at night and are attracted to light, notably *Aphodius rufipes* and *A. rufus*. Aphodiines can sometimes be found in large numbers among the debris left by winter and spring flooding.

Dung beetles are usually obtained by probing with the aid of a stick or spatula, though I find that a deft flick with a booted foot can be a quick way of checking whether anything is present! Specimens are often very dirty; filling your collecting tube with a twist of dead grass will give them a chance to clean themselves off. They may still require a final wash after killing and before mounting. If large quantities of dung are to be examined it may be collected in buckets or plastic bags and later dispersed in water or washed through sieves. Pitfall traps baited with dung have been used successfully and a number of complicated designs have been developed by researchers.

In the past much has been made of the mammalian source of the dung preferred by different scarabs, but the picture is not really that simple:

(1) Habitat preference. A species may utilise *any* dung within a restricted habitat, e.g. *Aphodius zenkeri* (deciduous woodland and parkland) and *A. nemoralis* (shady coniferous woodland).
(2) Availability of dung. Carnivore dung is often preferred, but is much less freely available than herbivore dung.
(3) The unit size of the dung. The large *A. fossor*, for instance, is usually confined to cow dung.

(4) The stage in the ageing process of the dung.
(5) Competition with other scarabs, etc.

It is evident from the above that one or two general habitats are particularly worth searching for scarabs. Many of the more local species are confined to sandy ground, especially among coastal sand dunes or on inland heaths. For several of these species the requirement for light, freely-drained soil is also satisfied on chalk downland. Generally, the southern coastal areas are the richest in species. On upland moors, in contrast, fewer species are present but they can occur in very large numbers.

Activity period

Most species may be found as adults in May and June, but it is worth noting those that must be sought earlier or later than this. The data of capture can sometimes be an aid to identification. Some species, e.g. *Euheptaulacus villosus*, are only active for a very short period and are easily missed. *Aphodius paykulli* is almost a winter species, rarely being found later than March. Those reaching a peak in March and April include *A. constans, A. nemoralis* and *Heptaulacus testudinarius*. Spring species such as *Typhaeus typhoeus* and *A. sphacelatus* reappear in smaller numbers in the autumn. Species confined to late summer and autumn (July to September) include *Geotrupes spiniger, A. foetens, A. ictericus, A. rufus, A. zenkeri*, and (from September) *A. contaminatus, A. obliteratus* and *A. porcus*.

Mounting

Species less than about 1cm in length are best mounted on card, with an occasional specimen gummed upside down to show the underside. Larger species are very robust and may be pinned. Care should be taken to display the antennae, mouthparts and legs; the hind tarsi in particular are frequently used in identification. Dissection of the aedeagus is not usually necessary, but can be helpful with such critical pairs as *Amphimallon solstitialis/ochraceus* and *Melolontha melolontha/hippocastani*, and with some of the more difficult species of *Aphodius* and *Onthophagus*. Scarabs are strong fliers, and a specimen mounted with the wings spread makes an interesting and attractive display – though a stag beetle so mounted takes up a great deal of space! Larvae are best preserved in 80% alcohol.

Identification

Fortunately an up-to-date key and check list is now available: *Dung beetles and chafers* by L. Jessop (*R.E.S. Handbooks* vol. **5**, part 11, £5). The species of *Aphodius* can be very variable, and some collectors in the past have been very keen on hunting for colour varieties. This is not so fashionable in this country now, but a current Italian book illustrates 65 varieties of *A. distinctus*, divided among 13 named forms! Jessop's book will also aid the identification of larvae as far as genus, but much more work is still needed on larvae of known species. Of allied interest is P.

Skidmore's *Insects of the cowdung community*, in the Field Studies Council's AIDGAP series.

The Status of Scarabs in Britain

Accounts of eight of the rarest species (*) are given in the British Red Data Books Insects. (Shirt, 1987). Most species become scarcer in the north.

1. Not recorded since c.1900 (presumed extinct). *Platycerus caraboides, Aphodius obscurus, A satellitius, A. scrofa, A sturmi, A. varians, Rhyssemus germanus, Pleurophorus caesus, Onthophagus taurus, Polyphylla fullo.*
2. Not recorded since c.1930 (possibly extinct). **Trox perlatus, *Brindalus (= Psammodius) porcicollis, Onthophagus nutans.*
3. Very rare (threatened). **Aegialia rufa, *Aphodius brevis, A. lividus, *A niger, A. quadrimaculatus, A. subterraneus, *Diastictus vulneratus, *Copris lunaris, *Gnorimus variabilis.*
4. Rare. *Odonteus armiger, Aphodius sordidus, Euheptaulacus (= Aphodius) sus, Heptaulacus (= Aphodius) testudinarius, Melolontha hippocastani, Gnorimus nobilis.*
5. Scarce or very local. *Trox sabulosus, Geotrupes mutator, G. pyrenaeus, Aphodius coenosus, A. conspurcatus, A. consputus, A. distinctus, A. fasciatus, A. nemoralis, A. obliteratus, A. paykulli, A. plagiatus, A. porcus, A. putridus, euheptaulacus (= Aphodius) villosus, Psammobius asper, Onthophagus nuchicornis, Omaloplia ruricola, Amphimallon ochraceus.*
6. Local to common in N+W, scarce to absent in SE. *Geotrupes stercorosus, G. vernalis, Aegialia sabuleti, Aphodius depressus, A. lapponum, Cetonia cuprea (not in W), Trichius fasciatus.*
7. Local to common in S, scarce to absent in N. *Lucanus cervus, Dorcus parallelipipedus, Trox scaber, Typhaeus typhoeus, Aphodius equestris, A. foetidus, A. granarius, A. haemorrhoidalis, A. ictericus, A. zenkeri, Oxyomus sylvestris, Onthophagus coenobita, O. joannae, O. similis, O. vacca, Hoplia philanthus, Amphimallon solstitialis, Anomala dubia, Cetonia aurata.*
8. Generally distributed, local to abundant. *Sinodendron cylindricum, Geotrupes spiniger, G. stercorarius, Aegialia arenaria, Aphodius ater, A. borealis, A. constans, A. contaminatus, A. erraticus, A. fimetarius, A. foetens, A. fosser, A. luridus, A. merdarius, A. prodromus, A. pusillus, A. rufipes, A. rufus. A. sphacelatus, Serica brunnea, Melolontha melolontha, Phyllopertha horticola.*
9. Established aliens. *Saprosites mendax, Psammobius caelatus.*
10. Doubtfully British, requiring confirmation. *Onthophagus fracticornis, Oxythyrea funesta, Trichius zonatus.*
11. Channel Islands only. *Onthophagus taurus,* and the four chafers, *Rhizotrogus aestivalis, R. maculicollis, Anisoplia villosa* and *Cetonia morio.*

CLAMBIDAE

BY J. COOTER

Another small family, both numerically and in the size of the beetles. In Britain there are nine species in two genera and they have been adequately dealt with by Johnson (1966).

REFERENCE

Johnson, C., 1966. Coleoptera, Clambidae. *Handbk. Ident. Br. Insects*, **4** (6(a)): 13pp. Royal Ent. Soc. London.

SCIRTIDAE

BY J. COOTER

All members of this family, of which there are twenty British species, are very delicate, poorly chitinised insects. To ensure perfect specimens in the cabinet, they are best kept individually in tubes, killed quickly and mounted at once, after dissecting the genitalia, using a very dilute glue. The Scirtidae possess a very complex genitalia not of the normal Coleopterous types. Kevan (1962) discusses the form and function of the male and female genitalia in genus *Cyphon*. The individual sclerites and structures being weakly chitinised are liable to distort upon drying and are therefore best preserved in Euparal on a celluloid/acetate strip.

Reference to the genitalia is a virtual necessity for the genera *Elodes* and *Cyphon*, for which Joy (1932) is quite out of date. For *Cyphon*, Kevan (1962) and Skidmore (1985) or Nyholm (1972) are necessary. Lohse (1979) gives a more modern and well illustrated key to the whole family, drawn from papers by Klausnitzer (1971) and Nyholm (1955).

All species have aquatic/semi-aquatic larvae and the adults can be found, often in some numbers, by beating and sweeping in damp places or near water. Adults of *Elodes*, *Microcara* and *Cyphon* can also be beaten from hawthorn blossom. *Prionocyphon serricornis* breeds in water-filled rot-holes in trees, particularly beech. *Hydrocyphon deflexicollis* is associated with running water and can be found on river shingle. *Scirtes* species have enlarged hind femora and are capable of jumping and might be unwittingly mistaken for halticine Chrysomelids. They are to be swept from lake- or stream-side vegetation.

REFERENCES

Kevan, D.K., 1962. The British Species of the Genus *Cyphon* Paykull (Col., Helodidae), Including Three New to the British List. *Entomologist's mon. Mag.*, **97**: pp. 114–121.

Klausnitzer, B., 1971a. Zur Kenntnis der Gattung *Helodes* Lart., *Ent Nachr*. **14**: 177pp.
Klausnitzer, B., 1971b. Beiträge zur Insektfauna der DDR: Col., Helodidae. *Beitr. Ent.*, **21**: 477pp.
Nyholm, T., 1955. Die mitteleuropäischen Arten der Gattung *Cyphon* Pk. *Ent. Arb. aus dem Museum G. Frey*, Tutzing 1955: 251pp.
Nyholm, T., 1972. Die nordeuropäischen Arten der Gattung *Cyphon* Paykull (Coleoptera). *Ent. Scand.*, Suppl. **3**: pp. 1–100.
Skidmore, P., 1985. *Cyphon kongsbergensis* Munster (Col., Scirtidae) in Scotland. *Entomologist's mon. Mag.*, **121**: pp. 249–252.

(*Eubria palustris*, a beetle which bears a strong superficial resemblance to *Cyphon* species and also frequents wet areas where it can be beaten or swept from shallows, *Equisetum*, and other vegetation inches above running water, is now placed in the family **Psephenidae** by most authorities, after long and unsettled taxonomic history).

BYRRHIDAE

BY J. COOTER

The thirteen British species of Byrrhidae are to be found under stones, by grubbing and by sifting litter and moss on moorland and heathland, especially in areas of established heather which is not managed for grouse and other game by cyclic burning. River banks in such places are favoured by Byrrhids as are sandy/chalky areas and vegetated shingle at the coast. In early spring some species are to be found walking in the open on moorland and generally, as the year progresses, the Byrrhids become abbraded.

The *Byrrhids* are commonly known as the "Pill Beetles", a name derived from their ovoid form which is enhanced when the beetles are disturbed or in death by tightly drawing their appendages against the ventral surface and flexing the head ventrally. Some, for example *Byrrhus* species, have tibiae furnished externally with a sulcus for reception of the tarsi. Such habit and structure makes it necessary to carefully tease out the appendages prior to mounting. This is best achieved by placing the beetle on its dorsal surface and holding it gently with a pair of forceps and arranging the appendages with the aid of a pin and/or fine forceps; the head should also be gently pulled forward. Care must be taken at this stage not to abrade the setae and scales of the dorsum. Dissection of *Byrrhus* species is advisable as the external characters are quite variable but the aedeagus offers reliable features.

With the passage of time, Joy (1932) has become out of date, but will enable identification to genus, bearing in mind *Syncalypta* has been divided into two genera known as *Chaetophora* and *Curimopsis*, see Johnson (1978). For *Byrrhus* Johnson 1965 is recommended. Alternatively Paulus (1979) or Mrozkowski (1958) present keys covering all the British species.

The most widespread and abundant Byrrhid is *Simplocaria semistriata* (F.), but the other member of this genus, *maculosa* Er., is excessively scarce with only three British specimens known. *Morychus aeneus* (F.) is a northern species known as far south as Lancashire, but the majority of specimens in collections originate from the Scottish Highlands. *Cytilus serriceus* (Forster) occurs in a wide range of habitats including woodlands and moors throughout Britain. *Byrrhus* species are typical of moorland and heaths with *arietinus* Steff., the most uncommon member of the genus, having a decidedly northern distribution. *Porcinolus murinus* (F.) is a rare heathland insect known from the Breck, New Forest and Surrey heaths; there appear to be no recent records. *Curimopsis nigrita* (Palm), recently added to our List, is known only from Thorne Waste, Yorkshire. *Curimopsis maritima* (Marsh.), as its name implies, is to be found near the coast in dry places such as vegetated shingle and sandy/chalky areas. It occurs in England and is not known from the south-west and west coasts, but there are in addition a few records from inland counties. *Curimopsis setigera* (Ill.) on the other hand is a western species found in dry sandy places from Dumfries to Hampshire and with one inland county record.

(In some classifications *Limnichus pygmaeus* (Sturm) is included in the Byrrhidae. In our current Check List it is placed in the **Limnichidae**. Externally very similar to a Byrrhid, it is found beside wet flushes and the like under stones and debris. It has a scattered distribution throughout England. As with the Byrrhids, *Limnichus* draws its appendages and head tightly in against the ventral surface; this, plus its small size (ca 1.75mm) and ovoid form, make it one of the most frustrating beetles to mount.)

References

Johnson, C., 1965. The British species of the genus *Byrrhus* L., including *B. arietinus* Steffahny (Col., Byrrhidae) new to the British List. *Entomologist's mon. Mag.*, **101** (1966): pp. 111–115.

Johnson, C., 1978. Notes on Byrrhidae (Col.); with Special Reference to, and a Species New to the British Fauna. *Ent. Rec. J. var.*, **90**: pp. 141–147.

Mroczkowski, M., 1958. Klueze do Oznaczania Owadow Polski. Coleoptera, zeszyt 50–51 Byrrhidae – Nosodendridae: 30pp. Warsaw.

Paulus, H.F., 1979. in Freude, H., Harde, K.W., & Lohse, G.A., *Die Käfer Mitteleuropas*, **6**: 47. Fam. Byrrhidae: pp. 328–350. Goeke & Evers, Krefeld.

HETEROCERIDAE

By J. Cooter

Our eight species of this family belong to the genus *Heterocerus* Fab. They are of characteristic appearance and habit and can often be found in numbers. Their identification is somewhat hampered by the beetles' dense pubescence, but they have been most ably covered by Clarke (1973) in the RES Handbook series. (see also p. 78)

Reference

Clarke, R.O.S., 1973. Coleoptera, Heteroceridae. *Handbk. Ident. Br. Insects*, 5(2)(c)): 14pp. Royal Ent. Soc. London.

BUPRESTIDAE (JEWEL BEETLES)

By Howard Mendel

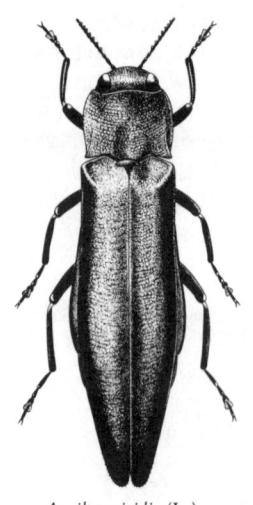

Agrilus viridis (L.)

The attractive coloration, unusual form and rarity of most of the British Buprestidae contribute towards the special fascination the group has always held for coleopterists. It is unfortunate therefore that only 12 of the approximately 200 European species are found in Britain. Most of these are restricted to the South and none is found in Ireland.

The larger species, *Melanophila acuminata* (Degeer), *Anthaxia nitidula* (L.) and *Agrilus* spp. may be taken by beating the appropriate host trees in the summer months or occasionally by general sweeping. The larvae develop in, or under, bark and are characterised by the greatly enlarged pro-thoracic segment of their bodies (Fig. 19). Adults appear to be short-lived and adult emergence is often closely co-ordinated. Some species, therefore, such as *Agrilus pannonicus* (Piller & Mitterpacher) and *A. sinuatus* (Olivier) may not be as rare as records suggest. The characteristic emergence holes of the former and the sinuate, subcortical larval burrows of the latter are found more frequently than the adults. Species with larvae that live sub-cortically or within the bark are not difficult to rear. The larval stage lasts between one and three years. In contrast, *Buprestis aurulenta* L., a large American species occasionally imported into Britain with timber, has a larval stage which may last over thirty years, and adults often emerge long after the wood supporting them has been processed.

M. acuminata adults are attracted to fire-damaged conifers and, when at rest, resemble small pieces of charred bark. *Anthaxia nitidula* is probably our rarest species and most of the specimens taken have been beaten from May blossom, *Crataegus* spp., or found on the flowers of Lesser Celandine, *Ranunculus ficaria* L. Of the species of *Agrilus*, *A. laticornis* and *angustulus* (Illiger) may be beaten from a variety of trees, but especially Oak, *Quercus* spp., and are the two species most likely to be found by the casual collector. *A. pannonicus* is also associated with Oak, *A. sinuatus* with Hawthorn, *Crataegus* spp. and *A. viridis* (L.) with Willows, *Salix* spp.

The life histories of the smaller species are less well known. The larvae of the two British species of *Aphanisticus* develop in the stems of rushes, *Juncus* spp. and possibly also sedges, *Carex* spp., but have not been described. The larvae of *Trachys* spp. are leaf miners. *T. minutus* (L.) is usually associated with Willows or Hazel, *Coryllus avellana* L.; *T. scrobiculatus* Kiesenwetter with Ground Ivy, *Glechoma hederacea* L., and *T. troglodytes* Gyllenhal with Devil's-bit Scabious, *Succisa pratensis* Moench. The larva of *T. troglodytes* has not been described. However, sites with this species may be most easily recognised by looking for the characteristic larval leaf mines or shiny black eggs on the foodplant (Alexander, K.N.A., 1989, *Br. J. Ent. Nat. Hist.*, **2**: 92) or the adult feeding damage in the form of notched leaf margins (Porter, K., 1988, *Dyfed Invertebrate Group Newsletter*, no. 11). *Aphanisticus* and *Trachys* adults may be found by sieving litter or moss beneath the larval foodplants, even in the autumn and winter. In suitable areas they are occasionally found in flood refuse.

All of the British species can be identified from the dorsal surface without resort to the genitalia. However, examination of the prosternal process helps with the separation of *Agrilus angustulus* and *A. laticornis* females and *Trachys scrobiculatus* and *T. troglodytes*. A small race of *T. troglodytes* from the Breckland of East Anglia may cause confusion (Allen, A.A., 1968, *Entomologist's mon. Mag.*, **104**: pp. 208–216).

Bibliography
Bily, S., 1982. *The Buprestidae (Coleoptera) of Fennoscandia and Denmark*. Klempenborg: Scandinavian Science Press Ltd. [written in English and includes all of the British species except *Agrilus sinuatus*].
Levey, B., 1977. Coleoptera Buprestidae, *Handbk Ident. Br. Insects*, **5(1b)**: 1–8.

ELATERIDAE (CLICK BEETLES)

By Howard Mendel

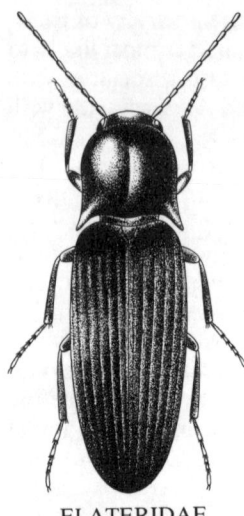

ELATERIDAE
(Typical shape)

These elongated beetles, with characteristic, pointed hind-angles to the pronotum are immediately recognised by their ability to "spring" into the air when disturbed; often with an audible "click". They are only likely to be confused with beetles belonging to the related families Throscidae and Eucnemidae.

The larvae of click beetles are collectively known as "wireworms", and are almost as well known as the adults. The majority are elongate with a tough cuticle so that the term "wireworm" is very appropriate (Fig. 20.). However, larvae of *Lacon querceus* and *Agrypnus murinus* (Agrypninae) are more fleshy and grub-like whilst those of *Cardiophorus* spp. and *Dicronychus equiseti* (Cardiophorinae) are incredibly long and thread-like with pseudo-segmentation and expandable, membranous cuticle. A few species can cause serious crop damage and it is largely because of this that they have been well studied. F.I. van Emden (1945), Larvae of British beetles. V. Elateridae. *Entomologist's mon. Mag.*, **81**: pp. 31–37) has written a useful key to the larvae of all of the British species known at the time, and this has been amended by B.A. Cooper (1945, Notes on certain Elaterid (Col.) larvae. *Entomologist's mon. Mag.*, **81**: pp. 128–130). There is also a detailed account of the larvae of *Ampedus* (Emden, H.F. van, 1956, Morphology and identification of the British larvae of the genus *Elater* (Col., Elateridae). *Entomologist's mon. Mag.*, **92**: pp. 167–188). Larvae of most species are now generally regarded as omnivorous, relying to a greater or lesser extent on other insect larvae, worms and various invertebrates, as well as frequently resorting to cannibalism. Superficially similar larvae of *Dryops* spp. and some of the Tenebrionidae lack the characteristic pygopodium of the 10th. abdominal segment of the Elateridae.

Although a few species of click beetles are extremely common, a large number are genuinely scarce and have precise habitat requirements. Such species are seldom encountered accidentally and to find them a knowledge of their life-histories and distributions (Mendel, H., 1988, *Provisional atlas of the click beetles (Coleoptera: Elateroidea) of the British Isles*. Institute of Terrestrial Ecology: Huntingdon) is most helpful. *Ampedus sanguineus, Cardiophorus gramineus, C. ruficollis* and *Selatosomus cruciatus* have not been recorded for over a century and should be regarded as extinct in the British Isles.

The 71 species on the British list are currently separated into 12 subfamilies but, for the purpose of this account, they are perhaps better divided by larval habitat. They divide nearly equally between those with larvae which inhabit the soil and those associated with dead wood.

It is amongst those inhabiting the soil that the pest species are found and the genus *Agriotes* is particularly well known in this respect. However, some of the species of *Athous*, *Ctenicera* and *Selatosomus* may, under certain conditions, become pests. Although there is considerable overlap, the various species tend to inhabit different soil types so that heavy lowland soils, thin peaty upland soils, heathlands, woodlands etc. will all have their own particular species assemblage. The more common and widespread species will be found in the normal course of collecting; under stones, on the beating-tray, in the sweep-net or in pit-fall traps.

Two particular soil habitats are of special note. *Cardiophorus assellus*, *Dicronychus*, *Melanotus punctolineatus* and *Cidnopus aeruginosus* are associated with sandy soils and dune systems. Searching suitable areas on fine days in the spring when the beetles may be found struggling over areas of bare sand or flying in the sunshine, can be very productive. *Negastrius* spp., *Zorochros minimus* and *Fleutiauxellus maritimus* are found amongst sand and shingle on river banks. These beetles are also best searched for in the spring and each species seems to have a preferred habitat. *F. maritimus*, for example, likes areas of "clean" shingle; *Zorochros* can tolerate shingle with sand and a certain amount of mud whereas the species of *Negastrius* prefer well-graded coarse sand or grit with sparse vegetation, often occupying a distinct zone some distance from the water.

The deadwood species have amongst their ranks some of the rarest of British beetles; relicts of the wildwood now confined to a very few localities. *Lacon*, *Ampedus nigerrimus*, *A. ruficeps* are known today only from Windsor Forest. *Limoniscus violaceus* christened the "Violet Click Beetle" has been added to Schedule 5 of the Wildlife and Countryside Act 1981, and should not be collected. Many of the deadwood species are associated with rot-holes in the living trees, developing in the red-rot (*Lacon*, *Ampedus cardinalis*) or black-rot (*Ischnodes*, *Elater*). Other species are commonly found under bark (*Melanotus villosus*, *Stenagostus*, *Denticollis*) or in the wood itself (*Ampedus* spp., *Procraerus*, *Megapenthes*). Larvae are often found in "colonies" of other wood-boring beetles (Lucanidae, Cerambycidae and Curculionidae) and no doubt prey on them. A strong knife, or even better a small pick, is needed to work suitable dead wood, but care must be exercised as a great deal of damage can be done in a short time by an eager collector. *Procraerus*, or even *Megapenthes*, may be found by the fortunate collector by beating the blossom of hawthorn in ancient woodland areas.

The length of larval life in the Elateridae is not fixed but rather determined by the conditions and the availability of suitable food. Species that may normally develop in two or three years may take five or six years if food is in short supply.

Two basic types of life-histories are found in the Elateridae. Some species emerge in July and August and remain in the pupal cell until the following spring. Searching for fully mature adults in their pupal cells, in the winter months, can be very rewarding. *Procraerus* and the species of *Ampedus* are most easily found in this way. In other species, adults emerge from the pupae in the summer and leave the pupal cell almost as soon as they are mature. It is surprising just how infrequently many of the species in this latter group are found as adults. The large black larvae of *Stenagostus*, for example, are not uncommon under loose bark in suitable localities across much of southern Britain, but adults in the open are rarely seen except perhaps at the lepidopterist's lamp. Cryptic behaviour and a short adult life combine to make some species elusive rather than rare.

The larvae of many species are more frequently encountered than the adults, and so rearing can be a very productive way of obtaining specimens. It is also a good way of learning more about the life-histories of Elateridae. Deadwood species, in particular, can easily be reared by placing sufficient pabulum in a jar and adding occasionally a few drops of water. The contents of the jar must not be allowed to become either too dry or sodden. Insect larvae may also have to be added as food, from time to time, but cheese (J.A. Owen, pers. comm.) or other high protein substitutes have been successfully used. However, patience will often be required.

Considering that click beetles form a well defined group which is generally popular with collectors, it is surprising how difficult some of the species can be to identify. The species of *Ampedus* are notorious. Fowler (vol. 4) and Joy are still the only readily available keys, in English, to the group, though Leseigneur (1972), *Coléoptères Elateridae de la faune de France continentale et de Corse*. Societe Linnéenne de Lyon: Lyon), which has numerous figures and interesting notes on biology, will prove more useful to coleopterists who can cope with a French text. In the majority of species the males have longer and differently formed antennae. Although the genitalia may be of value in the separation of certain species, overall the structure is of little use for identification. Unfortunately, comparative differences in shape and in punctuation still have to be relied upon.

The following references to additions and adjustments to the British list, since Joy (1932), are a good indication of the inadequacy of available keys.

References

Allen, A.A., 1936. *Adelocera quercea* Herbst (Col., Elateridae) established as British. *Entomologist's mon. Mag.*, **72**: pp. 267–269.

Allen, A.A., 1937. Limoniscus violaceus Mull. (Elateridae), a genus and species of Coleoptera new to Britain. *Entomologist's Rec. J. Var.*, **49**: pp. 110–111.

Allen, A.A., 1938. *Elater ruficeps* Muls.; a beetle new to Britain. *Entomologist's mon. Mag.*, **74**: p. 172.

Allen, A.A., 1966. The rarer Sternoxia (Col.) of Windsor Forest. *Entomologist's Rec. J. Var.*, **78**: pp. 14–23.
Allen, A.A., 1969. Notes on some British Serricorn Coleoptera with adjustments to the list. *Entomologist's mon. Mag.*, **104**: pp. 208–216.
Allen, A.A., 1990. Note on, and Key to, the often-confused British species of *Ampedus* Germ. (Col: Elateridae) with corrections to some erroneous records. *Entomologist's Rec. J. Var.* **102**. pp. 121–127.
Cooter, J., 1983. *Zorochros flavipes* (Aubé) (Col., Elateridae) new to Britain. *Entomologist's mon. Mag.*, **119**: pp. 233–236.
Hignett, J., 1940. *Corymbites angustulus* Kies.: an Elaterid new to the list of British Coleoptera. *Entomologist's mon. Mag.*, **76**: p. 14.
Mendel, H., 1990. The status of *Ampedus pomone* (Stephens) *A. praeustus* (F.) and *A. quercicola* (Buysson) (Col.: Elateridae) in the British Isles. *Entomologist's Gazette* **41**: pp. 23–30.
Owen, J.A., Allen, A.A., Carter, I.S., & Hayek, C.M.F. von, 1985. *Panspoeus guttatus* Sharp (Col., Elateridae) new to Britain. *Entomologist's mon. Mag.*, **121**: pp. 91–95.
Speight, M.C.D., 1986. *Asaphidion curtum, Dorylomorpha maculata, Selatosomus melancholicus* and *Syntormon miki*: insects new to Ireland. *Ir. Nat. J.*, **22**, pp. 20–23.

THROSCIDAE

By Howard Mendel

A small family of two genera, *Trixagus (Throscus* in older works) with four species and a single *Aulonothroscus*, together comprise most of the European fauna. There are some 200 species worldwide. The Throscidae are closely related to the true click beetles and are able to spring in the same way although this ability is less well developed and seldom observed. The species are superficially very similar; each less than 3.5mm in length, darker or lighter brown, elongate-oval with pointed pronotal hind angles and clubbed antennae.

When disturbed these beetles tuck their antennae and legs into grooves on the underside of their bodies and remain motionless assuming a seed-like appearance. They react in this way in the killing bottle and because of this are notoriously difficult to set. However, a little time spent extending the antennae and legs can save considerably more time when it comes to identification. Excess glue on the mounting card may also make identification difficult.

Although the male genitalia are quite distinct in each species it will seldom be necessary to resort to dissection. The British species can all be reliably separated on external characters and, with a little practice, careful examination of the head alone is all that is required for identification. The most useful characters are the extent of division of the eyes and the presence/absence and type of carinae between the eyes. Size, pronotal shape and puncturation of the pronotum and elytra provide additional characters for identification. There are also good characters on the underside which can be useful when identifying spirit preserved material.

Fowler or Joy will normally be used to identify the British species. However, *Aulonothroscus brevicollis* (de Bolvouloir) was only discovered "new to Britain" by Ashe (1942) and so is not included in these keys. It was recognised as an *Aulonothroscus* rather than a *Trixagus* by Burakowski (1975) and is the only British Throscid without distinctly emarginate eyes.

All of the British species show sexual dimorphism and this can provide a short-cut to identification. For example, only the males of *Trixagus elateroides* (Heer) have a distinct tooth near the base of the middle tibiae and the dense comb of pubescence on the elytral margin of male *T. obtusus* (Curtis) is diagnostic. Males of *T. dermestoides* (L.) and *T. carinifrons* (de Bonvouloir) have a distinctly expanded antennal club.

General collecting methods such as sweeping and beating (particularly in the late afternoon or evening) will produce Throscids in suitable localities. *Trixagus elateroides* and *T. obtusus* are most likely to be found in salt marsh localities in S.E. England and East Anglia. They may be swept in the summer or collected from grass tussocks or flood refuse in the winter months. *Aulonothroscus brevicollis* is a species associated with quality pasture woodland sites and is known only from a small number of scattered localities (Mendel, 1985). Where it occurs, the beetles may be beaten from old trees, usually oaks, found in wood mould or under bark. Throscids fly on warm still evenings and are often attracted to light. Mendel (1990) maps the distributions of the British species. Only our commonest, *Trixagus dermestoides*, has been recorded from either Scotland or Ireland.

Burakowski (1975) provides a detailed description of the larva and pupa of *T. dermestoides* and an excellent account of its life history. The larvae live in the soil and show little activity, feeding on ectotrophic mycorrhizas of various trees. The life cycle is usually completed in two years, the newly eclosed adults over-wintering within the pupal cell. Little is known about the life histories of the other British species.

Bibliography

Ashe, G.H., 1942. *Trixagus* (= *Throscus*) *brevicollis* Bonv. (Col., Trixagidae) a species new to Britain. *Entomologist's mon. Mag.*, **78**: p. 287.

Burakowski, B., 1975. Development, distribution and habits of *Trixagus dermestoides* (L.) with notes on the Throscidae and Lissomidae (Col., Elateroidea). *Annales Zoologici*, **32**: pp. 376–405.

Mendel, H., 1985. *Trixagus brevicollis* (de Bonvouloir) (Col., Throscidae) in Britain. *Entomologist's mon. Mag.*, **121**: p. 58.

Mendel, H., 1990. *Provisional atlas of the click beetles (Coleoptera: Elateroidea) of the British Isles.* Institute of Terrestrial Ecology.

EUCNEMIDAE (FALSE CLICK BEETLES)

BY HOWARD MENDEL

Of the six species on the British list only three were known prior to the 1950s, and it is quite possible that others await discovery. Eucnemids are secretive woodland species seldom found by the casual collector, but they may occasionally be seen in numbers at breeding sites.

Melasis buprestoides (L.) is the largest (up to 10mm) and most common species of the family in Britain. It is found as far north as Yorkshire and it is the only one of the group known from Ireland. *Melasis* larvae are not of the typical Eucnemid type but somewhat resemble those of Buprestids, having an enlarged pro-thoracic segment. They are most frequently found in dead beech, *Fagus sylvatica* L., less commonly in oak, *Quercus* spp., hornbeam, *Carpinus betulus* L. or other dead wood; often with dead adults. *Dirhagus pygmaeus* (F.) is also widespread, though very local, and found from southern England to the north of Scotland. Most specimens are swept in June or July beneath the ancient oaks in which the species breeds. *Eucnemis capucina* Ahrens, an ancient forest relict, is a great rarity. The larvae develop in dead wood, usually beech, and in warm days in June and July adults may be found crawling on the surface of the dead trunks.

The two species of *Hylis* are both relatively recent discoveries. *H. olexai* (Palm) is found sparingly in the south-east and is associated with beech, but *H. cariniceps* (Reitter) is only known, in Britain, from the original specimen swept in the New Forest near old beech trees, and another from Dorset. It is not known for certain whether or not our two species of *Hylis* are long overlooked forest relicts or recent arrivals. The former seems more likely which is probably not the case with our last species, *Epiphanis cornutus* Eschscholtz, an insect better known from North America. It is often associated with spruce, *Picea* spp., and is well established in southern England.

All of the British Eucnemids can be identified by external characters of the dorsal surface. A.A. Allen (1968, Notes on some British serricorn Coleoptera, with adjustments to the list. 1. Sternoxia, *Entomologist's mon. Mag.*, **104**: pp. 208–216) provides a very useful key to the British species. It also includes *Cerophytum elateroides* Latreille, now placed in a separate family, the Cerophytidae, as it is possible that this beetle was once found in Britain.

Bibliography

Allen, A.A., 1954. *Hypocoelus procerulus* Mannh. (Col., Eucnemidae, Anelastini) in Kent and Surrey: a tribe, genus and species new to Britain. *Entomologist's mon.*, 90: pp. 228–230.

Mendel, H., 1990. *Provisional atlas of the click beetles (Coleoptera: Elateroidea) of the British Isles.* Institute of Terrestrial Ecology.

Skidmore, P., 1966. *Epiphanis cornutus* Eschsch. (Col., Eucnemidae) new to the British list. *Entomologist*, **99**: pp. 137–139.

CANTHARIDAE

By Keith Alexander

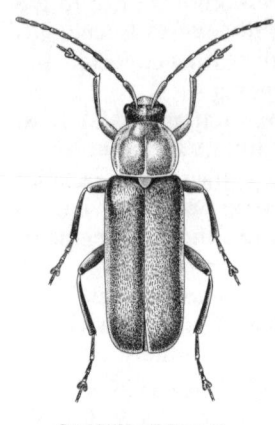

CANTHARIDAE
(Typical shape)

The Cantharidae are readily split in two. The larger, brightly coloured species form the *Cantharinae*, of which Britain has 25 species currently recognised. The other group are the *Malthininae*, which comprises 16 species of smaller dark species, usually having yellowish tips to their elytra. Although the bright colours and conspicuous behaviour of the *Cantharinae* make them an attractive group, their study is much neglected. Identification is not always easy as many species can be extremely variable in colour, and there may be considerable size variation. *Malthininae* also have their problems. In particular, female *Malthodes* can often only be taken down to two possibilities. The males of this genus are however very straightforward, with their characteristic terminal abdominal segments. Fowler and Joy are still the main identification guides, although there have been many changes in nomenclature – check Kloet and Hincks, one species split – *Cantharis cryptica* from *C. pallida* (G.H. Ashe, 1946 & 1947, *Ent. mon. Mag.*, **82**: pp. 138–139 & **83**: p. 59), and one synonymised – *C. darwiniana* is now regarded as an extreme form of *C. rufa*. There are other problems with these keys, but they should soon be resolved by the publication of a *R. ent. Soc. Handbook* (M.G. Fitton, in prep.). A key to larval genera is available (M.G. Fitton, 1975, *J. Ent.* (B), **44**: pp. 243–254).

The bright colours of the *Cantharinae* have attracted sufficient attention for them to be given English names: "soldier beetles" for the reddish and yellowish species, and "sailor beetles" for the black and bluish ones. However, the variation in colour possible in some species makes the distinction unhelpful, and all Cantharidae are generally known simply as "soldier beetles". The adult beetles are generally found amongst vegetation or on flowerheads, and are most readily captured by sweeping or beating. Tall, well-developed vegetation and trees and shrubs are the most productive sources, with the majority of species living in deciduous woodlands, hedgerows and scrub, or in marshland and wet meadows. Fewer species occur in dry grasslands or on heath and moor – although these have their specialists and should not be neglected. The velvety larvae are found in similar situations and are also present in the surface layers of the soil and in leaf litter. Both adults and larvae are mainly carnivorous, feeding on dead or injured invertebrates and also smaller and slower healthy ones. They are fluid feeders, and will also feed from plant material. The very abundant *Rhagonycha fulva*

– reddish–yellow with black elytral tips, so characteristic of flowerheads in July and August, presumably feeds extensively on nectar.

Larvae can be reared in captivity by feeding with inactive insect larvae, pupae, small worms, injured larvae and woodlice, or cut pieces of vegetables such as carrot or potato. Wheat grains will also be fed upon. A soft substrate of some kind should be provided as a pupation site – moist sterile sand will do; natural sites include dead wood and fungi.

Adult *Malthininae* are also found by sweeping and beating, although mainly in and around woodland or mature trees. Their larvae are mainly found under bark and within dead and decaying timber. The feeding habits are most probably very similar to Cantharinae.

Deciduous woodlands, parks and hedgerows are the most productive sites, producing *Podabrus alpinus, Cantharis cryptica, C. decipiens, C. livida, C. nigricans, C. pellucida, C. rufa, Rhagonycha limbata, R. lignosa* and *Malthodes marginatus* from about mid-May until early July. The very rare *M. brevicollis* and *M. crassicornis* are also early species. Other species occur from about mid-June to mid-August: *Malthinus balteatus, M. flaveolus, M. frontalis, M. seriepunctatus*, and *Rhagonycha translucida*. Damper sites are preferred by *Malthodes dispar* and *Rhagonycha testacea*. Other species are most readily found in the deciduous woodlands of the hill country of northern and western Britain: *Ancistronycha abdominalis, Cantharis obscura, Malthodes flavoguttatus, M. fuscus, M. guttifer* and *M. mysticus*. *Malthodes fibulatus* in most often found in the woods of chalk and limestone country, and *Rhagonycha elongata* is associated with the old pine forests of Scotland.

Marshes, wet meadows and similar places are the habitat of *Cantharis figurata, C. lateralis, C. nigra, C. pallida, C. thoracica* and *Silis ruficollis* – the latter mainly in East Anglia and sparingly from Kent to S. Wales. June and July are the best months. *Cantharis paludosa* is numerous on the peat moors and mosses of the North and West. *C. rustica* is a common species of dry grasslands in May and June, but also occurs widely in other situations. *Rhagonycha fulva* is ubiquitous in open flowery situations in July and August.

Some species may be commoner in southern Britain than in the north, although little information is currently available on such patterns of distribution. Likely species are *Cantharis lateralis, C. rustica, Rhagonycha lutea, Malthinus balteatus, M. seriepunctatus, Malthodes fibulatus* and *M. minimus*.

Care needs to be taken when mounting Malthininae to ensure that the male terminal abdominal segments remain free of gum. Pinning is a good alternative for the family since abdominal characters can be important in determination of the species. Antennae also may need to be examined and the minimum of gum should be used in their setting.

Distinguishing the sexes is not always straightforward. Male *Malthodes* are readily distinguished by the complex development of the terminal abdominal segments already referred to. In this and in other genera, the head of the male is often larger, especially broader, with the eyes more prominent. The antennal segments are often longer, and the middle segments have a fine groove or impressed line along their length in several *Cantharis – livida, pellucida,* and *nigricans.* The claws of *Ancistronycha abdominalis* are toothed in both sexes, but in the female the tooth is much longer and spine-like.

LAMPYRIDAE

BY J. COOTER

This family contains the "fire-flies" and "glow-worms" which are represented in the tropics by large number of species.

In Britain only one species, *Lampyris noctiluca* L., is likely to be encountered. The larviform female is strongly bioluminescent, the male only slightly less so; this ability to emit light has made the "glow-worm" one of our more familiar beetles, at least in name. It occurs from the south coast as far as Argyllshire in the west and Northumberland in the east. The female is totally devoid of wings and elytra, the male possesses functional wings and full elytra. The life-history of the "glow-worm" has been outlined by Wootton (1976), the main food being snails.

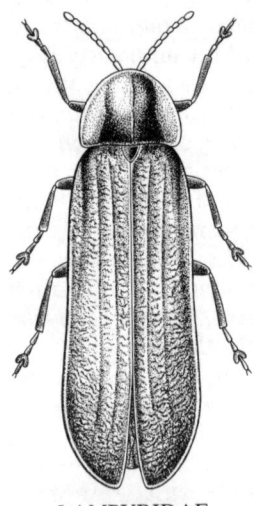

LAMPYRIDAE
Lampyris noctiluca (L.)

Phosphaenus hemipterus (Geoze) is known from East Sussex and single examples taken in 1894 and 1947 at Southampton, Hampshire; last recorded during July 1961, two males, East Sussex. The larviform female is excessively scarce and weakly luminescent, the male possesses abbreviated strongly dehiscent elytra leaving the bulk of the abdomen exposed.

A third species, *Lamprohiza splendidula* (L.) is known in Britain from two males specimens taken in Kent during 1884 (see Allen, 1989); their origin can only be speculated. The male bears a superficial resemblance to the widespread *Lampyris noctiluca.*

References

Allen, A.A., 1989. *Lamprohiza splendidula* (L.) (Col., Lampyridae) taken in Kent in 1884. *Entomologist's mon. Mag.*, **125**: p. 182.
Wootton, A., 1976. Rearing the Glow-worm, *Lampyris noctiluca* L. (Coleoptera: Lampyridae). *Ent. Rec. J. Var.*, **88(3)**: pp. 64–67.

DERMESTIDAE

BY J. COOTER

The dermestids are more familiar as pests of stored products (see Part III, p. 199) but a few are to be found out-of-doors. At the time of writing, an R.E.S. Handbook on the Dermestidae is nearing completion.

The antennae, especially in genus *Anthrenus*, exhibit specific characters and should be displayed during mounting. The pubescence and scales, as well as patterns formed by these, are all used in identification and should not be obscured by glue.

Dermestes murinus and *undulatus* can be shaken from dry carcases, the latter more so near to the coast; *lardarius* is more frequent indoors, but can also be found in dry carcases. *Attagenus pellio* frequents old houses, pigeon lofts/dove cotes and bird nests. *Megatoma undata* is to be found under dry bark as well as in old carcases and carrion debris associated with bird nests. *Ctesias serra* is a woodland beetle often encountered under thick dry bark of rotten trees in company with spiders webs. *Anthrenus* species can be swept from flowers; *museorum* and *verbasci* are more evident indoors where they are sometimes a pest damaging clothes, carpets and the like.

ANOBIIDAE

BY J. COOTER

Some of the twenty-eight British species of Anobiidae are of considerable economic importance; for example, *Anobium punctatum* (woodworm or furniture beetle), *Xestobium rufovillosum* (death watch beetle) and *Stegobium paniceum* (drug-store beetle). The latter and *Lasioderma serricorne* (cigarette beetle) are pests of a range of stored products (see Part III, p. 199) but the bulk of anobiids inhabit dead and dying timber including that used in buildings. *Dryophilus anobioides* frequents broom in the Breckland of East Anglia and *Ochina ptinoides* can be beaten from ivy; it has a scattered and disjointed distribution. *Ernobius* species may be beaten from pines and *Dorcatoma* from oak and other deciduous trees, though they breed in bracket fungi. *Caenocara affinis* and *bovistae* are scarce species, the former more so than the latter; they breed in puff-ball fungi. *Anitys rubens* is found in ancient woodland where it breeds in red-rotten oak; it is often found in company with *Mycetophagus piceus*.

With the exceptions noted below, Joy (1932) will be found quite adequate for identification purposes. *Ernobius*, Johnson 1966; *Gastrallus immarginatus*, Donisthorpe 1936; *Hemicoelus nitidus*, Mendel, 1982.

A good, well illustrated key to *Dorcatoma* which includes figures of the male genitalia of all the British species is given by Baranowski 1985.

References

Baranowski, R., 1985. Central and Northern European *Dorcatoma* (Coleoptera: Anobiidae), with a key and description of a New Species. *Ent. Scand.*, 16: pp. 203–207.

Donisthorpe, H.St.J., 1936. *Gastrallus laevigatus* 01. (Col., Anobiidae) A genus and Species of Coleoptera New to Britain. *Entomologist's mon. Mag.*, 72: p. 200.

Johnson, C., 1966. The Fennoscandian, Danish and British Species of the Genus *Ernobius* Thomson (Col., Anobiidae). *Opusc. Ent.*, 31: pp. 81–92.

Mendel, H., 1982. *Hemicoelus nitidus* (Hbst.) (Col., Anobiidae) New to Britain. *Entomologist's mon. Mag.*, 118: pp. 253–254.

PTINIDAE

BY J. COOTER

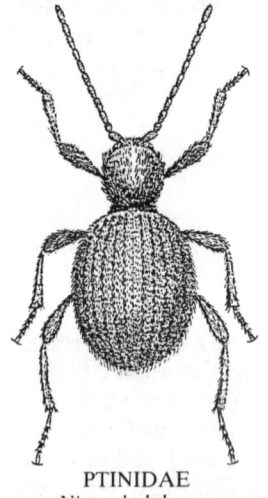

Like the Dermestidae (p. 117), the Ptinidae or "Spider Beetles" are more often associated with stored products (see Part III, p. 199) but a few are more typically found in out-door habitats.

Tipnus unicolor is recorded from old wood, bird nests and the like. *Ptinus palliatus* and *subpillosus* are associated with old oaks and rotten wood, the latter sometimes with ants, particularly *Lasius fuliginosus*; *P. lichenum*, *fur* and *sexpunctatus* in old wood, the latter two also frequently found indoors.

PTINIDAE
Niptus hololeucus
(Faldermann)

CLEROIDEA

BY J. COOTER

This superfamily comprises, in Britain, the families Phloiophilidae, Trogossitidae, Peltidae, Cleridae and Melyridae.

With a few exceptions their identification is straightforward and no special methods need be adopted in their preparation.

PHLOIOPHILIDAE: The sole British species, *Phloiophilus edwardsi* can be found during the autumn and winter; it is associated with the fungus *Phloebia* which grows on dead wood. Its ecology is the subject of an interesting paper by Crowson (1964).

TROGOSSITIDAE: Our two British representatives of this family have widely different habits. *Nemozoma elongatum*, a great rarity, is associated with the Scolytid *Acrantus vittatus* upon which it preys. *Tenebriodes mauritanicus*, the "Cadelle", is a pest of grain, flour, etc. in mills, warehouses and bakeries (see Part III, p. 199).

PELTIDAE: *Thymalus limbatus* is the most widespread species and occurs under bark of dead and dying conifers and hardwood trees. It bears a strong superficial resemblance to a "tortoise beetle" (*Cassida* spp.). *Ostoma ferrugineum* is a very scarce species, added to the British List in 1953 (see Lloyd, 1953). It is confined to ancient Caledonian forest remnants where it lives under the bark of fallen pine trees with the fungus *Phaeolus schweinitzii*. A third peltid occurs in Britain only under artificial conditions and is a stored product pest. This is *Lophocateres pusillus* (Klug) a cosmopolitan species spread by commerce; as yet it is not officially on the British List, though arguably has a stronger case for inclusion than some other species, *Thaneroclerus buqueti* for example (see below).

CLERIDAE: Of the 14 species on the British List, we are likely to encounter only 8; of the others, two are considered accidentally introduced species of doubtful status and four, *Tilloidea unifaciatus, Trichodes alvearius, T. apiarus* and *Tarsostenus univittatus* have not been recorded this century.

The three *Necrobia* species are associated with dry carcases and bones in old nests where carrion has been the food, around bone mills and in dried meat products in pet shops. *Korynetes caeruleus* is associated with timber and is a predator of anobiid and scolytid beetles. It is also well known from certain animal products such as dried meat and leather. *Tillus elongatus* is a predator of various anobiid beetles, particularly *Ptilinus pectinocornis* and *Anobium punctatum*; its larvae hunt nocturnally under bark and on the outside of the tree. The adult beetles are to be found under loose bark, particularly of beech. The female, with red pronotum and elytra expanded posteriorly, is more often noted than the wholly black, more parallel-sided male. *Thanasimus formicarius* is a widespread woodland species; its scarce relative, *T. rufipes*, is confined to N.E. Scotland and in particular the Caledonian Pine Forest remnants, but has also been recorded from plantations. Both are mimics of mutilid wasps. *Opilo mollis* is a scarce nocturnal predator of a variety of anobiid beetles, for example *Xestobium, Hadrobregmus* and *Xyletinus*. (*Thaneroclerus buqueti* as mentioned above, has a very dubious case for inclusion on the British List. It was found once in imported dried root ginger.)

MELYRIDAE: Included here are a number of species found by sweeping flower-rich meadows during early summer (*Dasytes*, and *Malachius*) or by beating trees and sweeping thereunder (*Hypebaeus* and *Aplocnemus*), the latter also occurring under bark. Several are predominantly coastal – *Psilothrix* can be found in numbers by sweeping dunes and grasslands in southern coastal areas of England and Wales; *Dolichosoma linearis* inhabits the East Anglian and Thames saltmarshes; *Dasytes puncticollis* is particularly fond of cliff grassland, *Malachius barnevillei* is recorded from saltmarshes in S.E. England, especially the Thames marshes of Essex and Kent. Wetland species include *Cerapheles terminatus* and *Anthocomus rufus*.

In the Malachiinae, the sexes often differ markedly, the males exhibiting a complex process at the elytral apices, or enlarged antennal segments; these parts, in the female, are simple.

For identification, Joy (1932) will be found adequate provided the user is aware of the fact that eleven species are not included and the key for *Dasytes* species does not work (it separates, *Dasytes niger* and *aeratus* only; the key in Freude, Harde & Lohse (1979) is essential for the separation of *puncticollis* and *plumbeus*). The Recording Scheme for the Cleroidea, Lymexyloidea and Heteromera, organised through the Biological Records Centre, Monks Wood, by Dr. Roger Key, from time to time issues Bulletins with useful guides to facilitate identification (see Part V, p. 254).

Species not in Joy

Ostoma ferrugineum (L.) – Lloyd, R.W., 1953. *Entomologist's mon. Mag.* **89**: p. 51.
Paratillus carus (New.) – Blair, K.G., 1944. *ibid.*, **80**: p. 132–135.
Trichodes alvearius (F.) – Allen, A.A., 1967, *ibid.*, **79**: p. 54 and 1969, *ibid.*, **81**: p. 272.
Thaneroclerus buqueti (Lef.) – Allen, A.A., 1969, *ibid.*, **81**: p. 272.
Aplocnemus pini Redt. – Allen, A.A., 1975, *ibid.*, **111**: p. 210.
Dasytes caerulenus (Dg.) – Johnson, C., 1975, *ibid.*, **111**: p. 179.
Ebaeus pedicularis (L.) – this species is included by Pope; (1977) there are three old specimens without data in the Stephens Collection, most likely taken at Windsor.
Hypebaeus flavipes (F.) – Donisthorpe, H.St.J., 1934. *ibid.*, **70**: p. 198, and Blair, K.G. & Donisthorpe, H.St.J., 1943, *ibid.*, **79**: p. 16.
Axinotarsus marginatus (Lap.) – Allen, A.A., 1971. *Ent. Rec. J. Var.*, **83**: p. 48.
Troglops cephalotes (Ol.) – Key, R.S., 1983. *Entomologist's mon. Mag.*, **119**: pp. 71–72. (A vagrant species).
Sphinginus lobatus (Ol.) – Allen, A.A., *Ent. Rec. J. Var.*, **96**: p. 243.

References

Crowson, R.A., 1964. Habitat and Life-cycle of *Phloiophilus edwardsi* Stph. (Col., Phloiophilidae). *Proc. Royal Entomological Soc. Lond.*, **39(a)** pp. 151–152.
Freude, H., Harde, K.W., & Lohse, G.A., 1979 *Die Kafer Mitteleuropas*, **6**. Krefeld.

KATERETIDAE

BY A.H. KIRK-SPRIGGS

The Kateretidae is a very small family which was formerly the subfamily Cateretinae of the Nitidulidae (Kloet & Hincks, 1977: p. 56). It has, however, been split off from the Nitidulidae based on male genital studies and is regarded as a distinct family (Audisio 1984 and Kirejtshuk, 1986).

The Kateretidae are small (1.5mm–4.2mm), dull-coloured beetles with clubbed antennae. The British fauna comprises only nine species in three genera: *Brachypterus* (2 spp.); *Kateretes* (3 spp.); and *Brachypterolus* (4 spp.). All have adults and larvae that develop on flowers, feeding on buds and pollen.

Brachypterus glaber (Stephens) and *B. urticae* (F.) are both very common beetles which have species of nettles (*Urtica* spp.) as their host-plants. They can be easily found on nettles in woodland, roadsides, field margins and waste ground etc., being commonest at the flowering period June to September. Both species are often found together on the same plant, with *B. urticae* being the more abundant.

The genus *Kateretes* is split into two subgenera, being the subgenus *Pulion* Des Gozis with only one British species *K. (Pulion) rufilabris* (Latreille) and *Kateretes* (*s.str.*), which contains the other two species *K. (s.str.) pedicularius* (L.) and *K. (s.str.) pusillus* Thunberg (= *K. bipustulatus* (Paykull)). They are rather flat beetles having characteristically developed basal antennal segments in males. All three species are to be found in wetland areas, particularly marshes and bogs, where adults and larvae feed on rushes and sedges, *K. pedicularius* and *K. pusillus* on *Carex* spp. and *K. rufilabris* on *Carex* and *Juncus* spp. *Carex* species flower from May to August and *Juncus* from May to September. By carefully examining flowering heads of these plants many examples can be collected. It does not appear to be known how many species of *Carex* these beetles utilise as hosts and this can only be ascertained by careful rearings on the individual sedge species.

Of the four species of *Brachypterolus* which now occur in the British Isles two are native species: *B. linariae* (Stephens) and *B. pulicarius* (Linnaeus). These two species were formerly both regarded as *B. pulicarius*, but a key for their separation has been published (Johnson, 1967: 143). Both these species are to be found feeding on flowers of *Linaria* species on roadsides, disused railway lines, woodlands, etc. The remaining two species *B. vestitus* (Kies.), and *B. antirrhini* (Murray) (= *B. villiger* (R.H.)), are introduced, *Brachypterolus vestitus* first being taken in Britain by Fryer (1929: pp. 101–102); and *B. villiger* by Williams (1926 pp. 262–263). They are to be found as pests on cultivated *Antirrhinum* flowers in parks and gardens. Many accounts of the damage they cause have been published, notably those of Jarvis (1944: p. 237); Henderson, (1946–47: p. 18); and Gimingham & Perkins, (1944:

p. 290). The larvae and damage caused have been described in detail by Tempère (1926).

All these genera can be collected by sweeping and by searching host-plants. *Kateretes* spp. have also been collected using pitfall traps in wetland areas.

Males of genus *Brachypterolus* should have their aedeagi dissected and the median lobe and tegmen carefully separated. These can be mounted, dorsal side uppermost, with a minimum of gum beside the beetle or kept in a small vial of glycerol attached to the mounting pin, see Part I, p. 57.

References

Audisio, P. 1984. Necessità di ridefinizione della Sottofamiglie nei Nitidulidae e nuove prospeltive per la ricostruzione filogenetica del gruppo (Coleoptera). – *Boll. Zool.*, **54** (suppl.): pp. 1–5.

Fryer, J.C.F. 1929. *Brachypterolus (Heterostomus) vestitus* Kiesenwetter, in Britain – *Entomologists' Mon. Mag.*, **65**: pp. 101–102.

Gimingham, C.T. & Perkins, J.F. 1944. *Brachypterolus (Heterostomus) vestitus* Kies. (Col., Nitidulidae) in Hertfordshire. – *Entomologists' Mon. Mag.*, **80**: p. 290.

Henderson, J.L. 1946–1947. *Brachypterolus vestitus* Kies. from his garden in Purley, Surrey [an Exhibit 10 July 1946]. – *Proc. S. Lond. ent. Nat. Hist. Soc.*,: p. 18.

Jarvis, C.M. 1944. *Brachypterolus (Heterostomus) vestitus* Kies. (Col., Nitidulidae). – *Entomologists' Mon. Mag.*, **80**: p. 237.

Johnson, C. 1967. The identity of *Brachypterolus linariae* (Stephens) (Col., Nitidulidae), with notes on its occurrence in Britain – *The Entomologist*, **100**: pp. 142–144.

Kirejtshuk, A.G. 1986. [An analysis of the genitalia morphology and its use in reconstructing the phylogeny and basis of the system of Nitidulidae (Coleoptera)]. – *Tr. Vses. Entomol. O–va*, **68**: pp. 22–28, [in Russian].

Tempère, G. 1926. Un Coléoptère nitidulide du muflier des jardins. – *Rev. Zool. agric. & appl. Bordeaux*, **25**: pp. 155–158.

Williams, B.S. 1926. *Brachypterolus (Heterostomus) villiger* Reitt., a clavicorn beetle new to Britain. – *Entomologists' Mon. Mag.*, **62**: pp. 262–263.

NITIDULIDAE

By A.H. Kirk-Spriggs

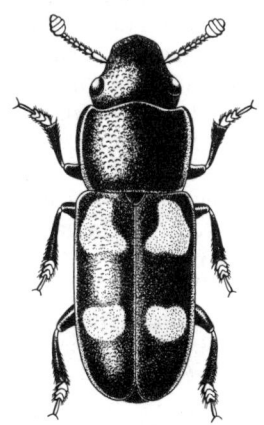

NITIDULIDAE
*Glischrochilus
quadripunctatus* (L.)

The classification of the family Nitidulidae has changed considerably since the publication of the check-list of British Insects (Kloet & Hincks, 1977: pp. 56–57). The nomenclature and higher classification used here are according to Audisio (1984) and Kirejtshuk (1986). The major change had been the removal of the subfamily Cateretinae from the Nitidulidae and its elevation to family status, the Kateretidae. This new family comprising the genera *Brachypterus*, *Brachypterolus* and *Kateretes* is dealt with in a separate section of this book.

The Nitidulidae are small- to medium-sized, obovate to oblong beetles, having 11-segmented antennae with a compact 3-segmented club, the elytra are foreshortened and one to three abdominal tergites are exposed, the tarsi are 5-segmented with the fourth segment always shorter than the others.

The British nitidulids are divided into four subfamilies, the Meligethinae (*Meligethes*, *Pria*); the Carpophilinae, with two tribes, the Carpophilini (*Carpophilus*) and Epuraeini (*Epuraea*); the Nitidulinae (*Nitidula*, *Omosita*, *Soronia*, *Amphotis*, *Pocadius*, *Thalycra*, *Cychramus*); and the Cryptarchrinae (*Cryptarcha*, *Pityophagus*, *Glischrochilus*).

The genus *Meligethes* is the largest of our British genera, with thirty-six species (Kirk-Spriggs in prep.). In Britain they are to be found feeding as larvae and adults on the unopened buds and flowers of the families Cistaceae, Rosaceae, Campanulaceae, Cruciferae, Labiatae, Papilionaceae and Boraginaceae. Each species is specific to an individual plant species, on which the larvae develop. In the case of our two commonest species *Meligethes aeneus* (Fabricius) and *M. viridescens* (Fabricius) the true British host plant, *Sinapis arvensis* L. (Cruciferae), is not the only larval host plant utilised. Unlike other British species they are both capable of completing there larval development on other species of yellow Cruciferae, notably oilseed rape, swede, turnip, cabbage and black mustard, as a result of which they have become very serious pests of these crops. Two other species of the genus are minor pests: *M. flavimanus* Stephens of cultivated roses and *M. nigrescens* Stephens of sweet pea flowers.

Although species are restricted to particular plants they are very commonly collected from a wide range of flowers, particularly before and after the flowering period of the larval host-plants. This has led to a great deal of confusion in the past, as to their true associations, for example the first edition of "A Coleopterist's Handbook" has an extensive list of plants with associated beetles but less than a quarter of these entries are correctly associated for this genus and should be ignored.

The adult beetles emerge from hibernation in the Spring, and on sunny days fly to their host-plants, or feed on pollen of other flowers prior to locating their flowering host-plants. Eggs are laid on developing buds, on which the larvae feed, larval development is usually very short, in the case of *M. aeneus* taking between 9 and 13 days under laboratory conditions (Osbourne, 1965: p. 748). After this period the mature larvae drop to the soil and bury themselves, forming an earthen cell in which to pupate. After a few weeks the adults emerge and feed on flowers, before seeking winter quarters, usually in the soil, in which to hibernate. There is usually only one generation per year. The larvae of *M. aeneus* and *M. viridescens* have been described by Osbourne (1965); *M. nanus* and *M. ruficornis* (as *M. flavipes*) by Perris (1873: p. 76); and *M. difficilis* by Rey (1866: pp. 174–175). Parasites have been discussed by Easton (1962); Askew (1979); Osbourne (1955; 1960). Recent papers on the control of *M. aeneus* include Hokkanen (1989); Hokkanen et. al. (1986; 1988).

The genus *Pria* is superficially similar to *Meligethes*, differing by the lack of arched impressions on the last abdominal sternite. Species of *Pria* also exhibit a high degree of sexual dimorphism, particularly in antennal structure, where the male has a four-segmented club and the female a three-segmented club. The genus has its centre of distribution in Africa and is represented by only one European species, *Pria dulcamarae* (Scopoli), which also occurs in the Yemen and East Africa. Seventy three species have been described so far and an excellent key is given by Cooper (1982).

Very little is known of the biology of *Pria*. The African species do not appear to be as host specific as *Meligethes*, but *P. dulcamarae* seems to be restricted to *Solanum dulcamara* L. and to a lesser extent *S. nigrum* L. (Solanaceae) in Britain. The life cycle has been described by Perris, (1875), and Norgaard, (1919). Eggs are laid amongst the stamens, on which the larvae develop, pupation takes place within the soil. The duration of the life cycle is not known, but Perris notes that adults are always found within flowers from May to September.

The genus *Carpophilus* (Carpophilini) includes generally cosmopolitan species which have been introduced from their tropical centres of distribution, being commonly associated with man as pests of dried fruit, particularly currants, raisins and figs. They only occur in large numbers as storage pests when the fruit has become slightly mouldy. Those species which have managed to establish themselves in non-synanthropic situations are encountered on fungi and mouldy fruits. The larvae are campodeiform with short feeble legs, the abdomen bears a pair of horns at the tip, with a pair of smaller horns just above, they are whitish to pale yellow in colour.

A figure of the rear end of a *Carpophilus* larva is given by Munro (1966: p. 101); a description of larvae is given by Wickham, (1894). Nikitskij (1980: p. 45) notes the genus as being predatory on Scolytidae;

nemotode parasites have been discussed by Remillett & Waerebeke (1975).

Ten species of *Carpophilus* are recorded from the British Isles. According to Kloet & Hincks (1977: p. 56), three species, *C. dimidiatus* (Fabricius); *C. freemani* Dobson and *C. ligneus* Murray, occurring in Britain only under artificial circumstances, while four species, *C. humeralis* (Fabricius) (as *Urophora*); *C. flavipes* Murray, and *C. maculatus* Murray are of doubtful occurrence, this leaves six species as occurring in natural situations in Britain.

A key to the species of economic importance as stored products pests is given by Mound (1989: pp. 22–23).

The second largest genus of nitidulids in Britain is the taxonomically difficult *Epuraea*, with twenty two species. The larvae and in many cases adult biology of the majority of species can only be guessed at or is completely unknown. Larvae are generally found in the galleries of boring beetles (Scolytidae), on flowing tree sap and in fungus, with adults also occurring in flowers. Spornraft (1967: p. 51) gives details of some species, upon which these notes are based. *Epuraea rufomarginata* (Stephens) has been recorded from the galleries of wood boring beetles, under the bark of spruce and on flowing sap of deciduous trees, particularly birch. It has also been recorded from the fungus *Daldinia concentrica*. 21 *Epuraea* species have been recorded in association with fungi in Europe, 2 tentatively associated, *E. silacea* (Herbst) and *E. limbata* (Fab.) and two have, with reservation, been associated, *E. deleta* Sturm and *E. unicolor* (Olivier). Spornraft (1967: p. 51) has suggested that *E. distincta* (Grimmer), may also be associated with fungi.

Saalas (1951) deals with eleven *Epuraea* species associated with spruce, boring beetle galleries under spruce bark being typical but not the only biotype utilised, these species being *E. boreella*, *E. angustula*, *E. pygmaea*, *E. thoracica*, *E. deubeli* and *E. abietina*. Three further species are common according to Saalas on other coniferous trees, *E. pusilla*, *E. oblonga* and *E. laeviuscula*. Not all of these species are British. *Epuraea laeviuscula* (Gyllenhal), a non-British species is known to be associated with the boring beetle *Xyloterus lineatus* (Olivier) (Scolytidae).

Epuraea depressa, *E. melina*, *E. longula* and *E. melanocephala* are all to be found on flowers. In the case of *E. depressa* it is known with certainty that it develops in subterranean nests, especially those of bumblebees, and the morphology of the immature stages has been described by Scott, (1920). Those British species not mentioned above are in many cases associated with tree sap, but the biology of these is almost completely unknown.

Larvae and adults of *Epuraea* species associated with wood-boring beetles have been seen to devour their hosts. In some cases these have been dead and mouldy boring beetle larva. The true associations remain a mystery. Further notes on scolytid associations are given by Nikitskij (1980: pp. 47 & 108); larval morphology is described by Pototskaya (1978).

Within the subfamily Nitidulinae are two genera found in association with bones, dried carrion and similar substances; the genus *Nitidula* with four British species, and the genus *Omosita* with three British species. Some of these species, particularly *N. bipunctata* (L.) and *O. discoidea* (F.), are occasionally encountered in larders and pantries, where meat or other substances have been allowed to dry-out in airy conditions. Some have been taken in birds' nests, particularly those of birds of prey, where animal remains are present in the nests. *Nitidula* species are rather convex, dark beetles often with small spots or other markings on the elytra, *Omosita*, on the other hand, are markedly flattened and have the elytra and pronotum with blotchy markings more extensively developed. The larva and pupa of *Omosita colon* L. has been described by Eichelbaum (1903).

The genus *Amphotis* is represented in Britain by only one species, *A. marginata* (F.). This species is myrmecophilous, living in the nests of the ant *Lasius fuliginosus*. Donisthorpe (1927) gives some very interesting notes on this species (pp. 25–27). He states that *A. marginata* is only very rarely collected away from the nests of this ant, and that the ants feed the beetles with honey as well as the beetles consuming flies and other prey given to their hosts. He gives an illustration of an ant feeding an adult beetle on page 27.

The genus *Soronia* has two British species, *S. grisea* (L.) and *S. punctatissima* (Illiger), which are to be found on tree sap, fermenting vegetable material, and especially in the tunnels and galleries of *Cossus* spp. (Lepidoptera:Cossidae). The remaining nitiduline genera, *Pocadius; Thalycra* and *Cychramus*, are all to be found in association with one sort of fungus or another. *Pocadius ferrugineus* (F.) is in powdery fungi, particularly puff balls, *Thalycra fervida* (Olivier) in soil-dwelling fungi such as truffles and puff balls, and *Cychramus luteus* (F.) is associated with fungi and also found in flowers, hawthorn in spring being particularly productive. Very little appears to be known of the biology of any of these species.

The fourth subfamily, the Cryptarchinae, comprises three British genera. *Cryptarcha* has two species, *C. strigata* (F.) and *C. undata* (Olivier), which are for the most part on running sap of deciduous trees. *Pityophagus ferrugineus* (L.) is frequently found under the bark of coniferous trees. *Glischrochilus*, has three species, *G. hortensis* (Fourcroy); *G. quadriguttatus* (F.) and *G. quadripunctatus* (L.), which can be collected on flowing tree sap, in the galleries of boring beetles and occasionally on rotting fruit, etc. in autumn.

The only complete key which includes the British nitidulid fauna currently available is written in German by Spornraft (1967).

Mounting: In the genera *Carpophilus, Meligethes, Epuraea, Nitidula, Soronia* and *Glischrochilus*, reference to the genitalia of the male or ovipositor of the female is often desirable. Because of its poorly sclerotised and tubular form, the ovipositor is best made up as a clove-oil preparation or kept in glycerol (see Part I, pp. 59). Dissection proceeds in the normal way (see Part I, p. 55) but with *Meligethes* it is important to remove only the apical three segments of the abdomen otherwise the caudal marginal line of the hind coxae which exhibits, in some cases, specific characters may be destroyed.

The aedeagus is dissected in the normal way, but after the careful removal of excess tissues, the median lobe and tegmen should be carefully separated and mounted dorsally separately. They can be mounted beside the beetle with a minimum of glue.

In order to appreciate certain diagnostic characters, *Meligethes* species require a special mounting technique. They can be card-pointed but specimens so treated are vulnerable to accidental damage and can easily be dislodged from the card-point unless adequate protection is afforded by the data labels. The preferred method is to mount the beetles on their right-hand side with the appendages of the left teased out and the right anterior leg removed and mounted dorsally alongside the beetle. In this way the teeth of the anterior tibiae on both sides of the specimen are visible and characters of the ventral and dorsal surfaces are easily observed.

Useful characters are displayed on the intermediate tibiae of male *Epuraea* species. The male genitalia in this genus require examination from above and in profile, for this reason they are best stored in glycerol vials pinned beneath the specimen (see Part I, p. 59) or glued upright, being attached to the card by their base.

The species of *Glischrochilus* are also best mounted on their sides as they exhibit diagnostic ventral characters; there is, however, no necessity to remove the right anterior leg.

References

Askew, R.R., 1979. The biology and larval morphology of *Chrysolampus thenae* (Walker) (Hym., Pteromalidae). – *Entomologists' Mon. Mag.*, **115**: pp. 155–159.

Audisio, P., 1984. Necessità di ridefinizione delle Sottofamiglie nei Nitidulidae e nuove prospeltive per la ricostruzione filogenetica del gruppo (Coleoptera). – *Boll. Zool.*, **54**: pp. 1–5 (suppl.).

Cooper, M.C. 1982. The species of the genus *Pria* Stephens (Coleoptera:Nitidulidae). – *Zool. J. Linn. Soc.*, **75 (4)**: pp. 327–390.

Donisthorpe, H. 1927. *The Guests of British Ants, their habits and life-histories*, George Routledge and Son Limited, London xiii + 244pp.

Easton, A.M. 1962. Mites parasitising *Meligethes* spp., (Col. Nitidulidae). – *Entomologists' Mon. Mag.*, **98**: p. 41.

Eichelbaum, F. 1903. Larve und Puppe von *Omosita colon* L. – *Allg. Z. Ent.*, **8(5)**: pp. 81–87.

Hokkanen, H.M.T. 1989. Biological and agrotechnical control of the rape blossom beetle *Meligethes aeneus* (Coleoptera, Nitidulidae) – *Acta Entomologica Fennica*, **53**: pp. 25–29.

Hokkanen, H: Granlund, H.; Husberg, G-B & Markkula, M. 1986. Trap crops used to control *Meligethes aeneus* (Col., Nitidulidae), the rape blossom beetle. – *Annales Entomological Fennici*, **52**: pp. 115–120.

Hokkanen, H.; Husberg, G-B; & Söderblom, M. 1988. Natural enemy conservation for the integrated control of the rape blossom beetle *Meligethes aeneus* F. – *Annales Agriculturae Fenniae*, **27**: pp. 281–294.

Kirejtshuk, A.G. 1986. [An analysis of the genitalia morphology and its use in reconstructing the phylogeny and basis of the system of Nitidulidae (Coleoptera)]. – *Tr. Vses. Entomol. O-va*, **68**: pp. 22–28 [in Russian].

Mound, L. (Ed.) 1989. *Common insect pests of stored food products a guide to their identification, Seventh Edition.* – publ. British Museum (Natural History), ix + 68pp.
Munro, J.W. 1966. *Pests of Stored Products*, The Rentokil Library publ. Hutchinson of London, 234pp.
Nikitskij, N.B. 1980. *[Insect predators of bark beetles and their ecology]*. Nauka, Moscow, 237pp. [in Russian].
Norgaard, A. 1919. Om *Pria dulcamarae* Scopoli og dens lavevis. – *Entomologiske Meddelelser*, **12**: pp. 128–136.
Osbourne, P. 1955. The occurrence of five hymenopterous parasites of *Meligethes aeneus* F. and *M. viridescens* F. (Col., Nitidulidae). – *Entomologists' Mon. Mag.*, **91**: p. 47.
Osbourne, P. 1960. Observations on the natural enemies of *Meligethes aeneus* (F.) and *M. viridescens* (F.) [Coleoptera:Nitidulidae]. – *Parasitology*, **50**: pp. 91–110.
Osbourne, P. 1965. Morphology of the immature stages of *Meligethes aeneus* (F.) and *M. viridescens* (F.) (Coleoptera:Nitidulidae). – *Bulletin of Entomological Research*, **55(4)**: pp. 747–759.
Perris, E. 1873. Résultats de quelques promenades entomologiques. – *Ann. Soc. Ent. Fr.*, **5(3)**: pp. 61–98; pp. 249–252.
Perris, E. 1875. Larves de Coléoptères. – *Ann. Soc. Linn. Lyon*, **22**: p. 289.
Potoskaya, V.A. 1978. Larval morphology and ecology of beetles of the genus *Epuraea* (Coleoptera:Nitidulidae), – *Enomologicheskoe Obozr.*, **57(3)**: pp. 570–577. [translated *Entomological Review*, **57(3)** 1978 [1979]: pp. 391–296].
Remillet, M. & Waerebeke, van D. 1975. Description et cycle biologique de *Howardula madescassa* n. sp. et *Howardula truncati* n. sp. (Nematoda:Sphaerulariidae) parasites de *Carpophilus* (Coleoptera: Nitidulidae). – *Nematologica*, **21(2)**: pp. 192–206.
Rey, C. 1866. Larve de Coléoptères. – *Ann. Soc. Linn. Lyon.*, **33**: pp. 131–254.
Saalas, U. 1951. Einiges über Charakterarten der Käferbestände an Fichten von verschiedener Beschaffenheit. – *Z. angew. Ent. Berlin*, **33**: pp. 12–18.
Scott, H. 1920. Notes on biology of some inquilines and parasites in nest of *Bombus derhamellus*. – *Trans. R. ent. Soc. Lond.*: pp. 105–124.
Sporncraft, K. von. 1967. 50. Familie: Nitidulidae, *(In) Die Käfer Mitteleuropas, Band 7. Clavicornia.* [Nitidulidae: 20–77], publ. Goecke & Evers, Krefeld, 310pp.
Walsh, G.B. & Dibb, J.R. 1954. *A. Coleopterist's Handbook*, Amateur Entomologists' Society, 120pp.
Wickham, H.F. 1894. Description of the larvae of *Tritoma, Carpophilus* and *Cyllodes.* – *Ent. News*, **5**: pp. 260–263.

RHIZOPHAGIDAE

BY J. COOTER

This small family includes twenty-two British species, divided almost equally between the subfamilies Rhizophaginae (13 species) and Monotominae (9 species). Members of both divisions are of characteristic appearance and easily recognised in the field. The majority of the Rhizophaginae are to be found under sappy bark of dead and dying trees, sometimes in numbers and occasionally two or three species together. *Rhizophagus parallelcollis* (the coffin beetle) and *perforatus* are subterranean and *cribratus* often occurs in vegetable refuse, fungi and the like. *Cyanostolus aeneus* occurs under bark near to or partially submerged in water. *Rhizophagus grandis* has recently been introduced to Britain in an effort to control the Scolytid beetle *Dendroctonus micans*; it is not included in the *R.E.S. Handbook* (Peacock, 1977) and to date has not been formally introduced as British via the entomological literature.

Species of the genus *Monotoma* are very slow moving beetles which inhabit rotting vegetable matter, such as manure and compost heaps. Two species *Monotoma conicicollis* and *angusticollis* are myrmecophilus with *Formica* species. *Monotoma* species, because of their habit of keeping still for a long time and subsequent sluggish movement, are easily passed over whilst examining siftings.

Reference

Peacock, E.R., 1977, Coleoptera, Rhizophagidae. *Handbk. Ident. Br. Insects*, **5(6)(2)**): 23pp. Royal Ent. Soc. London.

CRYPTOPHAGIDAE

BY J. COOTER

This large family is represented in Great Britain by about one hundred species, five of which comprise subfamily Telmatophilinae; the remainder are almost equally divided between the Cryptophaginae and Atomariinae.

TELMATOPHILINAE occur in marshy places and can be swept during late afternoon/early evening. *Telmatophilus typhae* (Fallen) is the commonest species and may be found quite easily by peeling the leaves off reed mace. (*Typha* spp.).

CRYPTOPHAGINAE occur in a wide variety of habitats including stored products and warehouses (see Part III) and other indoor biotopes, e.g. *Cryptophagus cellaris* (Scop.), *simplex* Miller, *saginatus* Sturm, *scutellatus* Newm. and *subfumatus* Kr. Although members of the genus do not seem to be associated with dung, decaying vegetable matter such as compost and manure heaps, bales of straw rotting in fields and mouldy hay are all worth sifting and a number of the more generally distributed species, such as *Cryptophagus acutangulus* (Gyll.), *pilosus* Gyll. and *pseudodentatus* Bruce will be found this way. Some of the less common species, such as *labilis* Er., *confusus* Bruce and *acuminatus* Coombs & Woodroffe inhabit old wood, particularly red-rotten oak and *Dorcus*-infested beech. Nests of birds, especially those in hollow trees, squirrel drays and bat-roosts also harbour rarities such as *Cryptophagus fallax* Balfour-Browne, *lapponicus* Gyll. and *badius* Sturm as well as more common species. *Cryptophagus setulosus* Sturm, *populi* Pk., and *pubescens* Sturm occur in the nests of bees and ground-nesting wasps, whereas *micaceus* Rey inhabits nests of tree-nesting wasps including the Hornet (*Vespa crabro* L.). Rotten fungi should also be searched and may produce *Cryptophagus ruficornis* Stph. (in *Daldinia concentricata* on Ash and Birch trees), *lycoperdi* Scop, and others. Apart from *Micrambe vini* Pz., which may be beaten in numbers from gorse (*Ulex*) and broom (*Cytisus*), other members of this genus, whose habits are not well known are quite scarce but have been taken by beating. *Antherophagus* species breed in bumble-bee nests (*Bombus* spp.), but the adults are more readily swept from flowers, *Caenoscelis* species have been recorded at light and may be found under bark of logs and at cut timber. *Henoticus* species can be beaten from dead branches or sought by searching cut timber.

ATOMARIINAE: A number of the more common and widely distributed species will be found by sifting a range of decaying vegetable matter of man-made origin such as compost heaps, rotting manure, decaying straw and hay and in flood refuse. The scarcer species seem more typical of more "natural" habitats including marsh and fen litter – piles of reed and sedge refuse generally repay investigation. At the coast *Atomaria rhenana* Kr. occurs in tidal and marsh refuse, under stones etc. Mouldy dung, especially horse and deer dung, should be sifted as should leaf-litter and the refuse that accumulates in hedge-row bottoms. *Atomaria fimetarii* (F.) is associated with the Shaggy cap fungus (*Coprinus comatus* (Fries) S.F. Gray), it occurs low down inside the stem. Evening sweeping in marshy areas may produce the brachypterous species *Atomaria atra* (Hb.), *mesomela* (Hb.) and *gutta* Stph. Piles of cut timber in woodlands and timber yards, especially the sawn ends, should be carefully examined as this is the characteristic haunt of *Atomaria pulchra* Er. and *strandi* Johnson.

Identification of cryptophagids is problematic and best attempted when several, or better still, a large number of specimens have been accumulated. This is especially true for *Cryptophagus* as differences are often comparative and there is a good degree of variation within species limits. Dissection of *Cryptophagus* species is recommended as a matter of course as the male genitalia offer reliable characters; the median lobe and parameres should be separated during dissection. Antennae should be displayed and care taken to ensure all the segments are flat on the mounting card. Glue should be kept off the dorsal surfaces as it will obscure or distort sculpture and pubescence, Joy (1932) is not satisfactory, the most up to date work available being Lohse (1967), but even this does not cover all the British species. Papers by Allen (1968) and Johnson (1975 and 1988) and the classic revision of *Cryptophagus* (which also includes *Micrambe*) by Coombs and Woodroffe (1955) are necessary reading.

References

Allen, A.A., 1968. Two additions to the British species of *Atomaria* Steph. (Col., Cryptophagidae), with notes on others of the genus in Britain. *Ent. Rec. J. var.*, **80**: pp. 318–326.
Coombs, C.W. & Woodroffe, G.E., 1955. A revision of the British species of *Cryptophagus* (Herbst) (Coleoptera:Cryptophagidae). *Trans. R. ent. Soc. Lond.*, **106(6)**: pp. 237–282.
Johnson, C., 1975. Nine species of Coleoptera New to Britain. *Entomologist's mon. Mag.*, **111** (1976): pp. 177–183.
Johnson, C., 1988. Notes on some British *Cryptophagus* Herbst (Coleoptera:Cryptophagidae), including *confusus* Bruce New to Britain. *Ent. Gazette*, **39**: pp. 329–335.
Lohse, G.A., 1967. in Freude, H., Harde, K.W., & Lohse, G.A., Die Kafer Mitteleuropas, **7**: pp. 110–158. 55 Fam. Cryptophagidae. Goeke & Evers, Krefeld.
Morris, M.G., 1990. Orthocerous Weevils. *Handbk. Ident. Br. Insects*. Vol. 5 part **16**. 108pp. Royal Ent. Soc. London.

PHALACRIDAE

BY J. COOTER

There are fifteen British species belonging to this distinctive family – *Phalacrus* 5 spp., *Olibrus* 7 spp., and *Stilbus* 3 spp.

They are commonly collected by sweeping, but sometimes it is necessary to search plants. *Phalacrus* species are associated with smutted grasses and sedges, *Olibrus* with Compositae and *Stilbus* with *Typha* and can be sifted from hay and other vegetable debris.

The beetles are of characteristic appearance, though separation to species level requires high magnification and dissection. There is an *R.E.S. Handbook* (Thompson, 1958) available for this family; it is well illustrated and straightforward to use. When preparing Phalacrids it is of prime importance to keep the dorsal surface clean as the microsculpture is easily obscured by glue and grease. The anterior tibiae of *Olibrus* and *Phalacrus* possess a varying number of spurs which must be kept free of glue. The ovipositor of the female often exhibits reliable specific characters and must be dissected with care and mounted in Euparal (after first passing through alcohol and xylene only) on a celluloid or acetate strip. The male genitalia should also be dissected and the median lobe separated from the tegmen; these parts can be glued to the card behind the beetle, but as it is sometimes necessary to view the tegmen from more than one angle, storage in a van Doesburg tube (see Part I, p. 59) is preferable; the ovipositor can be likewise treated.

Reference

Thompson, R.T., 1958. Coleoptera, Phalacridae. *Handbk. Ident. Br. Insects*, **5** (5(a)): 16pp. Royal Ent. Soc. London.

CORYLOPHIDAE

BY J. COOTER

These minute beetles rarely reach 1mm in length and to the naked eye are often strongly reminiscent of plant mites. They will be encountered by general sifting of vegetable debris and under bark.

Their identification is not easy and Joy (1932) will prove adequate for *Sericoderus* and *Corylophus* only (a total of three species!).

The British *Orthoperus* species have been ably revised by Allen (1970) who as well as presenting a key to all the British species, discusses each in turn and gives a commentary on the history of the genus. A more detailed account is given by Bruce (1948) including (pp. 12–15) advice on sexing and dissecting these diminutive beetles. *Rypobius ruficollis* was added to the British List by Hammond (1971).

References

Allen, A.A., 1970. Revisional Notes on the British species of *Orthoperus* Steph. (Col., Corylophidae). *Ent. Rec. J. var.*, **82**: pp. 112–120.
Bruce, N., 1948. The Scandinavian species of the genus *Orthoperus* Steph. (Coleoptera). *Opusc. ent.* Suppt. **9**: 1–34 with 4 plates.
Hammond, P.M., 1971. *Rypobius ruficollis* Jacqu. (Col., Corylophidae) A Genus and Species New to Britain. *Ent. Gazette*, **22**: pp. 241–243.

COCCINELLIDAE

BY J. COOTER

COCCINELLIDAE
Propylea 14–punctata (L.)

This is a family of world-wide distribution, numbering in excess of 3,500 species. Of the 43 generally considered British, 25 are normally thought of as "Ladybirds", familiar to the majority of people. The remaining 18 species are mostly small (under 3mm) and less brightly coloured and marked, generally more retiring and must be sought in the preferred habitats. While a considerable amount of information has been obtained concerning the biology, ecology and distribution of British Ladybirds, there are still many gaps in our knowledge and the biology and ecology of other coccinellids is poorly understood.

Coccinellids are of considerable economic importance because many species are predatory, feeding on aphids. However, not all are predatory and other prey types are favoured by some predatory ladybirds. The range of food types eaten by British coccinellids is considerable. Some species, such as *Adonia variegata, Coccinella 7-punctata, Adalia bipunctata, A.10–punctata* and *Calvia 14–guttata* have a wide range of prey species. Others are more specific; *Myzia oblongoguttata* and *Coccinella hieroglyphica* both appear to feed in the wild on a very restricted range of species. Indeed *C. hieroglyphica* may be monophagous. Not all predatory coccinellids feed on aphids; members of the Chilocorini, Hyperaspini and Coccidulini feed principally on coccids, although aphids and adelgids are also eaten. Members of genus *Nephus* feed almost exclusively upon coccids of the genus *Pseudococcus; Aphidecta obliterata* favours species of *Adelges; Stethorus punctillium* feeds on phytophagous mites. Of the non-predatory coccinellids, *Subcoccinella 24–punctata* is our only phytophagous species. Both larvae and adults of *Halyzia 16–guttata* and *Psyllobora 22–punctata* feed on powdery mildews of the Erysiphaceae, the former most commonly on Sycamore and the latter on Hogweed. *Micraspis 16–punctata* is apparently unspecific as to which of various mildew types it feeds upon.

We are fortunate in having our fauna well documented. The *R.E.S. Handbook* (Pope, 1953) is now out of print and seldom listed by dealers. It is somewhat difficult to use as it ignores colour pattern and concentrates on anatomical features, many of which are rather obscure and difficult for the novice or any one without a microscope to appreciate and are mostly displayed by the ventral surface. The recently published Naturalist' Handbook *Ladybirds* (Majerus & Kearns, 1989) is an excellent authoritative account of our ladybirds, with concise information about the other coccinellids. The genera *Nephus* and *Scymnus* were covered in considerable detail by Pope (1973); this and Majerus & Kearns (1989) are essential reading. The latter summarises all that is known about the biology and ecology of the ladybirds as well as giving field keys to adult and larval ladybirds, plus a more detailed key to all the British Coccinellidae. Alas, no British author illustrates the female genitalia of *Nephus* and *Scymnus*, but these are figured by Fursch (1967).

The colour patterns of the group are of an aposematic nature, that is they indicate to predators that the insect is distasteful or toxic. The family has the ability to "bleed" as a reflex action when endangered and so deters would-be predators.

A reason why coccinellid identification often proves tricky is that the colour patterns of many species are variable and in some cases subject to extreme genetic variation. The ground colour may vary as may the colour, size and number of spots and other marks which may become fused; some species have distinct melanic forms. Other species show more or less continuous variation in the number of spots – *Adonia variegata* may have from three to fifteen spots, though even numbers of spots are rare. *Harmonia quadripunctata* has either four or sixteen spots and only very rarely any other number. This last species was first recorded in 1937 in Britain and since then it has extended its range from Suffolk, arriving in Kent in 1978, Scotland in 1982 and Devon and the Welsh border in 1987. It appears to continue to widen its range.

Perhaps the most frequently encountered problem with mounting Coccinellidae is getting their appendages, especially the antennae, displayed. This is best achieved by placing the freshly killed beetle on its elytra on a small piece of damp tissue. It is held with a light downward pressure from fine-pointed forceps and the appendages are then carefully teased out with the aid of a fine pin. For several species, reference to underside characters will be required during identification, for which the beetle will have to be removed from its mounting card. Alternatively, identification can be carried out immediately prior to mounting, at which time it will be established if it is necessary to dissect the genitalia in order to confirm identity.

Many of our Coccinellidae are widespread and frequently encountered by beating and sweeping in marshes, meadows, woodlands and plantations. A small number are quite decidedly scarce and others restricted to certain areas of Britain; for an account of habitats see Majerus & Kearns (1989). *Clitostethus arcuatus* is a rare woodland species confined to south and central England and there are very few records. *Nephus bisignatus* is known as British from three old records from the south-east coast; *N. quadrimaculatus* is restricted to Kent and East Anglia where it is associated with coniferous trees, but occasionally deciduous trees and ivy. It has been taken in large numbers on a few occasions. *Hippodamia 13–punctata* has not been recorded in Britain since the 1950's and is thought now to be extinct. *Coccinella 5–punctata* is associated with unstable river shingles where it is most frequently encountered during the spring, roaming the shingle banks. During the summer it may be taken by sweeping river-side vegetation. It is known from the Spey valley, N.W. England, West Wales and the south coasts of Devon and Dorset. *Exochomus nigromaculatus* was recently reinstated on the British List after a specimen was captured near Doncaster in 1967 (Skidmore, 1985). Stephens (1831) records two other British specimens, one taken at Windsor in 1816 and the other, without mention of date, "in the vicinity of Bristol".

References

Fursch, H. 1967. Coccinellidae (Marienkafer) in Freude, H., Harde, K.W. & Lohse, G.A. *Die Kafer Mitteleuropas*, **7**: pp. 227–278. Goeke & Evers, Krefeld.
Pope, R.D., 1953. Coleoptera Coccinellidae. *Handbk. Ident. br. Insects* **5(7)**: 12pp. Royal Ent. Soc. London.
Pope, R.D., 1973. The Species of *Scymnus* (s.str.) *Scymnus (Pullus)* and *Nephus* (Col., Coccinellidae) occurring in the British Isles. *Ent. mon. Mag.*, **109**: pp. 3–39.
Majerus, M.E.N., & Kearns, P. 1989. Naturalists' Handbooks **10**, *Ladybirds*. Richmond Publishing, Slough.
Majerus, M.E.N., Forge, H. & Burch, L. 1990. The Geographical distribution of ladybirds in Britain (1984–1989) *British Journal of Entomology and Natural History*. Cambridge.
Majerus, M.E.N., Hostplant and habitat preferences of British ladybirds. *Ent. mon. Mag.* (in press).
Skidmore, P., 1985. *Exochomus nigromaculatus* (Goeze) (Col. Coccinellidae) in Britain. *Ent. mon. Mag.*, 121: pp. 239–240).
Stephens, J.S., 1831. *Illustrations of British Entomology*, Mandibulata **4**: 414pp. London.

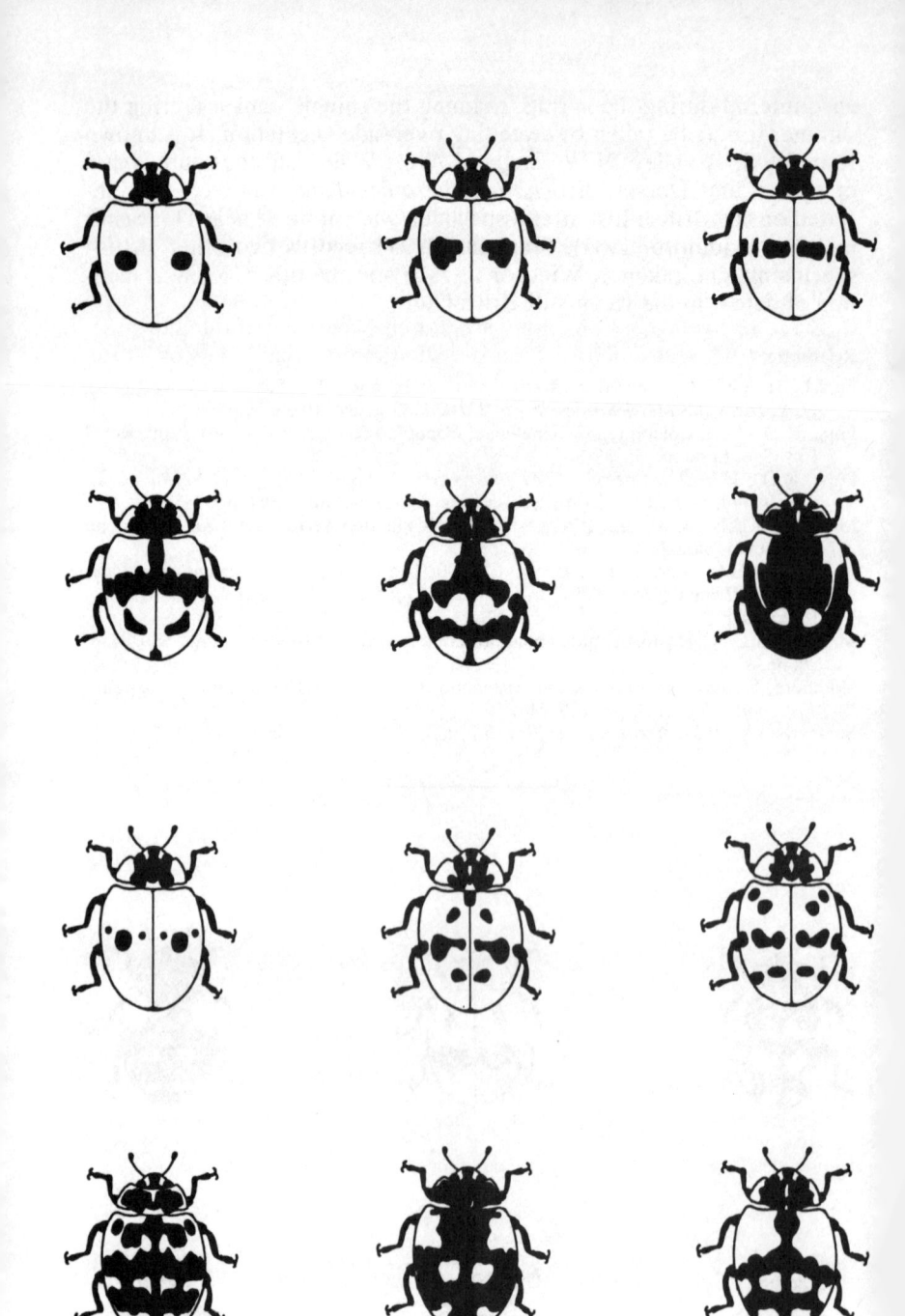

Forms of the Twin-spot Ladybird, *Adalia bipunctata*. (after Majerus)

LATHRIDIIDAE

BY J. COOTER

The fifty-one British species are roughly evenly divided between the subfamilies Lathridiinae (which includes the "Plaster Beetles") and the Corticariinae.

These beetles can be found in a wide range of habitats, though, with one or two exceptions, they are not characteristic of dung or marshy places. Humid mouldy places are the more typical haunt of lathridiids and such biotopes as garden and farm refuse, myxomycetes, mouldy bracket fungi that have fallen to the ground, flood and tidal refuse will all repay investigation. Some species will be found in old buildings, especially where there is dampness and mould. Beating trees, especially those in ancient woodlands may produce such scarce species as *Corticaria alleni* Johnson, and by grubbing at the roots of grasses and other plants at the coast one might turn up *Corticarina fulvipes* (Comolli) and *fuscula* (Gyll.). Some species are associated with stored products (see Part III); these have been dealt with in detail by Hinton (1941).

When preparing lathridiids it is very important that glue be kept off the dorsal surface, especially so with pubescent species. Antennae should be carefully displayed with all segments in the same plane flat on the card. It is necessary to dissect some species, particularly in genera *Enicmus* Thoms, *Cartodere* Thoms., *Corticarina* Reitt. and *Corticaria* Marsh. as the aedeagi exhibit reliable characters.

Within the family there are several difficult groups and as Joy (1932) is not adequate for determination purposes, Peez (1967) is recommended as the definitive modern key. Several useful papers on the British fauna have over the years appeared in the entomological press and are listed below with the references.

References

Allen, A.A., 1951. *Lathridius bifasciatus* Reitt. (Col., Lathridiidae) an Australian beetle found wild in Britain. *Entomologist's mon. Mag.*, **87**: pp. 114–115.
Allen, A.A., 1952. *Lathridius norvegicus* A. Strand (Col., Lathridiidae) Rediscovered: an addition to the British List. *Entomologist's mon. Mag.*, **88**: pp. 282–283.
Allen, A.A., 1966. A Clarification of the status of *Cartodere separanda* Reitt. (Col., Lathridiidae); and *C. schueppeli* Reitt. new to Britain. *Entomologist's mon. Mag.*, **102**: pp. 192–198.
Hinton, H.E., 1941. The Lathridiidae of Economic Importance. *Bull. ent. Res.*, **32**: pp. 191–247. 67 figs.
Johnson, C., 1974. Studies on the genus *Corticaria* Marsham (Col., Lathridiidae). Part 1. *Ann. Ent. Fenn.* **40(3)**: pp. 97–107.
Johnson, C., 1986. Notes on some Palaearctic *Melanophthalma* Motschulsky (Coleoptera: Lathridiidae), with special reference to *transversalis* auctt. *Ent. Gazette*, **37**: pp. 117–125.
Peez, A. von, 1967. *in* Freude, H., Harde, K.W. & Lohse, G.A., Die Kafer Mitteleuropas, **7**: pp. 169–190. 58 Fam. Lathriddiae. Goeke & Evers, Krefeld.
Tozer, E.R., 1972. On the British species of *Lathridius* Herbst (Col., Lathridiidae). *Entomologist's mon. Mag.*, **108** (1973): pp. 193–200.

CISIDAE

BY J. COOTER

Many of the twenty-two British species of this family, owing to their elongate cylindrical form, bear a superficial resemblance to certain ipinine scolytids. Members of the Cisidae are exclusively fungivorous and can be found in numbers in a variety of lignicolous bracket fungi on a wide range of trees as well as by beating dead twigs and branches in otherwise living trees, especially oak. Often more than one species will be found in the same fungus and it is good policy to sample as wide a variety of bracket fungi growing on a wide variety of host trees as possible.

Many species have a preferred host but are occasionally found in other fungi. *Cis pygmaeus* can be beaten from dead fungoid twigs and branches of oak; *C. bilamellatus* is common in *Piptoporus betulinus* and *Polystictus versicolor; C. lineatocribratus*, very scarce in England and a little less so in Scotland, frequents *Fomes* on birch; *C. jaquemarti* is another Scottish species found in *Piptoporus* on birch; *C. nitidulus* on *Ganoderma* on beech; *C. punctulatus* occurs in *Poltstictus abietinus* on branches of pines; *Rhopalodontus perforatus* occurs in *Fomes* in Scotland.

Many species show a fair range in size, scales easily abrade and immature specimens are frequently encountered which are paler than usual. Further, the male head and pronotal characters vary and, as differences between species are often comparative, determination is often tricky. It is, therefore, a good idea to amass a number of specimens before attempting to run them through a key. Joy (1932) will be found adequate, but Lohse (1967) presents a more modern and better illustrated key. The median lobe and parameres of the male aedeagus afford reliable characters for the separation of species. Dissection is best carried out on fresh specimens and parameres and median lobe separated. They can be lightly glued to the mounting card or, as it is often necessary to view from more than one direction, preserved in a van Doesburg tube (see Part I, p. 59).

References

Allen, A.A., 1990. Notes on the species-pair *Cis festivus* Panz. and *C. vestitus* Mell. (Col., Cicidae). *Ent. Rec. J. Var.*, **102**: pp. 177–180.

Aubrook, E.W., 1970. *Cis dentatus* Mell. (Col., Cisidae) an addition to the British List. *Entomologist*, **103**: pp. 250–251.

Kevan, D.K., 1967. On the apparent conspecificity of *Cis pygmaeus* (Marsh.) and *C. rhododactylus* (Marsh.) and on other closely allied species (Col., Ciidae). *Entomologist's mon. Mag.*, **102**: pp. 138–144.

Lohse, G.A., 1967. in Freude, H., Harde, K.W., & Lohse, G.A., *Die Käfer Mitteleuropas*, **Vol. 7**: pp. 280–295. Goeke & Evers, Krefeld.

"HETEROMERA"

By J. Cooter

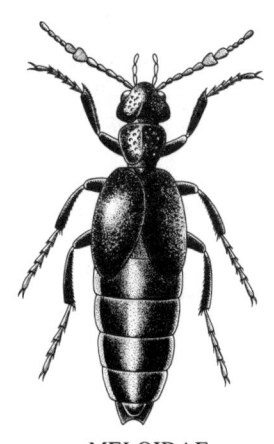

This grouping, in Britain, includes several numerically rather small families – Mycetophagidae, Colydiidae, Tenebrionidae, Tetratomidae, Salpingidae, Mycteridae, Pythidae, Pyrochroidae, Melandryidae, Scraptiidae, Mordellidae, Rhipiphoridae, Oedemeridae, Meloidae, Anthicidae and Aderidae.

MYCETOPHAGIDAE: *Typhaea stercorea* may be captured by sweeping in damp areas, but the other members of the family, in Britain, are associated with dead wood, especially fungoid and decaying timber. *Mycetophagus quadripustulatus* is often common in decaying vegetable matter, including wind-fall apples, etc.; *M. quadriguttatus* is to be found associated with certain "stored products" see Part III, p.199.

MELOIDAE
Meloe violaceus Marsham.
Oil beetle

COLYDIIDAE: Some Colydiids are excessively scarce, this is especially so of some timber-associated species such as *Teredus* and *Oxylaemus*. *Bitoma crenata* is another woodland species, widespread and often present in numbers, unlike other synchitinids it can be seen running over the bark of dead trees (especially beech) in sunlight. *Aglenus brunneus* is a stored-product pest (see Part III, p. 199) and *Orthocerus clavicornis* inhabits sand-dune areas where it may be found at large, trapped (see Part I, p. 8) or by grubbing at the roots of marram, etc. *Myrmechixenus* species inhabit "hot-beds", alas now but a rare feature of gardens, and have been sifted from pockets of fungoid hay contained within straw-rich manure.

All the British species are readily identified using Joy (1932) except *Cicones undata* Guerin (Mendel & Owen, 1987), *Synchita separanda* (Reitter) (Allen, 1964) and *Pycnomerus fuliginosus* Er. (Welch, 1964).

TENEBRIONIDAE: This is a very large family with in excess of 15,000 species world-wide exhibiting great diversity in form and structure. Many species live in hot arid regions and are adapted to survive extremes of temperature and prolonged drought.

In Britain, following the classification adopted by Pope (1977), we have 44 species (this includes the Lagriinae and Alleculinae, which are often regarded as separate families). Some tenebrionids are of economic importance (see Part III, p. 199) and are very rarely, if ever, encountered away from their adopted indoor environment. The remainder may be found in sandy areas, especially near the coast, beaten from blossoms or are associated with fungi and fungoid wood.

The student is directed to Buck (1954) and Brendell (1975) for identification and further information.

TETRATOMIDAE: Of the three British species, *Tetratoma fungorum* is the most widespread and abundant, often being found in large numbers during the winter in *Polyporus* fungi on birch. *T. desmaresti* and *ancora* are also woodland species, but much less widespread and more localised, the latter being associated with Hornbeam, though occasionally noted from other trees.

SALPINGIDAE: Another family of woodland beetles which are found by beating dead and dying branches, including burnt branches, under bark of dead and dying trees and under the loose flakes of thick bark on living trees.

MYCTERIDAE: Of *Mycterus curculioides* there are only a very few British records, none from the 20th century.

PYTHIDAE: The handsome violet-blue *Pytho depressus*, our sole representative of this family, frequents the older pine woodland of northern Scotland. Both adult and larva can be found under fungoid pine bark.

PYROCHROIDAE: The "Cardinal Beetles". *Pyrochroa coccinea* and *serraticornis* are widespread in England and Wales where they occur in established and older woodlands. *Schizotus pectinicornis* is confined to a few areas of northern Scotland and as scattered populations in the Welsh Borders (Herefordshire, Radnorshire, Breconshire notably); it is associated with mature/over mature birch.

MELANDRYIDAE: Another family of predominantly woodland beetles, associated with dead and dying timber attacked by fungi. They may be beaten from dead branches, bred from bracket fungi or found under fungoid bark. Some species, such as *Melandrya caraboides* and *Conopalpus testaceus*, may be beaten from blossoms in wooded areas.

SCRAPTIIDAE: In Britain this family (sometimes regarded as a subfamily of the Mordellidae) is confined to two genera – *Scraptia* with three species and *Anaspis* with fourteen.

Scraptia dubia is known as British from one or two ancient specimens captured in Dorset; it is a much larger beetle than either of its two congeners and despite being found not infrequently in France and the rest of the Continent, has not been taken in Britain for a great many years. *Scaptia fuscula* appears to be confined to Windsor Forest/Park and *S. testacea* to a very few ancient woodlands. They can be beaten from over mature oaks during June and early July. Great care must be taken when mounting as the beetles are very delicate and can easily be broken to pieces even with a fine hair brush; a minimum of very dilute glue is recommended.

Anaspis are also delicate and easily damaged. Females are difficult to identify, especially when captured in isolation. Fortunately, the bulk of the species occur in numbers. Males should be mounted ventrally after first removing the abdomen, this should be mounted dorsally alongside the beetle. The shape and form of the abdominal laciniae (or their absence) are of great diagnostic value. The antennae, palpi and legs should be set in the normal way.

As mentioned earlier, the bulk of *Anaspis* species can be obtained in numbers by beating blossoms – especially hawthorn and other spring flowering shrubs – and by sweeping umbelliferous plants. *Anaspis bohemica* is confined to the Scottish Highlands and is quite scarce; *A. septentrionalis* is known only from the Type specimens captured last century in Scotland; *A. schilskyana* is confined to Moccas and Blenheim Parks where it occurs on the most ancient oak trees during early June; *A. costai* occurs later in the summer than most other species, late July and early August typically; *A. melanostoma* is confined to Windsor. The others are not uncommon and are widely distributed.

For identification, both Joy (1932) and Buck (1954) are now out of date or otherwise unreliable. The excellent paper by Allen (1975) is necessary reading; it brings Buck's (1954) key up to date. Otherwise the key by Kaszab (1969) is recommended.

MORDELLIDAE: At the time of writing, this taxonomically difficult group, much better represented in warmer climates, is receiving attention from British coleopterists and several species New to Britain have been recognised, with the possibility of more as yet undetected.

Of characteristic form and appearance, they can be swept or taken individually from a range of pollen-rich flowers, notably *Heracleum, Daucus, Artemisia, Helianthemum, Sorbus* (*Mordellochroa abdominalis*) and *Crataegus* (*Mordellistena* species). They are especially active around mid-day in warm sunny windless weather during the summer.

When mounting it is important to keep gum off the femora and tibia as well as the upper surface especially in pubescent species. Killed with ethyl acetate, the beetles are transferred from the killing bottle to a sheet of white absorbent paper (kitchen tissue, filter paper etc.) and turned on their dorsal surface. The legs, head, mouth-parts, and antennae are frisked out with the aid of a fine pin, the beetle is held gently with the forceps. Male genitalia are dissected carefully with a fine pin and/or forceps via the terminal abdominal segment. The dissected genitalia are then transferred to a dissecting dish and the median lobe and minute parameres, paired and asymetrical, are painstakingly separated. They can be lightly glued to the mounting card alongside the beetle, but as the orientation of the parameres is critical, the van Doesburg method (see Part I, p. 59) might be preferable. The beetle is then carefully transferred to the mounting card, upon which waits a small elongate blob of glue. The head, mouth-parts, antennae and two anterior pairs of legs are then suitably arranged and lightly glued to the card. *The posterior pair of legs should not be gummed* but raised so that the tibia is level with the dorsal surface of the elytra. This not only greatly fascilitates identification, but makes a much more acceptable and more easily prepared mount.

For identification, the student is directed to the paper by Batten (1986), both Joy (1932) and Buck (1954) being long out of date. For collecting and preparing, full detail is contained in Batten (1988).

RHIPIPHORIDAE: *Meoticus paradoxus*, the sole British representative of the family, is a parasitoid of social wasps (Vespidae). Adults are to be found, rarely, at large during the autumn, but they are more frequently obtained by carefully working through a wasp nest (after first fumigating it and killing the wasps!).

OEDEMERIDAE: *Nacerdes melanura* is predominantly a coastal species, though there are some inland records. It inhabits dead wood and timber, including that used for coastal defences not exposed to the effects of tides. *Chrysanthia nigricornis* (Skidmore, 1973) is known in Britain from only one Scottish locality – Glen Tanar. *Oncomera femorata* is nocturnal and may be beaten from ivy, especially that growing on old and derelict buildings, and from hedges, during day time.

MELOIDAE: *Lytta vesicatoria*, the "Spanish Fly", is an occasional migrant which from time to time establishes itself in parts of East Anglia and Kent for a few years. The last such occasion being during the late 1940's early 1950's. It feeds on ash and privet. *Apalus muralis* was for many years known from an old wall at Cowley, Oxfordshire; the wall is no longer standing and there are only a very small handful of post-war records. It is a parasitoid of *Anthophora* bees.

Of the genus *Meloe*, several have not been recorded in Britain for many years and there are no 20th century records of *Meloe variegatus*; *M. cicatricosus* has not been recorded since 1906 and *autumnalis* and *brevicollis* since the 1940's/1950's. *M. rugosa* has been found recently during late winter/early spring in Gloucestershire, Worcestershire and Somerset. The two most frequent and widespread British species are *proscarabaeus* and *violaceus*; they occur throughout Britain, but *violaceus* has a more northerly and westerly distribution. The beetles are parasitoids of bees of the genera *Anthophora* and *Osmia*. Their life-history has been studied by many naturalists and is described by Fowler (1891).

ANTHICIDAE: *Notoxus monoceros* is typically a sand-dune beetle, but has been recorded from a number of sandy places well inland, unlike *Anthicus bimaculatus* which seems restricted to a small number of coastal sand-dune systems. *Anthicus floralis*, *formicarius* and *bifasciatus* can be found in compost and manure heaps, often in numbers, the former two being quite widespread and frequent, the latter more restricted.

The first British specimens of *Anthicus tobias* were taken on a municipal rubbish tip, and the other members of the genus may be sought by sifting tidal refuse, sea-weed or by grubbing near the sea on short turf or in salt-marshes.

ADERIDAE: Our three aderids are woodland beetles. *Aderus brevicornis* is very scarce and known from only a small handful of ancient woodland localities. *A. populneus* is likewise restricted, but not as scarce, though by no means common. *Aderus oculatus*, although restricted by habitat preference, can be found in quite large numbers by beating old oaks. In this genus there is a marked sexual dimorphism exhibited by the antennae which, in the male, are much longer.

They are very fragile beetles and should be mounted with a very dilute glue.

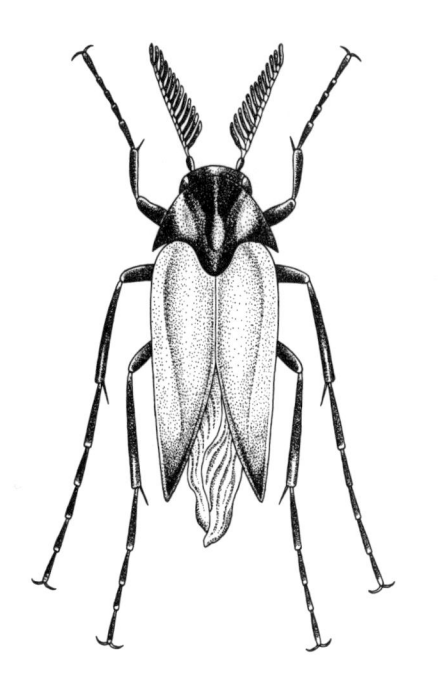

RHIPIPHORIDAE
Metoecus paradoxus (L.)

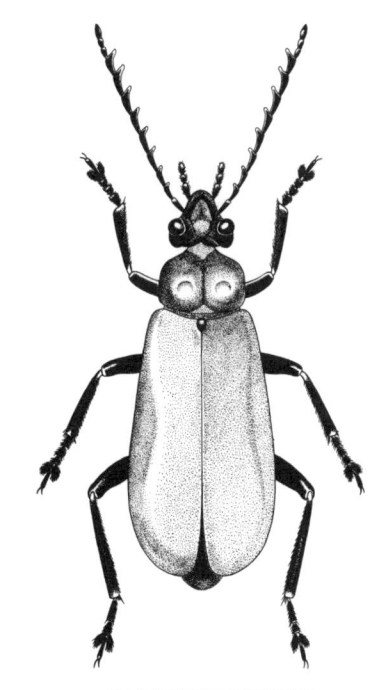

CARDINAL BEETLE
(PYROCHROIDAE)

References

Allen, A.A., 1964. The Genus *Synchita* Hellw. (Col., Colydiidae) in Britain; with an Addition to the fauna and a new synonymy. *Ent. mon. Mag.*, **100**: pp. 36–42.

Allen, A.A., 1975. Two Species of *Anaspis* (Col., Mordellidae) new to Britain, with a Consideration of the status of *A. hudsoni* Donis. etc. *Ent. Rec. J. var.*, **87**: pp. 269–274.

Allen, A.A., 1988. A fourth species of *Ischnomera* Steph. (Col., Oedemeridae) in Britain. *Ent. Rec. J. var.*, **100**: pp. 199–202.

Batten, R., 1986. A Review of the British Mordellidae (Coleoptera), *Ent. Gazette*, 37: pp. 225–235.

Batten, R., 1988. Mordellidae (Coleoptera): catching, preparing and mounting, *Nieuwsbrief European Invertebrate Survey – Nederland*, **18**: pp. 9–10.

Brendell, M.J.D., 1975. Coleoptera (Tenebrionidae) *Handbk. Ident. Br. Insects* **5 (10)**. Royal Ent. Soc. London.

Buck, F.D., 1954. Coleoptera (Largiidae, Alleculidae etc.) *Handbk. Ident. Br. Insects* **5(9)**. Royal Ent. Soc. London.

Kaszab, Z., 1969. in Freude, H., Harde, K.W., & Lohse, G.A., *Die Kafer Mitteleuropas* **8**: pp. 188–196 Goecke & Evers, Krefeld.

Mendel, H. & Owen, J.A., *Cicones undata* Guér. (Col., Colydiidae) New to Britain. *Ent. Rec. J. var.*, **99**: pp. 93–95.

Pope, R.D., 1977. in Kloet, G.S. & Hincks, W.D., A Check List of British Insects (Edn. 2). *Handbk. Ident. Br. Insects*, **11(3)**: xiv + 105pp.

Skidmore, P., 1973. *Chrysanthia nigricornis* Westh. (Col., Oedemeridae) in Scotland, a genus and species new to the British List. *The Entomologist*, **106**: pp. 234–237.

Skidmore, P., & Hunter, F.A., 1980. *Ischnomera cinerascens* Pand. (Col., Oedemeridae) new to Britain. *Ent. mon. Mag.*, **116**: pp. 129–132.

Welch, R.C., 1964. *Pycnomerus fuliginosus* Er. (Col., Colydiidae) new to Britain. *Ent. mon. Mag.*, **100**: pp. 57–60.

CERAMBYCIDAE (LONGHORN BEETLES) AND SCOLYTIDAE (BARK AND AMBROSIA BEETLES)

BY DR. DAVID LONSDALE

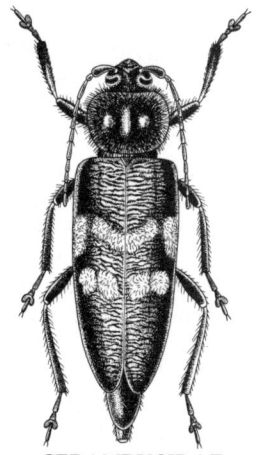

CERAMBYCIDAE
Hylotrupes bajulus (L.)

These two families have no close taxonomic affinity, but they share a strong association with trees. Thus, many species in both families are woodland beetles. This association with trees is by no means exclusive however, and there are species which feed only on herbaceous plants. Moreover, herbaceous plants provide an important adult food source for many of the wood-feeding longhorn beetles. Even among the tree-feeding members of the Scolytidae, there are species which derive most of their nutrition from fungi growing in the dead bark of wood, rather than on the plant material itself.

The cerambycids or "longhorns" are one of the most "popular" of beetle groups, especially amongst general naturalists, because many species are large, attractively shaped insects, and a good proportion are also brightly coloured and patterned. Indeed, some of the largest and most colourful species are protected by law from collection in some European countries, although none has been so far placed on the British protected list. There are just over 60 species in the British list, from among a world total of well over 20,000.

The adult body shape of nearly all longhorns is distinctive, being characterised by the long, tapering hind-body, with its "shoulders" wider than the adjoining edge of the prothorax, and the long antennae which can be reflexed over the body. The head and prothorax are often elegantly sculptured, and this combination of features makes it easy to distinguish longhorns from members of other families, with only a few exceptions among the Oedemeridae and Chrysomelidae. The antennae in some species are so long that they may exceed the length of the body; by several times in the case of the males of *Acanthocinus aedilis*, the "Timberman", which, apart from occasional importations, is confined to the Caledonian pine forests.

A few species are so distinctive in general appearance, e.g. *Strangalia maculata*, that they can be recognised by visual comparison with illustrations but, as with beetles in general, a key is necessary for the identification of most species. Some caution is needed in the use of keys in relation to the stated size ranges of longhorn species since, for some, it is not uncommon to find atypically small individuals. Two widely available keys to the British species are those of Joy (1931) and Duffy (1952). There are species which sometimes fail to "key out" correctly with either or both of these keys, and it may be useful to consult Fowler (Fowler 1887–1891. Fowler & Donisthorpe 1913) or the continental keys of Kuhnt (1912), Reitter (1908–1916) and other authors in difficult cases. Guides to particular genera and species may also be found in periodicals, such as Uhthoff-Kaufmann's (1988) paper on *Rhagium*.

The larvae of longhorns, like those of other wood-boring beetles, have poorly developed thoracic legs, or none at all. The body is stout, fleshy, somewhat flattened and tapering towards the tip of the abdomen. The head is dark, equipped with powerful mouthparts, and rather sunken into the prothorax which is broader than the more posterior segments. The body surface carries fleshy protuberances and these are used by the larva to help its progress along its gallery which, for many species is oval in cross-section. A key for the identification of all the British species, together with those that tend to be imported into the country, was provided by Duffy (1953a), who also added much useful information on their biology. There is also a slightly earlier key by van Emden (1948), which covers all the British genera and most of the species. If larvae are collected, they may be reared to the adult stage in captivity, and possibilities for this are discussed below.

Although the longhorns are very much woodland beetles in the main, they can more generally be regarded as plant stem borers, whether of woody stems as in most cases, or of herbaceous stems, as in the case of *A. villosoviridescens*, which feeds in thistles and hogweed. Some species may occur in roots. The wood-boring species are thought to obtain their food from the wood itself, rather than from wood-inhabiting fungi, unlike some of the scolytids. The condition of the wood chosen by adult females for egg-laying differs from species to species. Some choose sound, living trees, while the choice for others may range from recently dead trees to those that have begun to decay.

One of the advantages enjoyed by the collector of cerambycids is that the larvae of most wood species can be found all the year round. It is, however, very important to limit larval collection very strictly in order to avoid significant destruction of the habitat. Dismantling substantial quantities of dead wood can endanger populations of the invertebrates which depend on its continual presence in a locality. This practice should be avoided wherever there is not a superabundance of the food source. Dead-wood collecting is one of the few activities of the amateur which can pose a real threat to wildlife conservation. (Some members of the present two beetle families are under threat in Britain, and the cerambycid *Cerambyx cerdo* is now thought to be extinct.) With this important proviso, the following table may be used to aid the collection of particular species where there is a good reason, such as the need to survey a site when adults are not on the wing. The table is derived from a passage in the first edition of this Handbook, provided by the late Dr. E.A.J. Duffy, with more recent additions.

Type of woody habitat	Species of associated trees and cerambycids
Dead or slightly decayed stems or roots	*Prionus corarius* in both conifers and broadleaves *Arhopalus tristis* in pines (large emergence holes) *A. rusticus* in pines (large emergence holes) *Asemum striatum* in pines (small emergence holes)
Decaying logs or stumps	*Rhagium bifasciatum* in pine and other species *Strangalia quadrifasciata* in oak and alder
Decaying, attached branches	*Strangalia maculata* in sallow and birch *S. melanura* in oak, sycamore and spruces
Living stems, under bark or in wood	*Aromia moschata* in sallows *Saperda carcharias* in poplar and willow
In pith of saplings forming gall-like external swellings	*Saperda populnea* in sallow, aspen and birch
Under bark of logs or stumps	*Tetropium gabrieli* in larches and pines *Phymatodes testaceus* in oak and other broadleaved spp. *Rhagium mordax* in oak
Under bark of palings	*Callidium violaceum* in pine
Under bark of injured or decaying branches	*Clytus arietis* in various broadleaved spp. *Molorchus minor* in Scots pine and spruces *Leiopus nebulosus* in oak *Mesosa nebulosa* in oak *Rhagium inquisitor* in various broadleaved spp.
Dead twigs or stems of small woody spp.	*Molorchus umbellatarum* in crab apple and dogrose *Pogonerus hispidus* in apple, laurel and other plants
Roots of small, woody plants	*Strangalia melanura* in *Cytisus*
Herbaceous plant tissue	*Agapanthia villosoviridescens* in thistles and hogweed *Phytoecia cylindrica* in umbellifers (early spring)

The adults of many longhorn species, including wood-feeders, spend some of their time on the flowers of herbaceous plants where they feed on pollen, and this makes it very easy to observe, photograph and, if necessary, collect them. The only disadvantage of searching flowers for the adults, as compared with breaking up logs or tree stumps for larvae and pupae, is that it can only be done during a limited part of the year but, on the other hand, it does not destroy the often scarce deadwood habitat.

Members of the family differ widely in their degree of specialisation towards particular foodplants, and this is true of both larval and adult feeding. It is not possible here to give an exhaustive list of the adult foodplants of the British Cerambycidae, but some examples can be given.

The flowers of the Umbelliferae are a good starting point for the observer because the adults of many longhorn species frequent them.

Examples include the common but strikingly coloured *Strangalia maculata* and other members of the genus such as the relatively small and sombre *S. melanura*. Hogweed (*Heracleum sphondylium*) is by far the commonest umbellifer used by longhorns. Angelica (*Angelica sylvestris*) and ground elder (*Aegopodium podagraria*) are also commonly chosen. The flowers of rosaceous plants are also prominent as an adult food source, and are used exclusively by some, such as *Rhagium mordax* which seems to occur mainly on bramble. Hawthorn, meadowsweet and dogrose are also well worth inspecting. Adults of some species (e.g. *Leptura sexguttata* and *Obrium brunneum*) are best found by sweeping or beating vegetation where they may be resting or feeding on the flowers of shrubs and small trees. In the case of *O. brunneum*, the lower, dead branches of conifers may also be beaten (Uhthoff-Kaufmann, 1985).

Adult scolytids are very different from cerambycids in appearance, being more closely related to the weevils (Curculionidae), but they are distinctive in their own right, with their rather "hunched" appearance which is produced by the wholly or partly concealed head and the usually rather large and broad pronotum. The antennae are rather short and clubbed and the legs stout. In some species the apical portion of the elytra are modified to form a shovel-like declivity which is used to help move frass during burrowing and which has elaborate sculpturing which differs between otherwise similar species and may perhaps be concerned with species-matching prior to copulation. This sculpturing is of value in identification and is used in the key provided by Duffy (1953b). Duffy's key is currently under revision, and there is an improved version available for the genus *Hylastes* (Nash 1983). No keys are available for identification of the larvae of the British species. These resemble the larvae of weevils, being legless, white, wrinkled and fleshy.

Unlike cerambycids, scolytids are almost exclusively small beetles, with few exceeding 4mm in length. Among the 60-or-so British species, *Ips sexdentatus* at about 8mm is something of a giant, although not quite equalling the recently introduced continental species, *Dendroctonus micans*, which reaches 9mm. Their colours are as sombre as those of the Cerambycidae are striking and varied. The small size and unprepossessing appearance of scolytids does not make them very attractive to the general naturalist, but many of them are fascinating biologically and some are of great economic importance as we shall see at the end of this chapter.

As in the case of cerambycids, different members of this family choose a very wide range of plant material in which to breed. Many spend their larval life in the main stems of trees, while others are found in small twigs or in roots. As mentioned at the beginning of the chapter, yet others occur in herbaceous plants, as do some of the cerambycids. The tree-feeding species have a number of different feeding patterns (Linssen, 1959), but these can be divided into two main types from the point of view of the field coleopterist. These are (a) feeding within the inner bark (phloem and cortex) and (b) feeding within the wood.

The species which feed within the bark are called, simply, bark beetles. The adult female bores into the bark of a tree or log and constructs a gallery in which eggs are laid. Pairing may take place on the bark surface, in the gallery entrance, or – in the case of polygamous species – in a special nuptial chamber constructed by the male. The larvae burrow outwards from the mother gallery, forming feeding galleries of their own. Each larval gallery terminates in a pupal chamber with an exit hole, unless the larva has succumbed to unfavourable conditions or to predators or parasites. The shape of the system of mother gallery and larval galleries is characteristic of the species concerned, and can easily be seen on the underside of bark if it is carefully peeled away from the wood. The pattern can, with most species, also be seen on the wood surface itself, since the galleries extend partly into the outermost sapwood. The pattern is intricate and indeed beautiful in the case of some *Scolytus* species.

A rather unusual larval feeding habit which has been adopted by *Dendroctonus micans* (the Great spruce bark beetle), a species introduced to Britain, is the production of a communal feeding chamber, rather than a series of individual larval galleries. This mass feeding is thought to enable the larvae to overcome the natural resistance of the tree; indeed most other species confine their attacks to weakened or dead trees and logs. The extent to which they can nevertheless cause economic loss is discussed at the end of this chapter.

The adults of some bark beetles have feeding habitats which are separate from the larval breeding substrates, a lifestyle which parallels that of the longhorns but, unlike longhorns, these adult bark beetles feed on woody plant tissues. Some feed on living bark in the twig crotches of the tree species in which the larvae breed (e.g. *Scolytus* spp. on elm), while others tunnel in the shoots (e.g. *Tomicus piniperda* in pine). Thus, from the collecting point of view, the adults cannot be found in a conspicuous situation like the flower feeding sites of many cerambycid species. Nevertheless, adult bark beetles can sometimes be found in great numbers when they are attracted to dying or physiologically stressed trees, suppressed branches on otherwise healthy trees, or recently felled logs. This aggregation phase is of course soon followed by the disappearance of the beetles into the egg-laying galleries, but the numerous entrance holes and frass-ejection holes betray their presence beneath the bark. One way of collecting adults at the aggregation phase is to put out "decoy" logs, recently cut from living trees. Ring-barking trees, though generally not environmentally acceptable and always requiring permission if they are either someone else's property or if they come under a tree preservation order, can achieve the same effect.

Collection of adult bark beetles at aggregation or egg-laying time can only be done over a relatively short period of the year; in late spring or early summer for most species, but larvae and pupae can of course be found over a longer period. Thus trees and logs which have entry holes in the bark can be sampled by removing patches of bark, thus exposing the feeding and pupal chambers underneath. As with all forms of deadwood collecting, restraint should be exercised unless there is a great deal of habitat material present in the area. Logs which have been colonised in this way can be taken home (again seeking permission where appropriate) and kept inside a cage until the adults (plus any interesting parasitoids) emerge. General notes on breeding and on caging colonised wood are given later in this chapter.

The remaining scolytids, and the related family Platypodidae, are wood-borers. They go under various names, such as "pinhole borers", "shothole borers" and "ambrosia beetles". The small size of the beetles and thus of their boreholes gives rise to the first two of these names. The third name describes the way of life adopted by most of the wood-boring scolytids, for these beetles feed not so much on the wood itself as on the "ambrosia fungus" which lines the walls of their galleries and grows extensively in the surrounding sapwood. There are a number of fungi which form "ambrosia", and they have an intimate relationship with the beetles, being carried within special cavities within the body called mycetangia. Mycetangia, which may be derived from a variety of body parts, also occur among some of the bark beetles, although many of the latter have no such organs and are true plant-feeders.

One ambrosia beetle, *Platypus cylindrus*, has been included on the British Red Data Book list. It is also quite an elegant beetle, being very elongated, and indeed belonging to the Platypodidae rather than the Scolytidae. It is interesting that it can occasionally attain the status of a pest when its numbers build up enough to riddle the timber with their galleries.

The wood-boring scolytids are, like their bark-boring cousins, attracted to damaged or unhealthy trees or parts of trees. In particular, some seem to be attracted to areas of dead bark which are undergoing fermentation due to the growth of yeasts and other micro-organisms. They bore into the wood underlying such bark to construct their egg-laying galleries, expelling a fine sawdust-like frass which is lighter in colour than the frass of bark beetles. As in the case of bark beetles, the adults can be collected over the fairly short period when the adults are on the bark surface, but collection of larvae from then on would involve a major operation; i.e. dismantling the hard, unrotted wood of a standing tree. This can rarely be attempted by the amateur, and it is wiser to collect emerging adults by caging areas of the tree stem either *in situ* or as sawn lengths of stem.

Most scolytids feed on a range of host plants, but few will attack both conifers and broadleaved trees. In general, conifers are more prone to attack, as discussed at the end of this chapter. The following table shows a selection of associations between scolytid and plant species.

Host Plants	Associated Scolytids
Pines	*Tomicus piniperda, T. minor* (also spruce and fir), *Polygraphus poligraphus* (also fir), *Hylastes ater* (branches, roots and stumps), *H. brunneus* (N. Britain), *H. attenuatus, H. angustatus, H. opacus* (also elm and ash), *Pityophthorus pubescens* (in twigs), *P. lichtensteini, Pityogenes bidentatus, P. trepanatus* (Scotland, E. Anglia), *P. quadridens* (Scotland; also in firs), *Ips acuminatus, I. sexdentatus, Orthotomicus suturalis, O. laricis, Crypturgus cinereus* (following attack by other scolytids; also in fir; not native in Britain, occurring on imported timber), *Xyleborinus saxeseni* (also in broadleaved trees)
Spruces	*Polygraphus poligraphus, Hylastes angustatus, H. cunicularis, Crypturgus subcribrosus* (also fir and sometimes pine and larch; uses old galleries of other spp), *C. cinereus* (also, see pine, above), *Cryphalus asperatus* (also fir), *C. piceae, Pityophthorus pubescens, Pityogenes chalcographus*
Larches	*Ips cembrae, I. acuminatus, Orthotomicus laricis*
Juniper and *Cupressus*	*Phloeosinus thujae*
Various conifers	*Hylurgops palliatus* (especially on cut logs), *Dryocoetes autographus* (especially on dead or felled Silver fir), *Trypodendron lineatum, Pityogenes chalcrographus*
Oak	*Dryocoetinus villosus* (also beech, sweet chestnut; egg gallery boring produce a very fine brown dust), *Scolytus intricatus* (in oaks with root rot; sometimes also in other broadleaved trees), *Trypodendron domesticum and T. signatum* (also in beech), *Xyleborinus saxeseni* (also maples and conifers), *Ernoporus fagi* (young wood on mature trees; also in beech)
Ash	*Pteleobius vitattus* (more typically in elm), *Hylesinus crenatus* (also other broadleaved trees), *H. oleiperda* (in twigs), *H. orni* (in slender branches), *H. fraxini* (also other broadleaved trees)
Elm	*Pteleobius vittatus* (also occasionally in ash), *Xyleborus dryophagus* (also other broadleaved trees), *X. saxeseni* (also maples; see also pines above), *Scolytus scolytus* (also in other broadleaved trees), *S. laevis, S. multistriatus, S. mali* (also on rosaceous trees)
Rosaceous trees and shrubs	*Scolytus mali* (also in elm), *S. rugulosus*
Leguminous plants	*Phloeotribus rhododactylus, Hylastinus obscurus* (in dead stems of both woody and herbaceous spp.)
Alder, hazel and birch family	*Dryocoetes alni* (alder, birch, hazel), *Scolytus ratzeburgi* (birch only; Scotland),
Poplar and willow families	*Trypophloeus binodulus* (in dry branches), *T. granulatus* (poplars only)

Ivy	*Kissophagus hederae*
Old man's beard	*Xylocleptes bispinus*
Various broadleaved trees and shrubs	*Xyleborus dispar, Scolytus multistriatus, Xyleborus dryographus* (especially oak), *Taphrorychus bicolor*

Mention has been made of the rearing of scolytids in captivity through the simple expedient of removing naturally colonised logs from the natural habitat to a more convenient location. The simplest method of retaining the emergent adults (as well as any predators and parasitoids which may have been attacking them) is to sew up the logs or other breeding substrate in fine net curtain fabric. The same technique can be used for cerambycids and for other wood-inhabiting insects with free-flying adult stages. This method has the disadvantages that it is difficult to examine the material during the rearing period, and that emerging insects are likely to get caught in the folds of the netting and cannot easily be removed for study. A much more refined device has been described by Owen (1989); this is essentially a pyramidal mesh tent with the aperture of a trapping bottle inserted near its top.

Rearing wood or bark-inhabiting insects in artificial habitats is also possible, but it is much more difficult to provide the right conditions of moisture and nutrition. Different species have different requirements in these respects, and too few examples are known to give any worthwhile guidance here. However, the requirements of any one species can be to some extent determined by observing the nature of the natural food material. One of the main problems associated with excessive moisture is that it encourages over-abundant fungal growth. Fungi are, on the other hand, essential for the larval development of those scolytids which normally infect the food material with their own fungal flora. It is not generally feasible to rear these in substrates which are sterile or which are colonised by the wrong type of fungi. Such species (e.g. the ambrosia beetles) are best reared only from naturally colonised substrates.

Wood-boring cerambycids can be reared by simply removing colonised wood from the wild and keeping it in containers which can be kept at the right moisture content. The containers traditionally used were metal biscuit boxes, but a modern alternative can be provided by various items of plastic ware if these can be equipped with small perforations to provide ventilation and to prevent build-up of condensation. Large ice cream tubs can be used for the smaller species; dustbins can accommodate the larger ones which may need to have substantial pieces of wood in which to complete their tunnelling.

The collection of naturally colonised wood may deplete the wild population, and it is often difficult to know whether a particular piece of wood has been colonised. Another method (which also unfortunately involves destruction of part of the dead wood habitat) is to search for individual larvae and to provide them with pieces of food material in the rearing containers. The larvae may have difficulty in establishing new galleries in such material, and it is advisable to drill a number of suitably-sized "priming" holes in the pieces of wood to which they are introduced. Some species may also feed in small pieces of wood and sawdust, and it is advisable to provide heaps of such material amongst the larger lumps of wood.

A number of wood-boring beetles, mainly lucanids, have been reared on unconventional plant materials such as apples and beetroots. Such foods, rich in nutrients, may actually produce adults in a shorter time than the natural food, and they can be provided without depleting the supply of dead wood in the wild. It is well worth trying such foods on cerambycid larvae, and since there are many which have not yet been reared in this way, there is much scope for publishing the results. Bark beetles, on the other hand, are most easily reared in the more natural conditions provided by logs of the host tree. However it is also possible to rear some species on pieces of detached bark, if they are clamped between two pieces of glass. This method, which has been used for the gregariously feeding species *Dendroctonus micans*, is excellent for observation of the insects. An even more artificial method, which has been used for *D. micans* and which could probably work for other bark beetles, is to use powdered bark, kept at a carefully regulated moisture content, instead of entire sheets of bark. Such material can be kept in a variety of containers such as Petri dishes.

As has been mentioned in this chapter, certain wood or bark-inhabiting beetles can become forest pests and there may be a need to practise some degree of forest hygiene by removing fallen trees in certain situations where they could harbour beetle species capable of attacking standing trees or log stores in the forest area. On the other hand, it is extremely important that all those who use or manage forests should realise that dead wood habitats are under severe threat in many areas, and need urgent protection. Many foresters now realise this, but there are still some who feel that any dead wood in any forest is at worst a tree health hazard and at best a source of firewood. To provide the coleopterist with some information on the important balance between conservation, forestry and collecting, this chapter ends with a brief account of the economic importance of some of the British Scolytidae and Cerambycidae.

A number of cerambycids can damage felled produce or timber products, while a few, such as *Saperda* spp. can attack living trees. Nevertheless, none is considered a serious pest in British forestry. In some tropical countries, however, some of the species which form galleries under the bark of trees can kill them by destroying the layer of cells which lays down new bark and wood, the cambium, and this can cause serious losses in plantations. Exotic cerambycids are frequently imported into Britain, and a few have become established here. There is no suggestion that this poses a direct threat to British forestry, but some species of *Monochamus* are vectors of the potentially lethal pine wood nematode, which is causing devastation in some parts of the world, notably southern Japan. It has been claimed that damage from this nematode is only severe in climates warmer than almost anywhere in

Europe, but importation controls are being attempted in case this relationship is not water-tight.

The Scolytidae include some of the most serious forest pests, although the threat which they pose is almost entirely confined to coniferous forests. Some of them, such as *Tomicus piniperda*, can attack logs *en masse*, allowing the rapid invasion of the wood by blue-stain fungi which greatly reduce its saleability. Some, like species of *Ips*, are able to attack the bark of living trees if the trees are under stress, as commonly occurs in areas disturbed by forest thinning, drought or windthrow of adjacent trees. A typical situation of risk is a partly windthrown stand, where the fallen trees are a breeding ground for large beetle populations, and the surviving trees are still in a stressed condition.

Some scolytids such as *T. piniperda* can cause an additional type of damage by killing the young shoots of conifers when feeding as adults prior to mating. The Great spruce bark beetle, *Dendroctonus micans*, which is thought to have entered Britain on timber imported from the Continent, can attack the main stems of apparently healthy trees, and the law has been invoked to designate the outbreak areas as a controlled zone, beyond which it is illegal to transport living *D. micans*, or any produce which might carry the beetle. A biological control agent, the beetle *Rhizophagus grandis*, has been released by the Forestry Commission to help control populations within these areas.

A chapter on the scolytids cannot end without some mention of Dutch elm disease. The fungus which causes this lethal disease exists as a number of "aggressive" and "non-aggressive" races, spread by various scolytids in the different regions of the world where the disease occurs. In Britain the chief vector is *Scolytus scolytus*, which does not have any special association with the fungus other than to carry spores from disease victims to fresh host trees. Contrary to popular misconception, the infection does not take place when the beetles go to the bark for breeding, but at the stage of maturation feeding, when the recently emerged adult flies up into the crown of the tree and gnaws grooves into the twig crotches. The tree is killed by the fungal attack in the sapwood, and the fungus later grows out into the bark where the next generation of beetles are breeding and where they will pick up spores to continue the next disease cycle.

Dutch elm disease has implications for conservation in relation not only to the loss of habitat for the insects and other animals which have depended on elm trees, but also the conservation of the dead wood habitat which the disease generates. This dead wood habitat is largely a thing of the past in southern England, where the epidemic is long past and where it was generally destroyed for firewood or in the supposed interests of sanitation. However, there are regions where it is still present or being produced and, for some, advice on sanitation may help to protect this valuable, if transient, wildlife resource. Sanitation is not considered appropriate in areas where the disease is out of control, but even where it is practised, the dead wood habitat need not be entirely ruined. The object of sanitation is to prevent the breeding of the scolytid vectors, and this requires not the disposal of the timber, but the removal of any live bark from the main stem and the burning of branches too small to be easily debarked.

Further information on the conservation of dead wood habitats, and its compatibility with forestry is given in the AES handbook No. 21 (*Habitat Conservation for Insects*). By carefully defining what is necessary for sanitation and for conservation, the status of our dead-wood habitats can perhaps be improved, not only for scolytids and cerambycids, but also for the many other insects which depend on this much neglected wildlife resource.

References

Duffy, E.A.J., 1952. Cerambycidae: *Handbooks for the Identification of British Insects.* Royal Entomological Society, London. **5(12)**, 18pp.
Duffy, E.A.J., 1953a. A monograph of the immature stages of British and imported timber beetles. British Museum (Natural History) Monograph, 350pp.
Duffy, E.A.J., 1953b. Scolytidae and Platypodidae: *ibid.* **5(15)**, 20pp.
van Emden, F.I., 1948. A key to the genera and most of the species of British cerambycid larvae. *Entomologists' Monthly Magazine* **75**, pp. 257–273; **76**, pp. 7–13.
Kuhnt, P., 1912. Illustrierte Bestimmmungs-Tabellen der Käfer Deutschlands. Schweizerbart (Nägele & Sproesser), Stuttgart. 1138pp.
Nash, D.R., 1983. The genus *Hylastes* Erichson (Col.: Scolytidae). *The Coleopterists' Newsletter* **14**, pp. 5–7.
Owen, J.A., 1989. An emergence trap for insects breeding in the dead wood. *British Journal of Entomology* **2**, pp. 65–67.
Reitter, E., 1908–16. *Fauna Germanica.* Die Käfer der Deutschen Reiches. Lutz, Stuttgart, 5 vols.
Uhthoff-Kaufmann, R.R., 1985. The genus *Obrium* (Col.: Cerambycidae) in Great Britain; a re-appraisal. *Entomologist's Record* **97**, pp. 216–223.
Uhthoff-Kaufmann, R.R., 1988. A revised key to the varieties of *Rhagium bifasciatum* F. (Col.: Cerambycidae) *Entomologist's Record* **100**, pp. 217–225.

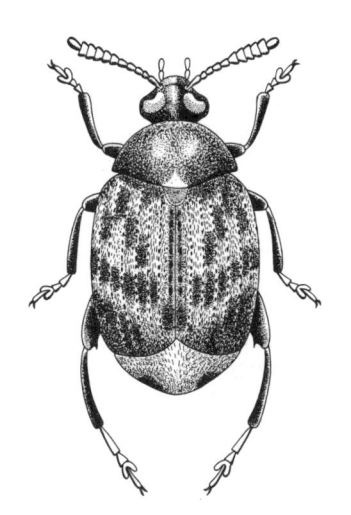

BRUCHIDAE
Bruchus pisorum (L.)

CHRYSOMELIDAE

By M.L. Cox

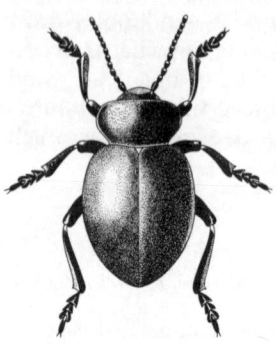

Twelve subfamilies (listed in taxonomic order), totalling at present 257 species, are recognised within the British Chrysomelidae: Donaciinae – 20 species; Orsodacninae – 2; Zeugophorinae – 3; Criocerinae – 7; Clytrinae – 4; Cryptocephalinae – 19; Lamprostomatinae – 1; Eumolpinae – 1; Chrysomelinae – 41; Galerucinae – 19; Alticinae – 124 and Cassidinae – 14. Some of these are still given full family status by some authorities.

They are surprisingly not a particularly popular group with the amateur, considering the frequency with which they are encountered in the sweep-net or on the beating-tray and many are large and showy. However, about half the British fauna is represented by the flea-beetles which are small, often difficult to catch and often uniform in colour and thus present the most difficult problems as regards accurate identification. The lack of good up-to-date keys also partly explains why this still remains an unpopular family.

CHRYSOMELIDAE
Timarcha tenebricosa (F.)

Specific identification in the majority of genera is relatively straightforward since many display good external characters. However, there are still "tricky" groups in most subfamilies since colour in these is variable and often external characters are only relative and unreliable. In such cases resort has to be made to dissection of the genitalia. Notably difficult genera include *Phyllodecta*, *Phaedon*, *Galerucella*, *Longitarsus*, *Altica*, *Phyllotreta*, *Aphthona*, *Chaetocnema*, *Psylliodes* and *Cassida*.

DONACIINAE. These usually brightly metallic-coloured beetles frequent the foliage and flowers of a variety of aquatic plants during the summer. They eat strips from the leaves and also feed on pollen.

The ova are usually laid in groups, e.g. on the undersides of water-lily leaves, and are covered by an opaque gelatinous substance. The larvae are aquatic and feed on the rhizomes and roots of their host plant and obtain oxygen through a pair of hollow caudal spines arising from the 8th abdominal segment which are inserted into the host plants tissues. Pupation occurs in tough brown cocoons attached to the roots of the host plant. Larvae and pupae may overwinter but adults of the new generation may emerge in the autumn, either remaining in the cocoon or emerging above water. They are univoltine.

Adults may be collected by sweeping vegetation in the littoral zone of lakes, lochs and small ponds or from the margins of streams and rivers. Larvae and pupae are obtained by pulling up clumps of uprooted water plants using a grappling iron.

Although the adults in this subfamily are moderately large and sometimes with distinctive coloration and sculpture they still prove quite difficult. This is because the body colour is sometimes very variable intraspecifically and also in those species with the hind femora armed, the sexes are often dimorphic as regards the size and position of the teeth. These teeth are usually a good specific character but are sometimes variable as in females of *versicolorea* where they may be large or greatly reduced (virtually absent). Attempts have been made to use the elytral depressions in species separation in *Donacia* but these are variable and not a valid character. The aedeagus is useful in separating males of *Macroplea appendiculata* and *M. mutica* since keys based on external characters are unreliable.

The sexes are usually easy to distinguish since the males have the anterior tarsi dilated; the size and arrangement of teeth on the hind femora are usually different and rarely the body colour is dimorphic (as in *Plateumaris affinis* in which the females are coppery and the males black with a purple or violet reflection).

Since ventral characters are not used in specific determination, specimens are best mounted dorsally with the legs (especially the anterior tarsi and hind femora and tibiae) well displayed.

ORSODACNINAE. The pale-brown or black adults of *Orsodacne* species frequent the flowers of various herbaceous and woody plants from May until July. They feed upon the anthers and pollen and probably oviposit from June to July. The ova are white and very elongate with rounded ends. Incubation requires about two weeks and larvae are probably present in the field from mid-June. The position of feeding of *Orsodacne* larvae is not known with certainty. Morphological evidence suggests that they are external or internal root feeders (Cox, 1981) and not miners in leaf petioles as proposed by Balachowsky (1963). In Britain it is possible that *Orsodacne* species overwinter as first instar larvae after a lengthy summer diapause. They emerge from hibernation in early spring, feed up and the adults emerge from May until July. It is not definitely known where the larvae hibernate and when and where they commence feeding. It is possible that hibernation is above ground. *O. lineola* larvae were most commonly dissected from gizzards of bluetits, which obtain most of their food from twigs, buds and leaves, but rarely were they found in gizzards of ground-feeding great-tits (Betts, 1955, 1956).

The adults are best collected from areas in and around mature broad-leaved woodland by beating, especially rowan or hawthorn in flower, or by sweeping vegetation rich in flowering Umbelliferae. *Orsodacne lineola* and *cerasi* are easily separated using external characters and should be mounted dorsally.

ZEUGOPHORINAE. Zeugophora adults are dull, either unicolorous orange–brown or bicolorous yellow–orange and black. They occur on the foliage of young poplars and aspens from May to October. Their damage is characteristic since they cause netting of the leaves which become completely perforated by small holes and only the sclerotised veins remain intact.

The yellow oval ova are laid singly in small concavities eaten by the females in the upper epidermis and covered by a secretion. Oviposition probably commences from mid-May and the larvae mine the parenchyma of the leaves during June and July causing characteristic black blotch mines. Pupation occurs within a cell in the soil. New generation adults emerge from about mid-August, feed and move below into their overwintering quarters from September. The *Zeugophora* species are univoltine.

Adults may be collected by beating young aspens and poplars especially those showing the characteristic damage of the adults and larvae.

The three British species are readily separated on head and body colour and hence it is unnecessary to dissect their genitalia. The sexes are difficult to distinguish and apparently can only be separated on the shape of the apex of the final abdominal sternite.

CRIOCERINAE. The adults are typically narrow, elongate, with the pronotum much narrower at the base than, and often differently coloured from, the elytra which are rectangular and often with a bright metallic coloration. The adults, which usually stridulate using an elytra/abdominal apparatus, may be seen from April to September feeding on the leaves of their hosts (Gramineae, Compositae, Liliaceae). Those feeding on grasses and cereals (*Oulema* species) cause characteristic damage since they eat short longitudinal strips in parallel lines 1mm wide, completely perforating the leaf blade.

The ova, white (*Lema cyanella*), greenish–grey (*Crioceris asparagi*), yellow (*Oulema*) and red–brown (*Lilioceris lilii*) are struck by an adhesive substance in one's and two's on the undersurface of the leaves or sometimes deposited in grooves eaten by the adults. In *C. asparagi* they are fixed by their extremities, while in *L. lilii* they are stuck laterally. The larvae are strongly convex and cover themselves dorsally with a viscous layer of excrement which seems to protect them against predators. They feed openly, eating elongate strips 1mm wide from the upper layer of the epidermis, leaving the lower epidermis intact as a fine transparent membrane, especially in those feeding on Gramineae. There are four instars and pupation occurs in white cocoons constructed by the larvae either on the host plant or in the soil. New generation adults usually emerge from July to October and enter hibernation in grass tussocks and other sites from September onwards. *O. melanopus*, *O. gallaeciana* (*lichenis*) and *L. lilii* are probably univoltine whilst *L. cyanella* and *C. asparagi* may have two generations per year.

Adults are best obtained by sweeping their known hosts, especially if showing characteristic damage. *O. melanopa* and *O. gallaeciana* also occur in grass tussocks during the winter. *Lema cyanella* occurs in broad-leaved woodland sites, clearings and edges, grassland, scrub etc.

Oulema erichsoni occurs in coastal areas and grassy places besides dykes and ditches; *C. asparagi* is to be found on arable land, gardens and allotments on Asparagus. *Lilioceris lilii* is associated with lilies in gardens, garden centres and the like.

Most criocerines are distinct and easily identified, except *Oulema gallaeciana*, *erichsoni* and *septentrionis*. *O. gallaeciana* is the smaller and more widespread and with elytra much less elongate than in *erichsoni*. *O. septentrionis* can be separated by the shape of the elytra and black pronotum (which in *erichsoni* is concolorous with the elytra) as well as the strength of elytral puncturation, especially in the apical half.

Sexes are difficult to recognise, except of *C. asparagi*, the males of which have longer more strongly curved anterior tibiae. Species should be mounted with the dorsal surface uppermost.

CLYTRINAE. The adults which may be partly metallic are usually encountered nibbling the leaves or young shoots of various bushes, including *Salix*, *Betula* and *Prunus*, or around the nests of ants, especially *Formica rufa* from April to October.

The eggs are covered with excrement by the females and either dropped to the ground (*Clytra*, *Gynandrophthalma*) near the nests of *Formica rufa* or attached together by long pedicels of excrement to the leaves of bushes (*Labidostomis*). The ants move the eggs into their ant-hills possibly mistaking them for seeds or other plant debris. The larvae, after hatching, build a larval case onto the egg case and feed upon plant material in the ant-hill. The larval cases are very variable in form and furnish excellent generic characters. The larvae may be met with in ant-hills (*Clytra*, *Gynandrophthalma*) whilst *Labidostomis* larvae occur under stones near ant-hills, but not at the interior. Pupation occurs within the case but the entrance is completely closed by the final instar larva. According to Donisthorpe (1902) the cycle in *C. quadripunctata* is yearly and as eggs have an incubation period of about 20 days, the first instar larvae are likely to occur within the nest the third week of June. The larvae would probably feed during the summer, autumn and winter becoming fully developed the following spring, giving rise to a new generation during May and June.

Adults are best collected by beating bushes but those of *Clytra quadripunctata* have been taken in light traps.

The adults of British clytrines are easily separated as they belong to three distinct genera. *Gynandrophthalma affinis* is confined to the Wychwood area of Oxfordshire where it has been beaten from hazels in dediucuous forest (Holland, 1910; Atty, 1970). *Labidostomis tridentata* has been collected from young birches, usually in rough open ground in woodland during May, June and July (Massee, 1945); recorded mostly from eastern England, the most recent being the mid 1950's. *Clytra quadripunctata* is widespread and sometimes common, especially in woodland throughout Britain. *Clytra laeviuscula* is doubtfully British but there are possible old records for the Black Wood, Rannoch where it is possibly associated with *Formica sanguinea* and also chalk hills

(Streatley, Berkshire, Chilterns) where it may be associated with *Lasius niger* and *L. alienus* (Allen, 1976).

Sexing is fairly straightforward, especially in *Labidostomis* which has serrated antennal segments 5–10 broader in males, the mandibles, are longer and have a dorsal prominence, the anterior tibiae are longer and more curved in relation to the mid- and posterior tibiae, and tarsal segments 1 and 2 are longer in relation to the same segments on other legs. The females, as in *Clytra* and *Gynandrophthalma*, have a deep median circular pit in the apical abdominal sternite. There is sometimes only a shallow broad smooth depression in the males of *Clytra*.

CRYPTOCEPHALINAE. Cryptocephalus adults are variable in colour and may be almost entirely metallic green or blue; or entirely yellowish, black or brownish or a combination of these. They occur on the foliage of either bushes and trees or low growing herbaceous plants from June until September. During oviposition the female retains each egg between the posterior tarsi, turning them around and depositing on each a regular layer of excrement. They are then dropped to the ground where they may be collected by ants. Incubation of the eggs takes about three weeks and the larvae never abandon their cases but enlarge them by adding new pieces made from excrement. There is much controversy over the larval food of *Cryptocephalus* species, but in captivity *C. labiatus, pusillus* and *parvulus* larvae took *Betula* leaves although *parvulus* seemed to prefer leaves which had turned brown and had a fungus infection. There are four instars and before pupation the larvae close up their cases and turn to face the opposite end of the case. Pupation usually occurs with the case attached to the bark of trees or on the ground. Adults emerge by removing a small cap from the larval case about 1mm from the anterior end.

Cryptocephalus species overwinter as larvae and may complete their development the year following oviposition (e.g. *C. pusillus*). However, some species (e.g. *C. parvulus* and *C. labiatus*) probably require two years to achieve development to the adult. Larvae (probably of *C. pusillus*) may feed on leaf litter as three larvae (one penultimate and two final instar) were extracted from oak leaf litter on March 25th 1980. Donisthorpe (1908) suggested that they feed on lichens on trees. On April 16th, 1910 he found a larva of *C. fulvus* in a nest of *Lasius fuliginosus*, the adult emerged on June 8th–9th and was eaten by ants. *C sexpunctatus* has been recorded near a nest of *Formica rufa*, and it was suggested that all *Cryptocephalus* species pupate within ant nests.

Cryptocephalus species occur in the following general habitat types (N.B. some species occur in more than one type of habitat):

Calcareous grassland – *aureolus, bilineatus, bipunctatus, hypochaeridis, labiatus, moraei, nitidulus, primarius, pusillus*.

Broad-leaved woodland – *aureolus, bipunctatus, coryli, decemmaculatus, frontalis, fulvus, hypochaeridis, labiatus, moraei, nitidulus, parvulus, punctiger, pusillus, querceti, sexpunctatus*.

Wetland (bog, fen, carr) – *biguttatus, exiguus, frontalis, labiatus, pusillus, ?fulvus*.
Heathland – *biguttatus, bipunctatus, fulvus, labiatus, moraei, parvulus, pusillus*.
Coastal sand dunes – *aureolus, bilineatus, fulvus, labiatus, pusillus*.
More specific habitats of some rare species are as follows:

- *C. biguttatus* – swept from cross-leaved heath (*Erica tetralix*) and heather (*Calluna vulgaris*) on boggy heaths and moors.
- *C. coryli* – beaten from young birch and oak.
- *C. decemmaculatus* – beaten from dwarf sallows and birch
- *C. exiguus* – beaten from *Betula* and *Salix cinerea*.
- *C. nitidulus* – beaten from young birches.
- *C. primarius* – swept from warm sheltered dry hillsides with grasses, *Hieraceum* and *Helianthemum*.
- *C. querceti* – oaks and hawthorn.
- *C. sexpunctatus* – from hazel, birch, aspen and crack-willow.

Sexing is readily achieved as all female *Cryptocephalus* have a deep median circular depression in the apical abdominal sternite whilst this is absent in the males. In addition the antennal segments are usually more elongate in the males, especially notable in *pusillus*. The males of this species often have the elytra dark-brown or black, whereas in the females they are yellow. Colour is also sexually dimorphic in *coryli* in which the pronotum of the males is black whilst in the females it is orange. There are also modifications of the abdomen in some species. In males of *sexpunctatus* abdominal sternite 2 bears a pair of lateral processes.

Cryptocephalus species are easily identified using colour, elytral plus pronotal puncturation, antennal and leg characters making resort to genital dissection unnecessary. Specimens are best mounted with the dorsal surface uppermost and with legs and antennae extended.

LAMPROSOMATINAE. The small brassy adults of *Lamprosoma concolor* are seldom seen openly on vegetation. Little is known concerning the biology of this species. Adults have been collected in all months of the year except March, with the greatest number of records for June. They probably overwinter, reappearing in April and ovipositing during May, June and July.

Kolbe (1898) demonstrated that in this species hibernation is also undertaken in the larval stage. The adult fed upon the leaves of *Astrantia* and *Aegopodium* (Umbelliferae) whereas the larva was polyphagous, feeding nocturnally on the stem and leaves of different plants on the ground and usually preferring the petioles. This work was substantiated by the observations of Kasap & Crowson (1976) who collected fully grown larvae at Morroch Bay, Scotland on 11th September 1974. The case-bearing larva fed nocturnally on the fresh green leaves and stems of *Hedera helix* (Araliaceae).

Adults are best collected by evening sweeping, by grubbing at the roots of low plants, in water nets (October, November, December) or by beating in old ivy hedges. Adults and larvae are also extracted by collecting leaf litter and often detritus from *Hedera* and other vegetation. This species occurs chiefly in broadleaved woodland throughout England and Wales, but only rarely in Scotland.

There are no obvious differences between the sexes.

EUMOLPINAE. This subfamily is represented in Britain solely by the rare *Bromius obscurus* (L.). The dull black or brownish, pubescent adults occur on various willowherbs, *Epilobium* spp. and especially the Rosebay Willowherb *Chamaenerion angustifolium* (L.) Scop. On the Continent it was recognised as a pest of the grape vine *Vitis vinifera* L. as early as the 15th century.

According to Balachowsky (1963) this species is parthenogenetic and only the female is known. However, this is not true for N. American populations, in which males commonly occur (pers. observation). Adults start to emerge from the soil during May. Oviposition commences in early June and continues up to August and adults may survive for nearly three months. The bright yellow, 1.0×0.5mm ova are laid in batches of 20–40 either in the soil near the host plants or at the base of the stem, slightly above the root neck under the old sloughed-off epidermis. The white slightly "C"-shaped larvae feed in groups on the roots, decorticating them and thus removing the epidermis and even sometimes the superficial wood. They eat either linear or irregular incisions, the latter resembling the damage caused by certain scarabaeid larvae. The larvae develop during the summer and early autumn and penetrate deep into the soil to overwinter. Pupation occurs in an earthen cocoon during March or early April and the adults emerge 20–30 days later. The adults feed on the foliage of the host plant.

This species seems to prefer light sandy soil alongside rivers. It is known as a single colony from only one site in Cheshire where it was first discovered during 1970. It is possibly a relict species since it was apparently very common in Britain during the mild phase about 11,000–12,000 B.P. (Before Present). Stephens (1831) is perhaps the only British author to include this amongst the reputed British species. He refers to a specimen in the BM(NH) collection which was said to have been taken in Lincolnshire, but no subsequent record has since been reported.

The males and females are extremely difficult to distinguish with only a slight difference in the shape and puncturation of the apical abdominal sternite.

CHRYSOMELINAE. The majority of this subfamily are small to medium-sized, usually brightly coloured beetles frequently encountered in the spring and summer on the foliage of their host plants. The larvae and adults feed openly on the foliage of low herbaceous plants in various families (*Timarcha, Chrysolina, Gastrophysa, Phaedon, Hydrothassa,* and *Prasocuris*) or on bushes and trees, especially of *Salix* and *Populus* (*Plagiodera, Chrysomela, Phytodecta* and *Phyllodecta*).

The white, yellow or reddish, oval or elongate-oval ova are usually laid singly or in batches on the undersides of leaves but sometimes covered with faeces and regurgitated food and deposited at the base of their host-plants or at a shallow depth in the soil as in *Timarcha* spp. In some *Phaedon* and *Hydrothassa* species they are laid in depressions cut into the leaves or actually laid inside the hollow stems of their host-plants. The larvae have three or four instars and usually have well-developed dark dorsal sclerotised tubercles. They also sometimes have paired eversible defensive glands on the thorax and abdomen. Pupation usually occurs in the soil but in *Plagiodera* and *Chrysomela* it takes place on the leaves and stems of their host plants. Most of the species are probably univoltine, the adults usually overwinter, rarely as diapausing eggs (*Timarcha*), non-diapausing eggs or possibly early instar larvae (*Chrysolina hyperici* and *brunsvicensis*) or diapausing late instar larvae (*Chrysolina menthastri* for example).

Adults and larvae are best collected by sweeping or beating vegetation; some species may also be collected in grass tussocks during the winter (for example *Chrysolina staphylaea, Chrysomela aenea, Hydrothassa glabra, Phaedon cochleariae* and *tumidulus* and *Prasocuris junci*).

Specific habitats and distributions of some of the rarer species are as follows:

Chrysolina graminis – sweeping *Tanacetum vulgare; Mentha* spp.

Chrysolina haemoptera – sweeping maritime cliffs, coastal shingle, sand dunes and possibly salt marshes, also sometimes downland inland.

Chrysolina marginata – sweeping well-grazed short-turf grassland, river margins and dry sandy habitats, for example sandpits.

Chrysolina menthastri – sweeping wetland mostly, but also grassland, wood edges, commons, gardens and disturbed ground.

Chrysolina oricalcia – sweeping woodland rides, clearings and woodland edges, also calcareous grassland, disturbed ground and possibly wetland.

Chrysolina sanguinolenta – sweeping broad-leaved woodland rides, clearings and wood-edge, grassland and disturbed ground.

Chrysolina violacea – sweeping calcareous grassland, possibly also scrub and disturbed ground on base-rich soil, also chalk quarries.

Chrysolina brunsvicensis – broad-leaved woodland rides, clearings and wood-edge, grassland and disturbed ground, also on disused railway lines and grassy places beside rivers.

Chrysolina cerealis – Very rare and still only known from Snowdonia though apparently well established there. Adults occur on the host plant in montane grassland between mid-April and September. They may also be found by grubbing at the base of the host plant or by searching under stones.

Chrysolina latecincta – maritime cliffs and probably dry sandy grasslands near the coast. N.W. and S.W. Scotland, Orkney and Shetland.

Phaedon concinnus – saltmarshes, by sweeping or grubbing at plant roots.
Hydrothassa glabra – sweeping grassland, wetland, quarries, downland, heathland, possibly also woodland edges and disturbed ground. Also by grubbing, tussocking and flood refuse.
Hydrothassa hannoveriana – sweeping wetland and semi-aquatic habitats.
Chrysomela aenea – by beating alder. Under loose bark and in tussocks during winter. England except S.E. and E. Anglia, Wales, Scotland.
Chrysomela populi – beating in broad-leaved woodland, wetland, particularly willow carr. Also recorded from wet flushes on heathland, dunes and cliff top.
Chrysomela tremula – Formerly widespread in southern England but last recorded during the 1950's. Broad-leaved woodland.
Phytodecta decemnotata – beating in broad-leaved woodland.
Phytodecta viminalis – beating in broad-leaved woodland.
Phyllodecta polaris – a high altitude species usually collected by shaking out moss from below the host plants. So far only recorded from two vice-counties in N.W. Scotland but probably occurring in other similar montane localities in Scotland above 950m.

Sexing is usually accomplished by examining the anterior tarsi, the males *usually* having one or the first three segments dilated. In those species where this obvious difference is not exhibited, then the apical abdominal sternite should be examined. The apex of this is entire and usually rounded in females, but slightly indented or sinuous in males.

Specific determination in this subfamily usually relies upon colour or puncturation, antennal and leg (tarsal) characters. Specimens are best mounted with the dorsal surface showing and legs and antennae extended. However, in *Chrysolina* ventral characters are sometimes used so at least one specimen should be thus mounted.

GALERUCINAE. The soft-bodied adults are usually small to medium sized, dull brown, yellow or black, rarely metallic. They occur during the spring, summer and autumn on the foliage of low growing herbaceous plants (some *Galerucella, Galeruca, Phyllobrotica,* and *Sermylassa*) or shrubs and trees (some *Galerucella, Pyrrhalta, Luperus, Calomicrus* and *Agelastica*).

Galerucine ova are usually yellow or brown, approximately spherical, hemispherical or oval and are usually deposited singly or in batches on foliage or in the soil. Those laid in the soil often have the chorion with strong hexagonal microsculpture. Sometimes eggs are laid inside stems of the hostplant (*Pyrrhalta viburni*) or in oothecae attached to low vegetation (*Galeruca*). The ova in some cases diapause; in *Sermylassa halensis, Pyrrhalta viburni* and *Galeruca tanaceti* this is probably obligatory, whilst in *Phyllobrotica quadrimaculata* it is faculative. These species oviposit in late summer/autumn whilst the other species oviposit in early or mid-summer.

First instar larvae lack egg bursters and hatch by biting their way from the egg. The larvae either feed openly exposed on foliage (*Galerucella, Pyrrhalta, Galeruca, Agelastica, Sermylassa, ?Phyllobrotica*) or at the roots of their host plants (*Luperus, Calomicrus, ?Phyllobrotica*). The adults and larvae of the former group feed on the same hostplant whilst the hosts of the adults and larvae of the latter group may differ. For example, adults of *Luperus* species feed on the foliage of *Betula* whilst their larvae probably feed on the roots of grasses (*?Molinia*). The larvae of *Lochmaea crataegi* are unusual since they feed within the pulp of *Crataegus* haws whilst their adults feed on pollen from flowering hawthorn. Foliage frequenting larvae usually have dark dorsal tubercles on thorax and abdomen and dark heavily sclerotised heads, whilst root-feeders have pale weakly sclerotised body tubercles and pale heads. Larvae of *Sermylassa* and *Agelastica* have paired segmental openings of defensive glands on the abdomen. Larvae have three instars, except *P. viburni* which have four. Pupation usually occurs in earthen cells in the soil, except in *Galerucella nymphaeae* and *sagittariae* which pupate on the leaves of their aquatic and semiaquatic hostplants.

The adults of *G. tanaceti* have a period of summer aestivation (June to August) when they stop feeding and remain inactive in grass-tussocks (Donia, 1958; Siew, 1966). *Agelastica alni* also apparently aestivates from July to August.

The adults are best collected by beating or sweeping vegetation; *L. crataegi* especially from hawthorn, and *G. tanaceti* by tussocking from June to August.

Specific habitats and distributions of some of the local or rare species are as follows:

Galeruca interrupta – last recorded in Britain in 1919 (Sherborne, Dorset). Dry heathland/grassland, areas of dry hillsides.

Galeruca tanaceti – grassland, maritime cliff, disturbed ground, heath, commons, quarries.

Phyllobrotica quadrimaculata – wetland and semi-aquatic habitats; also wetland areas within woodland, alder carr.

Luperus flavipes – beating bushes (birch, willow) in broadleaved woodland, parkland, scrub, heath etc.

Calomicrus circumfusus – beating *Ulex* spp., *Genista* spp., broom on heathland, grassland, scrub, maritime cliff.

Agelastica alni – beating alders in wetland, particularly old carr, also along river margins and in wet flushes. Presumed extinct in Britain since last recorded in 1900, Deal, Kent.

The sexes in the galerucines are usually distinct and easily recognisable. In *Luperus* the antennae of the males are as long or longer than the body and the apical segment is at least twice as long as the first. In the females the antennae are much shorter than the body and the apical segment is at most only slightly longer than the first. In addition males have large, more prominent eyes, and the head broader than the pronotum, whilst in the female the head is narrower.

In *Calomicrus* the sexes can only be distinguished by the apical abdominal sternite, which in males has a deep apical notch on both sides and a large median depression; the apex is entire in the female and has no depression.

The sexes of *Sermylassa*, *Agelastica* and *Galeruca* can only be distinguished by the shape of the apex of the final abdominal sternite. In the males of the first two genera this is sinuous with a median lobe and in the females it is evenly rounded. In *Galeruca* the apex is deeply emarginate in the male and evenly rounded in females. In addition females of this genus when gravid show physogastry.

The sexes in *Phyllobrotica* differ in the following respects. First tarsal segment of the fore-legs is dilated and as broad as segment three in males, whereas it is slender and narrower in the females. There are also distinct differences in the shape of the abdominal sternites 3, 4, and 5.

In *Galerucella tenella*, *lineola*, *pusilla*, *Pyrrhalta* and *Lochmaea* the mid-tibiae of the males bear a short spur whereas these are absent in the females. Males of *Galerucella calmariensis* have the hind tibiae armed with a spur. In this species the apical abdominal sternite is also sexually dimorphic. In addition in *Lochmaea* the males usually have the first segment of the metatarsus strongly dilated and the hind tibiae stouter and more sinuous than in the females. In *G. sagittariae* tibial spurs are absent in both sexes and the only reliable difference in the sexes is the shape of the apex of the final abdominal sternite.

Galerucine species are easily separated except for those in *Galerucella*. Here resort has to be made to the dissection of the male aedeagi for accurate identification. Specimens are best mounted with the dorsal surface uppermost except in *Galerucella* where at least one specimen should be mounted on its back since ventral characters are of value.

ALTICINAE. The "flea-beetles" are small, dull or brightly coloured beetles which have the ability of jumping. They often cause characteristic damage by shot-holing or completely perforating with numerous small holes the foliage on which they feed. This is particularly true of the genus *Phyllotreta* when they attack young cruciferous plants. They usually occur in the spring, summer and autumn on low vegetation since the majority feed on herbaceous plants with the exception of *Altica brevicollis* and *Chalcoides* which are beaten from bushes and trees of *Corylus*, *Salix* or *Populus*.

The oval or elongate-oval, whitish, yellow or orange eggs are usually laid in the soil with the exception of *Altica*, *Phyllotreta nemorum*, *Hippuriphila modeeri* and certain *Psylliodes*, *Mantura*, *Sphaeroderma* and *Apteropeda*, where they are deposited on the foliage. Those laid in the soil often have distinct hexagonal microsculpture on the chorion. Larvae are elongate and hatch by biting an escape hole in the egg or cut slits in the chorion using paired egg bursters, usually situated on the meso- and metathorax. The larvae of the majority of species feed in concealed positions within their host plants or in the soil, except for those of *Altica* which feed openly on foliage. Some mine leaves, others stems, but the majority live inside roots or at roots in the soil. There are usually three larval instars and pupation always occurs in cells in the soil.

Alticines are usually univoltine and the eggs apparently do not diapause. The overwintering stage is usually the adult but occasionally the larva may also overwinter (*Sphaeroderma* – leaf-miner in thistles; *Crepidodera* – stem-miner in various plants; *Derocrepis* – root or stem-miner; *Podagrica* – root feeder; *Longitarsus gracilis, flavicornis* and *luridus*–root feeders). There is usually only one period of oviposition annually, either in Spring/early summer but in some species there is also oviposition in late summer/autumn (e.g. some *Longitarsus*). Brachypterous genera in the Alticinae include *Derocrepis, Podagrica* and *Hermaeophaga* whereas the apterous genera are *Mniophila, Apteropeda* and *Batophila; Longitarsus* and *Aphthona* include apterous and brachypterous species.

Adults are usually collected by sweeping and beating vegetation and secured with a pooter before they leap away! They may also be collected in a Malaise trap (*Altica, Longitarsus*); suction traps (*Chaetocnema concinna*); yellow water traps (*Phyllotreta*); in tussocks during winter (*Chalcoides, Chaetocnema, Sphaeroderma, Altica, Psylliodes* and *Longitarsus*) or by grubbing at the base of plants and shaking out short turf and moss (*Chaetocnema*).

Specific habitats of some of the more local/rare species are as follows:

A. Calcareous grassland (chalk or limestone, downland).
Aphthona atrovirens, herbigrada.
Longitarsus agilis, curtus, anchusae, dorsalis, nigrofasciatus, abliteratus, quadriguttatus, tabidus.
Batophila aerata.
Mantura matthewsi.

B. Grassland.
Aphthona nigriceps.
Longitarsus anchusae, ochroleucus, parvulus, pellucidus.
Psylliodes chalcomera, sophiae.

C. Broadleaved woodland.
Altica brevicollis.
Chalcoides nitidula.
Epitrix atropae.
Chaetocnema conducta (N.E. Yorkshire only).
Mniophila muscorum.

D. Saltmarsh.
Longitarsus absynthii, plantagomaritimus.
Crepidodera impressa.
Chaetocnema sahlbergi.
Podagrica fuscipes.

E. Wetland.
Phyllotreta tetrastigma.
Longitarsus brunneus, ferrugineus, holsaticus, nigerrimus (S. England), *rutilus.*
Lythraria salicariae (England).
Chaetocnema aerosa (England S. & S.E.), *arida, subcoerulea.*
Apteropeda globosa.

F. Moorland.
Altica britteni, ericeti.

G. Arable land, disturbed ground, gardens, allotments.
Phyllotreta aerea, cruciferae, flexuosa, vittata
Psylliodes attenuata, luteola.

The specific identification in many genera is relatively straightforward, except in the large, intraspecifically variable *Longitarsus* and the small genus *Altica* (in which there are no reliable external characters). The spermatheca, notably the number of twists in the spermathecal duct, is especially important in *Longitarsus* for species recognition and in *Altica* for species group recognition. The aedeagus is an extremely important specific character in both of these genera. Males are needed for accurate determination in *Altica*.

Sexing of beetles in these two genera, especially *Altica*, is important. In *Altica*, males usually with first segment of anterior tarsi wider than second; females with tarsal segments one and two sub-equal in width. In addition the males have the apical abdominal sternite with the posterior margin laterally incised, females with posterior margin entire, not laterally incised. In *Longitarsus* the first segment of the anterior tarsi is usually more strongly dilated in the male. However, sometimes this difference is not very obvious and if both sexes are not available for comparison it is difficult to decide which sex you have (especially in *L. succineus, ochroleucus, membranaceus* and *parvulus*). In these cases the difference in the shape of the apical abdominal sternite is particularly useful. In addition the elytral shoulders in the males are often less strongly marked than in the females.

The antennae are modified in the males of several *Phyllotreta*. The fifth segment is very elongate and dilated in *P. ochripes* and *exclamationis*, the fourth and fifth are moderately dilated in *nemorum*, the third and fourth are strongly dilated in *nodicornis*, whilst *consobrina* has the fifth segment elongate and slightly dilated.

CASSIDINAE. The adults of the "tortoise beetles" are usually greenish dorsally and some have metallic golden bands (*vittae*) on the elytra. These are chemical colours which fade soon after death of the beetle but can be maintained if the beetle is preserved in 70% alcohol. They frequent the foliage of various low growing plants during the spring and summer. Their green colour provides good camouflage against the foliage of their host plants.

The elongate-oval ova are laid either singly, in pairs or in groups of up to 10 within brown oothecae, usually on the under-surface of leaves. They do not diapause, but hatch in from one to two weeks depending upon the prevailing temperature. There are five larval instars and the larvae are characterised by long lateral forked processes on the thorax and abdomen and a pair of usually long caudal processes which are bent forwards over the abdomen. Exuviae and faeces are sometimes retained on this caudal fork, which when held over the body acts to camouflage the larvae since they resemble droppings of birds. Pupation always occurs on the stem or foliage of the host plant. They are usually univoltine with the new generation adults emerging from the end of June until mid-September and overwinter in this stage.

Adults are best collected by sweeping vegetation in the summer.

Specific habitats of some of the more local/rare species are as follows:

Pilemostoma fastuosa – sweeping on grassland (downs), disturbed ground and probably also coastal habitats.

Cassida hemispherica – sweeping on grassland, woodland rides, clearings and also coastal habitats.

Cassida murraea – sweeping wetland, grassland, scrub and possibly disturbed ground. Formerly widespread in England (except north), Wales (except north). This species appears to have contracted its range to S.W. England, S. Wales and Dyfed/Powys where it is still common.

Cassida nebulosa – sweeping on disturbed ground (arable farmland) probably also grassland and scrub.

Cassida nobilis – sweeping in coastal habitats (England, except N.W.; Wales except N., S.W. Scotland).

Cassida prasina – sweeping on grassland, disturbed ground and probably scrub.

Cassida vittata – sweeping coastal shingle, maritime cliff, saltmarsh and probably sand dunes.

Since dorsal (colour, puncturation, etc.) and ventral (colour) characters are used in specific determination, it is best to mount one specimen dorsally and one ventrally or to micro-pin a singleton. Resort does not have to be made to genital dissection and sexing is very difficult, the only apparent difference being the shape of the apex of the final abdominal sternite.

References

Allen, A.A., 1976. Notes on some British Chrysomelidae including amendments and additions to the list. *Entomologist's Rec. J. Var.*, **88**: pp. 220–225 & pp. 294–299.

Atty, D.B., 1970. Gloucestershire beetles: a few records and an appeal. *Entomologist's mon. Mag.*, **105**: p. 199.

Balachowsky, A., 1963. Entomologie Appliquée a L'Agriculture. Tome I Coléoptères. 2nd vol. 1,391pp. Paris.

Betts, M.M., 1955. The food of titmice in oak woodland. *J. anim. Ecol.*, **24**: pp. 282–323.

Betts, M.M., 1956. A list of insects taken by titmice in the Forest of Dean (Gloucestershire). *Entomologist's mon. Mag.*, **92**: pp. 68–71.
Champion, G.C. & Lloyd, R.W., 1910. Some interesting British Insects (ii). *Entomologist's mon. Mag.*, **46**: p. 2.
Cox, M.L., 1981. Notes on the biology of *Orsodacne* Latreille with a subfamily key to the larvae of the British Chrysomelidae (Coleoptera). *Entomologist's Gaz.*, **32**: pp. 123–135.
Donia, A.R., 1958. Reproduction and reproductive organs in some Chrysomelidae (Coleoptera). Ph.D. Thesis, London University.
Donisthorpe, H.St.J.K., 1902. The life history of *Clytra quadripunctata* L. *Trans. ent. Soc. Lond.*, **35**: pp. 11–24. 1 pl.
Donisthorpe, H.St.J.K. 1908. A few notes on Cryptocephali. *Entomologist's Rec. J. Var.*, **20**: pp. 208–209.
Kasap, H. & Crowson, R.A., 1976. On systematic relations of *Oomorphus concolor* (Sturm), with descriptions of its larva from Australia. *J. nat. Hist.*, 1976. **10**: pp. 99–112.
Kolbe, W. 1898. *Lamprosoma concolor* Sturm in biologischer Beziehung. *Z. Ent.* **22**: pp. 22–29.
Massee, A.M. 1945. Abundance of *Labidostomis tridentata* L. (Col., Chrysomelidae) in Kent. *Entomologist's mon. Mag.*, **81**: pp. 164–165.
Siew, Y.S. 1965. The endocrine control of adult reproductive diapause in the chrysomelid beetle *Galeruca tanaceti* L. *J. Insect Physiol.*, **11**: pp. 1–10.

Additional British Key Works to Adult Identification

Kevan, D.K., 1962. The British species of the genus *Haltica* Geoffr. *Entomologist's mon. Mag.*, **98**: pp. 189–196 (43 figs).
Kevan, D.K. 1967. The British species of the genus *Longitarsus* Latreille. *Entomologist's mon. Mag.*, **103**: pp. 83–110.
Shute, S.L. 1975a. *Longitarsus jacobaeae* Waterhouse (Col., Chrysomelidae): identity and distribution. *Entomologist's mon. Mag.*, **111**: pp. 33–39.
Shute, S.L. 1975b. The specific status of *Psylliodes luridipennis* Kuts. *Entomologist's mon. Mag.*, **111**: pp. 123–127.

British Key Works to Larval Identification

Subfamily Keys

Cox, M.L., 1981. Notes on the biology of *Orsodacne* Latreille with a subfamily key to the larvae of the British Chrysomelidae (Coleoptera). *Entomologist's Gaz.*, **32**: pp. 123–135.
Mann, J.S., & Crowson, R.A., 1981. The systematic positions of *Orsodacne* Latreille and *Syneta* Lac. (Col., Chrysomelidae), in relation to characters of larvae, internal anatomy and tarsal vestiture. *J. nat. Hist.*, **15(5)**, pp. 727–749.

Generic and Specific Keys

Cox, M.L., 1982. Larvae of the British genera of chrysomeline beetles (Coleoptera, Chrysomelidae). *Syst. Ent.*, **7**: pp. 297–310.
Marshall, J.E., 1979. The larvae of the British species of *Chrysolina* (Chrysomelidae). *Syst. Ent.*, **4**: pp. 409–417.
Emden, F.I. van-. 1962. Key to species of British Cassidinae larvae. *Entomologist's mon. Mag.*, **98**: pp. 33–36 + 10 figs.

CURCULIONIDAE (WEEVILS)

BY DR. M.G. MORRIS

CURCULIONIDAE
Hypera rumicis (L.)

To the coleopterist, though not to the 19th century seafarer, weevils are members of the superfamily Curculionoidea, which is placed at the end of most classifications of beetles, if the Strepsiptera are regarded as a separate order. Members of the Curculionidae most commonly come to mind as "the weevils", but recognition of the families Attelabidae and Apionidae may suggest that "weevil" is a convenient common name for any member of the Curculionoidea, including Anthribidae and Scolytidae. However, the terms bark-beetles and ambrosia beetles are probably too well established to be superseded. Seven families of Curculionoidea are recognised by Kloet and Hincks (2nd edition) and, in this account, *Bruchela rufipes* is often placed in a separate family, the Urodontidae, but is here included in the Anthribidae.

Although there are a few exceptions (such as *Brachytarsus*), nearly all weevils feed on plant material and very many are stenophagous, i.e. they are associated with one species, genus or family of plants. General methods of collecting, particularly in the three levels of vegetation represented by the ground zone, field (or herbaceous) layer, and the arboreal zone (trees and bushes), will yield many species of weevils. Grubbing (which goes under other names), sweeping and beating are collecting methods the coleopterist learns in the cradle, metaphorically speaking. However, the collector or student of weevils, and other phytophagous beetles (for example *Meligethes* and Chrysomelidae) will find it much more rewarding if he (or she) can develop a good working knowledge of field botany. Not only is working particular plants the most effective way of collecting weevils, but the foodplant can often be a useful guide to identification. For instance, a series of *Apion* shaken from Dyer's greenweed. *Genista tinctoria*, will almost certainly be *A. (Exapion) difficile*. This knowledge could save long poring over the unsatisfactory keys in Joy, or struggling with Fowler's 4.A.b.a*.b ** ## style of notation.

Grubbing, sweeping and beating, together with a systematic searching for and working of known (and occasionally only suspected) foodplants are the bread-and-butter collecting methods of the weevil specialist. Special methods for particular species are described in due course.

Our single species of Nemonychidae, *Cimberis attelaboides*, is associated with Scots Pine, *Pinus sylvestris*, and will most frequently be taken by beating that tree. It is local, but widely distributed.

The British Anthribidae are all uncommon species and not easy to work for. The fungus known imaginatively as "King Alfred's Cakes", *Daldinia concentrica*, should be searched, particularly in spring and where it occurs abundantly (on dead and dying trunks, particularly of Ash), for the large *Platyrhinus resinosus*, one of the less rare species. Faggot traps, of dead sticks and branches wired together in bundles, can be set in deciduous woodland and beaten vigorously over a sheet. Some of our rare anthribids have been taken this way. *Brachytarsus* spp. can be beaten from both conifers and deciduous trees and from old hedgerows but are most usually come upon casually, as are most of the other species. It is difficult to work specifically for anthribids, though our only synanthropic species, *Araecerus fasciculatus*, can be looked for in abandoned or unhygienic grain warehouses and the like; it feeds on stored foodstuffs.

Our small fauna of Attelabidae includes some particularly attractive and interesting species. All are fairly specific in their plant hosts, but the feeding habits of the larvae and behaviour of egg-laying females are very varied. One well-known method of larval feeding, unique to the attelabids in our fauna, is for the ovipositing female to construct a leaf roll by rolling up a part of the leaf lamina of a host plant (usually a tree) after laying an egg in it. The conical, pendant rolls of the common *Deporaus betulae* are often abundant on Birch, Alder or Hazel. The adults of this and other species of attelabids are usually collected by beating or sweeping. Curiously, our other *Deporaus*, *D. mannerheimi* has a different method of nutrition, the larvae mining the petioles of Birch leaves. Other leaf-rollers include *Attelabus nitens*, on Oak, *Apoderus coryli*, on Hazel and our two *Byctiscus* spp., both rarer weevils (*B. populi* on poplars, particularly Aspen, and *B. betulae* on various trees, perhaps most frequently on Hazel). Although most usually obtained by beating their hosts these species may often also be "spotted" as they sit on foliage, as all are conspicuous and brightly coloured.

Several of the Attelabidae have rather short seasons, unlike many weevils which can be taken during most months of the year. For example *Rhynchites aeneovirens* can be beaten, often plentifully, from Oak, but only in April and May; at least it is very seldom collected at other times. Though in general less closely restricted in time, many other attelabids are taken mainly in spring or early summer, as they tend not to overwinter in the adult state. There are exceptions, however; *Rhynchites caeruleus*, for example, is often beaten from rosaceous shrubs such as Hawthorn and fruit trees in autumn as well as spring. This species is a twig-cutting weevil. The females lay eggs in fresh shoots of the hosts and partially sever them, thus providing a source of dying and dead tissue for the developing larvae.

Not all attelabids are found on trees and shrubs. The commonest species associated with herbaceous vegetation is *Rhynchites germanicus* which feeds on a variety of Rosaceae. It has behaviour similar to that of *R. caeruleus* and lays eggs in shoots, runners and stolons and is a minor pest of strawberries and other soft fruit.

Many coleopterists are put off by the large number of species of *Apion* and the lack of good keys to members of the family, which includes our two species of *Nanophyes*. Despite the many species, *Apion* are not difficult to identify provided it is recognised that the two sexes of one species may be very different in structure and proportions. As mentioned earlier, the host plant is often a valuable aid to identification, whilst a reasonable working knowledge of botany will be helpful in collecting the different species. By working various clovers and vetches (Papilionaceae), docks and knotgrasses (Polygonaceae), mallows (Malvaceae) and selected Compositae (= Asteraceae) the beginner will accumulate many of the commoner *Apion* and perhaps some more local species as well.

Some hosts in other plant families support only one species of *Apion*. *A. sedi* feeds on species of stonecrop, particularly *Sedum anglicum* and *S. acre*. It is a tiny species and the host cannot be swept because of its recumbent growth form. Occassionally plants growing on a wall can be "brushed" or shaken into a sweep net, but more usually the weevil must be sought by grubbing, particularly on light soil such as sands, and on rocky outcrops where the host grows.

Each of our species of Mercury (Euphorbiaceae) is host to a species of *Apion*. The widely distributed but local *A. pallipes* feeds in the stems of Dog's Mercury and *A. semivittatum* has the same larval feeding site on Annual Mercury. This species is restricted to SE and Southern England, but is often common where found. A closely related third species, *A. urticarium*, feeds on common Stinging Nettle (Urticaceae) but is much more local than its host and tends to occur mainly on light soils such as sands, peats and limestones.

A small group of *Apion* species feeds on Labiatae; an example is the minute *A. atomarium*, associated with species of wild Thyme.

Two of our *Apion* are arboreal, associated with trees or bushes, an unusual habit in the family. *A. simile* feeds as a larva in the catkins of Birch and is of interest as being one of our few Holarctic weevils (occurring in North America as well as Europe). The species is generally common where it occurs but although widely distributed it is not found everywhere. *A. minimum* is much rarer and has particularly interesting larval habits. It occurs as an inquiline (harmless guest) in the galls of sawfly larvae on sallows (Salicaceae). Both these arboreal *Apion* are most easily collected by beating.

The best way of collecting a good number of species of *Apion* is to study the association of weevils with plant species and carefully work the known hosts. Some of the less common hosts, such as Wild Liquorice (*Astragalus glycyphyllos*) or our *Genista* species support uncommon and local species. There are few specialised techniques for collecting *Apion*, although some species may be reared from their hosts (try doing this with collected clover seed heads as a start). Once he or she begins to find a way about the group the coleopterist will find *Apion* particularly rewarding, and many of the species are quite common. Before passing on to the family Curculionidae proper, mention must be made of our two

Nanophyes. *N. marmoratus* is a generally common species on Purple Loosestrife (Lythraceae) and easily collectable, but *N. gracilis* is rare and often difficult to find. As its host, Water Purslane, is insignificant, creeping, and found in wet places, collecting from it can be problematical and uncomfortable. However, *N. gracilis* is a pretty and variable species.

The family Curculionidae, which contains the bulk of the weevils, is divided into two great series, the short-nosed group, Adelognatha, which feed on roots as larvae, and the Phanerognatha, which in general have the well-developed rostra (beaks) of "typical" weevils. Phanerognatha tend to be more restricted to particular plants than Adelognatha and they mostly feed inside plant structures as larvae, for instance in fruits, buds and stems. The root-feeding Adelognatha usually live free and exposed in the soil. Although they feed below ground as larvae the adults of many adelognathous species may be found by beating or sweeping, though there are also many ground-living species. The genus *Otiorhynchus* contains arboreal, field-layer and ground zone species, though it is sometimes difficult to distinguish between the first two categories. Moreover, although many *Otiorhynchus* (and other weevils, especially Adelognatha) appear to us, as diurnal animals, to be ground-living, collecting them after dark tells a different story. This is when the entomologist/gardener finds *O. sulcatus* devouring his prize cyclamen, or *O. porcatus* eating his primulas or saxifrages. In fact, very few *Otiorhynchus* are restricted to particular plants. Collecting the rarer species (several are very common) depends more on geography than botany. *O. articus*, *O. nodosus*, *O. scaber* and two rarities, *O. morio* and *O. auropunctatus* (which is common enough in Ireland) are all Scottish, or at least northern species. *O. scaber* is a good example of an arboreal species, beatable from Birch in particular, whilst *O. arcticus* is exclusively ground living. Incidentally, it was common in England during "the Ice Age" though not found there now.

Caenopsis, *Trachyphloeus*, *Cathormiocerus* and their allies are all ground-living weevils with many features of interest. Most of our *Trachyphloeus* are parthenogenetic, males being unknown, very rare, or restricted to a few small areas of Europe; *T. scabriculus* is an exception. *Cathormiocerus* are not only among our rarest weevils, they are restricted to the extreme West of Europe, their "headquarters" being the Iberian Peninsula. You will look in vain for any mention of the genus in Freude, Harde & Lohse. All our species are very local and restricted to the south coast. The Lizard Peninsular is a favoured collecting area, and the home of *C. britannicus*, one of our very few endemic beetles. Grubbing and searching on hands and knees are required to collect these species, but more specialised methods can also be employed. Over a long stay in a suitable area, moss-traps can be used. Damp moss, bundled up and secured by chicken wire or string, can be placed in rabbit burrows, hollows, under heather or in other situations, to be taken up and sifted at a later date. Vegetation can be collected and sieved, or turf and moss "extracted" by heating under a lamp until the weevils crawl out. Stone walls are often favoured by *Cathormiocerus* and other short-nosed weevils. Grubbing on top of a stone wall can be precarious, but is often profitable.

Adelognathous weevils with very different habits include species of *Phyllobius* and *Polydrusus*, which despite their similarity of facies and biology are usually placed in the subfamilies Otiorhynchinae and Brachyderinae respectively. Although the larvae of these species are root-feeders, the adults are arboreal and often very abundant on deciduous trees. In fact, there can hardly be said to be an uncommon *Phyllobius*, though *P. vespertinus, P. calcaratus* and *P. viridicollis* are not found absolutely everywhere. The last species belongs to a small group which is found on herbaceous vegetation rather than trees; its members are therefore more frequently swept than beaten. The genus *Polydrusus* contains a few genuine rarities as well as some species which are temporary colonists or of doubtful provenance. Some species, such as *P. cervinus* and *P. pterygomalis*, are abundant and often collected with *Phyllobius* species. *Polydrusus pulchellus*, however, is a salt-marsh species, thought its text-book association specifically with Sea Wormwood is erroneous or at least exaggerated. All our other species are arboreal, though *P. confluens*, as well as being beaten from Common Gorse can also be found by grubbing under Dwarf Gorse (wear gloves). All our *Phyllobius* and *Polydrusus* have limited seasons in spring and early summer. Although some, such as *Polydrusus pterygomalis*, may persist into August, or even September, the covering of bright metallic scales will have become worn by then and the collector will want to take specimens earlier in the year.

Between *Polydrusus* and *Sitona* (in most modern lists and classifications) comes a group of disparate genera, each with few species. Some are ground-living, others arboreal. *Philopedon plagiatus* is a well-known inhabitant of coastal dunes and other open sandy areas. *Liophloeus tessulatus* has a frequently-recorded association with Ivy, despite the fact that the larvae feed in the roots of Umbelliferae. Our three species of *Tropiphorus* are all uncommon and each has been associated with Dog's Mercury, though again this seems not to be true very often. On the other hand, *Barynotus moerens* certainly does seem to feed, perhaps exclusively, on that plant.

The genus *Sitona*, which concludes the adelognathous weevils, is a specialised one. As far as is known all the species feed as larvae on the nitrifying root-nodules of Papilionaceae. They are much more host-specific than most other Adelognatha (though a few exceptions have been mentioned). Because many of the hosts are forage crops for man's domesticated farm animals, several species of *Sitona* are notorious pests. Consequently, they have been well studied and we probably know more about the biology of some species of *Sitona* than of any other British weevils. For the collector, however, the group may seem a relatively dull one. Most of the species are common and easily collected by sweeping and grubbing, or, in the case of the Broom and Gorse species, by beating. However, there are exceptions, as there usually are in any reasonably large group. *S. cambricus* is one of the less abundant species and it has two close relatives which are even rarer still. It may often be detected by finding the marks of its feeding on its host plant, in this case Large Birdsfoot Trefoil. Adults of *Sitona* feed very characteristically by gnawing semicircular notches on the edges of the leaves of their hosts. These notches are readily observable, for instance by gardeners on their peas and beans, but are also useful signs for detecting some of the rarer and more desirable species which make them. Most *Sitona* species are found underneath their food-plants, or in the soil, by day, as they are nocturnally active, like *Otiorhynchus*.

The series of phanerognathus weevils begins with the Cleoninae and Lixinae. This is appropriate because in general habitus these weevils are a half-way house between the Adelognatha and "typical" weevils with long thin snouts or rostra. Almost all of our few species in this group are rare. Many more species are known from the Mediterranean region and the species are also much more abundant there. One of the least uncommon species is *Cleonus piger*, associated with Creeping Thistle, but occurring only in sandy places. Under favourable circumstances it may be quite abundant and can be found sitting on bare sand near its food-plants. Its mottled appearance tones in well with the substrate, but does not make it particularly difficult to see. Also associated with thistles (though probably with a wider spectrum of hosts) are *Larinus planus* and *Rhinocyllus conicus*. Both can be quite common in the extreme south of England, but they are rare elsewhere. Tapping thistles over a sweep net, or just spotting the weevils on the plants, are the best methods of collecting them. Probably our commonest *Lixus* at the present time is one added to the British list only five years ago. Like many other insect colonists, *L. scabricollis* is abundant at its only known site, perhaps because its specific parasites have not yet caught up with it. In contrast, *L. paraplecticus*, which was once so abundant that it was a danger to cattle through getting stuck in their throats when eaten with vegetation, is now very rare, if not extinct. It feeds on waterside Umbelliferae, the larvae inhabiting the stems (like other *Lixus*) and is perhaps our best-documented case of a decline brought about by intensive agriculture and land-use change.

Fortunately, most of our species of *Hypera* have been little affected by such changes. Most of the species are fairly common. They feed on various Papilionaceae, Carophyllaceae, Umbelliferae and Geraniaceae, and are most easily collected by working the appropriate foodplants. Searching, tapping and sweeping are all effective. However, *Hypera* species, unlike most weevils can also be reared from collected larvae. Again, unlike most weevils, the larvae feed externally on host-plant foliage (occassionally on flowers etc.). Weevil larvae are usually white or yellowish, inactive, comma-shaped grubs, but those of *Hypera* are colourful (green, pink or purplish, often striped) and they crawl actively over vegetation by means of body-lobes (pedal discs). The larger larvae are easily reared, rather like lepidopterous caterpillars. Indeed, like some moths, *Hypera* larvae spin cocoons in which to pupate. After emergence from the pupa the callow adults take some days to mature, and do so more effectively if allowed to feed. This is a good way to get specimens, particularly of the less common species if an expedition to collect them is mistimed so that larvae, but no adults, are present. This is likely to happen most often in midsummer. Our two *Limobius* species are both uncommon, but have larval habits like those of *Hypera*.

Another group of weevils with exophagous larvae (those feeding exposed on the footplant) is the familiar genus *Cionus*, associated with species of Scrophulariaceae (Figworts and Mulleins). The slug-like larvae of several species may be found and reared from the paper-like globular cocoons, which contrast with the open network ones of *Hypera* species. However, many *Cionus* are common and hardly need to be reared, particularly as some bred specimens do not always mature or colour up well. Like many weevils, *Cionus* (together with the closely related *Cleopus pulchellus*) can be found on their foodplants for much of the summer and are readily collected by shaking them into a net or onto a sheet.

The subfamily Hylobiinae is a small group of mainly large weevils. *Hylobius* itself is represented by the common and notorious Pine Weevil, *H. abietis*, but a second, much rarer species has the very different habit of feeding in the roots of Purple Loosestrife. *H. transversovittatus* may be found by searching the ground beneath well-grown loosestrife plants in the few areas where the weevil occurs. *Liparus germanus* is one of our largest species. Occurring only in Kent and Surrey, it is another species which is often given away by the feeding defoliation of its foodplant, Hogweed. Because its smaller and commoner congener *L. coronatus* feeds on Cow Parsley (also known as Keck and Queen Anne's Lace), a plant with finely divided leaves, its feeding is much less readily detected. Both species may occasionally be found walking along country lanes and tracks. *Leiosoma* species are like miniature *Liparus*. *L. deflexum* is a common species, the presence of which, again, can easily be ascertained from its feeding damage on buttercup and Windflower foliage. It is ground-living and is not frequently swept. The exceedingly local *Anchonidium unguiculare*, which is only found in woods about Gweek in Cornwall, is a hylobiine which is most readily collected by sieving moss and leaf litter from its habitat. It is often overlooked, even when very small quantities of litter are painstakingly examined.

Magdalinae and Cossoninae are two subfamilies of wood-boring weevils which require special collecting methods, at least in the case of the latter. Species of *Magdalis* can be beaten like other aboreal weevils, despite the fact that the larvae feed in cells under the bark of their hosts. However, the adults are on the trees for a very short time compared with other weevils. They should be sought in May and early June in southern England, later in the north. There are species on Elm (a common one), various roseaceous shrubs and fruit trees, Birch, and Scots Pine.

In contrast, cossonine weevils are seldom beaten and have to be extracted from dead wood, particularly when this is not too dry and hard. *Pselactus spadix* has the specialised habitat of dead wood which is periodically inundated by the sea; it is often plentiful in old groynes. The Holly weevil, *Mesites tardii*, actually feeds in many kinds of wood. It is a comparatively large species found mainly in western Britain. Several cossonines are semi-synanthropic, occurring in old floorboards and similar situations indoors; an example is *Euophryum confine*. Also with specialised habits are our three species of *Sitophilus*, which feed on grain and stored food products. They are less common under today's hygenic warehouse and factory conditions than formerly.

Several groups of weevils are aquatic or semi-aquatic. The little *Tanysphyrus lemnae* feeds on the floating plants of species of duckweed (Lemnaceae). However, the weevil is unable to swim. The introduced *Stenopelmus rufinasus* has rather similar habits, feeding on the Water Fern (*Azolla*). Both species are found in ponds and ditches and can be collected by examining the foodplants or with a water net. The interesting genus *Bagous* is one of the Erirhininae, like *Stenopelmus*. All the species are semi-aquatic and feed on water plants. Most of the species are rare. They can be taken by collecting water weed, wringing it out to expel most of the water, and sieving it over a waterproof sheet. Alternatively, the weed can be heat-extracted, using an ordinary desk lamp or other source of warmth. In many of the areas where *Bagous* occur, particularly in drainage ditches, the vegetation is periodically cleared from the ditch, or "roded". The cleared vegetation can be worked for the weevils and on hot sunny days in spring they may be seen crawling back to the water on the muddy sides of the ditch. Some species of *Bagous* seem to be sporadic or uncertain in appearing and much patience is needed to collect more than a few of the less rare species.

All the species of *Dorytomus* are associated with poplars and sallows, most of them feeding as larvae in the catkins; consequently they are easily reared. Timing of the life-history depends on the appearance of the catkins of the host tree, which is often early in the year. In winter many of the species can be found hibernating under bark or in hollow trees. Species of *Notaris* and *Thryogenes* are waterside erirhinines, feeding on bur-reeds and sedges. They can be found in marsh litter and in tussocks on marshy ground as well as by sweeping. *Grypus equiseti* feeds on horsetails and is most usually found by grubbing at the roots. However, at evening, particularly after wet weather, it comes up on its host plants and is conspicuous.

Our remaining Erirhininae are dry land species. *Orthochaetes* are leaf-miners as larvae and conspicuously ground-living insects. *O. setiger* is quite a common weevil but seldom collected if beating and sweeping only are used. *Smicronyx* species are associated with dodders (Convolvulaceae) and Centuary (Gentianaceae). The dodder species are rarely collected by general methods and virtually the only way to find them is by recognising the plant and working it.

The subfamily Ceutorhynchinae is species-rich and has representatives with varied habits. However, nearly all the species are closely tied to particular plants. This is especially true of *Ceutorhynchus* itself. The distinctive *Mononychus punctumalbum* feeds in the seedpods of *Iris* species and is most readily found by examining and collecting these in late September and October. It is a local species of southern England.

Few ceutorhynchines are arboreal. *Rutidosoma globulus*, a rare species, is associated with poplars, especially Aspen, while three of our five *Coeliodes* are found on Oak and two on Birch. Many of our *Zacladus, Stenocarus, Micrelus, Cidnorhinus, Ceuthorhynchidius* and *Ceutorhynchus* species can be found by general sweeping. However, this is a poor substitute for working the known food-plants of the species. Many *Ceutorhynchus* feed on only one species of Cruciferae (although there are a lot of exceptions) and a botanical knowledge of this family of plants will be found particularly valuable.

The same principle, though not applied specifically to the Cruciferae, is relevant to other ceutothynchines, but many of these are waterside or aquatic species. Most aquatic of all is *Eubrychius velutus*, which can spend its whole life under water. It is beautifully adapted for this existence. It should not be passed over when netting water beetles. It can also be found by sieving roded water-weed, particularly if its foodplant, Water Milfoil, is present in quantity. Many of the semi-aquatic Ceutorhynchinae are able to swim in the surface of water and can occasionally be found this way. Most can be obtained, under favourable circumstances, by working their known food-plants in the normal way, for example species of *Litodactylus, Rhinoncus, Phytobius, Amalorrhynchus, Drupenatus, Tapinotus* and *Poophagus*. However, a few species require special methods. Thus *R. albicinctus*, recently added to the British list, can only readily be found by examining the floating leaves of its foodplant, Amphibious Bistort, on which the weevils sit.

One problem encountered occasionally when working foodplants for weevils is that the insects drop from the plants when approached. Certainly care needs to be exercised and any rough or sudden movements avoided. One species with a reputation for dropping from its foodplant at the slightest provocation is *Tapinotus sellatus*, an uncommon species in southern England which feeds on Yellow Loosestrife. In the author's experience, however this ability has been exaggerated. With *Orobitis cyaneus* the Ceutorhynchinae end with a fully terrestrial species once again. This weevil feeds in the seed capsules of various violets (Violaceae) and in appearance is so like a violet seed that it can easily be passed over.

Baridinae are small, elongate weevils, often markedly metallic. Only five *Baris* are reliably recorded from Britain, but the genus is rich in species in other parts of the world. In contrast to our two *Limnobaris*, which are associated with sedges. *Baris* feed on a varied range of dicotyledonous plants, usually in the base of the stem. They can be found underneath their foodplants but many species can also be swept.

The subfamily Anthonominae is represented in Britain by 12 species of *Anthonomus* and single species of *Furcipus* and *Brachonyx*. Most are arboreal, exceptions being the common *A. rubi* and the mainly northern *A. brunnipennis* which feed on herbaceous Rosaceae. Hawthorn, Blackthorn, fruit trees, Elm and Scots Pine are the main hosts, the larvae feeding chiefly in the flower buds and "capped" blossoms. Because of this, and unlike most weevils, many species are active in winter. For example, *A. bituberculatus* is common by beating the bare branches of Hawthorn from late November to March and is less frequently taken at other times. Care is needed in the identification of *Anthonomus*, the host trees once again being a useful guide. In the older literature (Fowler and Joy) keys to the genus are not satisfactory and some of the species were omitted, or included as "varieties".

One of the species which many naturalists would quote as the typical weevil is *Curculio nucum*, one of 8 *Curculio* which constitute the subfamily Curculioninae in Britain. All the species have particularly long and slender rostra, used for boring into nuts, acorns and other structures in which the larvae develop. These structures are galls induced on Oak (by gall wasps) and Sallow (by sawflies) in the case of *C. pyrrhoceras* and *C. salicivorus* respectively. Both species of weevil are common but their habits and biology are known only in outline.

Tychiinae are very small weevils, most of which are seed feeders as larvae. Species of *Tychius* itself are all associated with various Papilionacae, while *Sibinia* affect Carophyllaceae. On the other hand our single species of *Ellescus* and *Acalyptus* are arboreal, feeding as larvae in the buds of sallows.

Most of our Mecininae are also seed feeders. Species of *Miarus* are associated with Campanulaceae. Because plants of this family often flower late in the summer *Miarus* may frequently be found on other vegetation earlier in the year; many seem to have a preference for yellow-flowered Composites and Rockrose, of which they may eat pollen. Species of *Mecinus*, a genus of small elongate weevils, are associated mainly with plantains and feed as larvae in stems rather than fruits. *M. collaris* may be reared from inconspicuous galls below the flowerheads of Sea Plantain, but the larvae sustain high rates of parasitism. Plantains are also the foodplants of some of our *Gymnetron* species, including the commonest, *G. pascuorum*, but most are associated with Scrophulariacae. The British list is in disarray and currently includes species with no right to be there.

Most coleopterists know species of Rhynchaeninae because of their ability to jump. This is also the group in which stridulation (by means of a file on the underside of the elytra and a scraper on the abdomen) has been most studied. Most British species are arboreal, and we have ones associated with Oak, Elm, Hazel, Birch, Alder and Sallow. By far the commonest is the Beech weevil, *Rhynchaenus fagi*. Rhynchaeninae are leaf miners as larvae and the mines of *R. fagi* can be very numerous and conspicuous. Only one British species *R. pratensis*, is found on herbs; it feeds on Hardheads (Compositae) and is rather local. Our two species of *Ramphus* are minute jumping insects which need looking out for, not because they are rare but because of their small size and active habits. *R. oxyacanthae* is associated mainly with Hawthorn and *R. pulicaris* with Sallow and Birch.

Bark and Ambrosia beetles are often included nowadays with the Curculionidae, but it is more convenient to regard them as a good family, Scolytidae. All are associated with dead wood, or at least dead stems of woody plants, and many contruct complicated larval galleries in which the larvae feed. The Ambrosia beetles, in particular, have a close symbiotic relationship with fungi that are part of the decomposition cycle of dead wood.

Bark beetles are only occasionally swept or beaten, though abundant species such as *Leperisimus varius*, *Phloeophthorus* and *Scolytus multistriatus* may be found in this way on their respective foodplants, Ash, Gorse (and Broom) and Elm. Similarly, *Xyclocleptes bispinus* may be beaten from Clematis. Several scolytids engage in maturation feeding in which the newly emerged adults eat green shoots and foliage; they can be collected more easily at this time than when in dead wood. If the timing is right peeling back the bark of suitable trees may result in the discovery of mature adults of different species (nor normally in the same tree). Though many species make superficial galleries underneath the bark others penetrate more deeply into the wood, for instance *Xyleborus dispar*. Careful use of the chisel is needed to extract this species, which is most frequently found in the wood of Oak and fruit trees. Puffing smoke into the exit holes has sometimes been effective in inducing the beetles to emerge. Many scolytids are associated with conifers rather than deciduous trees, in contrast to Cossoninae, which predominantly attack broad-leaved trees. Some species are forest pests and much of the biological information on this group has been gathered by forest entomologists.

Platypus is the only British genus in the family Platypodidae. These beetles are especially associated with tree stumps and are usually considered uncommon. They bore into solid wood and can sometimes be detected from the fine sawdust they leave by their exit holes. Two species are on the British list; they are elongate, cylindrical insects, the pronotum in particular being very long.

Curating

The details of forming a collection of weevils, or of beetles more generally are largely a matter of building up personal preferences. However, some advice may be acceptable to beginners. Weevils are best kept alive until reaching home and not killed in the field. In my view nothing beats ethyl acetate as a killing agent.

Unlike many beetles, weevils do not relax well after killing. If it is intended to make a tidy, neat and aesthetically pleasing collection well-relaxed specimens will be needed. The best plan is to stun the weevils in ethyl acetate, set them, and then return them to ethyl acetate vapour to complete the killing process. A good alternative to this procedure, particularly if time is short, or facilities limited (i.e. on a collecting trip or expedition) is to kill the weevils and store them (with adequate labels!) in tubes of 2% acetic acid solution. This leaves the insects nicely relaxed. They should not be kept too long in the solution, though weevils can last for several years in it; the exact period depends on the size and robustness of the species.

Nowadays considerable reliance is placed on the aedeagus of weevils as an aid to determination. This is particularly true in the Curculionidae-Phanerognatha. The student should get used to dissecting at least a specimen or two of any new species taken, in order to use this important characteristic for determination. It is much more difficult to extract aedeagi from dry and long-dead specimens, though not impossible of course; museum taxonomists do it every day.

Once "hooked" on weevils the coleopterist may find his interest in other beetles waning. But that is a risk worth taking!

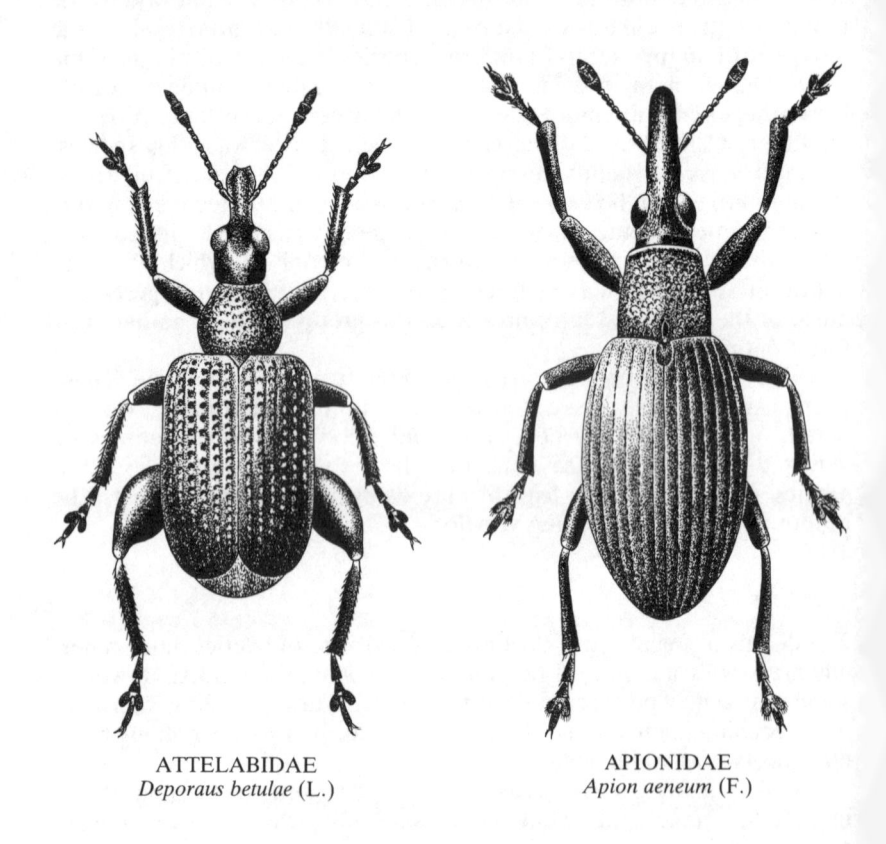

ATTELABIDAE
Deporaus betulae (L.)

APIONIDAE
Apion aeneum (F.)

Reference

Morris, M.G., 1990. Handbooks for the Identification of British Insects, Vol. **5**. Part 16. *Curculionoidea. Orthocerous Weevils*. Royal Ent. Soc., London.

Part III – Beetle Associations

VASCULAR PLANTS AND THE BEETLES ASSOCIATED WITH THEM

BY ERIC G. PHILP

Many species of beetle are closely associated with certain flowering plants and it will be of great help to the coleopterist to gain a working knowledge of the wild flowers of the countryside in order to search out some of the more elusive species of his quest. Beetles will be associated with certain plants because they or their larvae feed upon the plant (or parts of the plant such as the root, fruits, stem etc., each of which might have to be in a certain condition, young, ripe, dying etc.); or feed upon other insects or fungi that are associated with that plant.

The following list of some of the British beetles, in Check-list order, also gives the major host-plants on which these beetles are found. A large number of "common" species of beetle feed upon a wide range of plants or are attracted to almost any plant in flower and these will be found by general beating or sweeping and are not dealt with here. By first seeking out the host-plant as listed here the reader will have a much better chance of finding the specific beetle they seek.

CARABIDAE

Most species of Carabidae feed upon a mix of both animal and vegetable matter, and many are often found resting around the base of various plants, i.e. mats of *Stellaria media*, although this is more often as for cover than for food. Most species of **Amara** and **Harpalus** will often climb herbaceous plants in search of pollen or seeds.

Calosoma inquisitor; *Quercus* spp. **Amara intima**; *Calluna vulgaris*. **Amara aulica**; Mayweeds, *Matricaria*, *Chamomilla*, etc. **Amara convexiuscula**; *Atriplex*, *Chenopodium*. **Zabrus tenebrioides**, in cultivated corn fields on *Triticum*, *Secale* etc. **Harpalus rufipes**; *Fragaria*. **Trichocellus cognatus**; *Calluna vulgaris*. **Bradycellus collaris**; *Calluna vulgaris*. **Odacantha melanura**; *Phragmites*, *Typha*. **Demetrias monostigma**; *Ammophila arenaria*, *Leymus arenarius* and occasionally other grasses or sedges. **D. imperialis**; *Typha* spp.

HYDROPHILIDAE

Helophorus nubilis; *Triticum* and the roots of various other plants. **H. porculus** and **H. rufipes**; at the roots of *Brassica rapa* and other plants.

HISTERIDAE

Plegaderus dissectus; in wet decaying wood of *Ulmus* and *Quercus*. **P. vulneratus**; under bark of *Pinus sylvestris*. **Kissister minimus**; at the roots of *Rumex acetosella*. **Paromalus flavicornis**; under bark of *Fagus*, *Quercus* etc.

SILPHIDAE
Dendroxena quadrimaculata; on *Quercus* and occasionally other trees where it feeds upon Lepidopterous and other larvae.

LUCANIDAE
All three species feed upon dead or rotten wood of various trees, particularly those listed.
    Lucanus cervus; *Quercus, Ulmus*. **Dorcus parallelipipedus**; *Fagus sylvatica, Fraxinus excelsior, Salix* spp., *Ulmus* spp. **Sinodendron cylindricum**; *Fraxinus, Fagus, Salix, Betula, Malus, Tilia, Castanea.*

SCARABAEIDAE
Most of the Chafers develop underground, feeding upon the roots of various plants. The adults are frequently found at flower heads particularly those of *Carduus, Cirsium, Heracleum* and *Rubus*. The larvae of **Gnorimus nobilis** develop in wood mould in stumps of *Betula* and *Prunus*, and those of **G. variabilis** in *Quercus*.

BUPRESTIDAE
Melanophila acuminata; in and under bark of scorched and burnt *Pinus* spp., *Picea* spp. and occasionally *Betula* spp. **Anthaxia nitidula**; larvae under bark of *Prunus spinosa* and other *Prunus* spp., adults on *Crataegus* spp., *Viburnum opulus*. **Agrilus angustulus**; *Quercus* spp., *Corylus avellana*. **A. laticornis**; *Quercus* spp., *Salix* spp., *Corylus avellana*. **A. pannonicus**; *Quercus* spp. **A. sinuatus**; mature specimens of *Crataegus*. **A. viridis**; mature or coppiced *Salix cinerea* and other *Salix* spp., stunted or damaged *Quercus* spp. **Aphanisticus emarginatus**; *Juncus articulatus* and other *Juncus* spp. **Aphanisticus pusillus**; *Schoenus nigricans, Juncus* spp. **Trachys minutus**; *Salix* spp., *Carpinus betulus, Corylus avellana*. **T. troglodytes**; *Succisa pratensis*.

ELATERIDAE
The wire-worm lavae of the click beetles feed upon the roots of herbaceous plants or in dead or decaying wood in trees (although some are predaceous upon other larvae in these situations).
    Lacon quercus; *Quercus* spp. **Ampedus balteatus**; *Pinus* spp., occasionally *Quercus* spp. and *Betula* spp. **A. cardinalis**; *Quercus* spp. **A. cinnabarinus**; *Fagus sylvatica, Betula* spp., *Quercus* spp. **A. elongantulus**; *Quercus* spp., *Pinus* spp. **A. nigerrimus**; *Quercus* spp. **A. nigrinus**; *Pinus* spp., *Picea* spp. **A. quercicola**; *Betula* spp., *Fagus sylvatica, Quercus* spp. **A. pomorum**; *Betula* spp. and occasionally other trees. **A. ruficeps**; *Quercus* spp. **A. rufipennis**; *Fagus sylvatica*. **A. sanguinolentus**; *Pinus* spp., *Betula* spp., adults often at roots of *Calluna vulgaris*. **A. tristis**; *Picea* spp. **Ischnodes sanguinicollis**; *Ulmus* spp., *Quercus* spp., *Acer* spp. **Procraerus tibialis**; *Quercus* spp., *Fagus sylvatica*. **Megapenthes lugens**; *Ulmus* spp., occasionally *Fagus sylvatica*. **Dicronychus equiseti**; *Ammophila arenaria*. **Melanotus villosus**; *Ammophila arenaria*. **Limoniscus violaceus**; *Fagus sylvatica*. **Harminius undulatus**; *Betula* spp., *Pinus* spp. **Selatosomus bipustulatus**; *Salix* spp., *Alnus* spp., *Quercus* spp. **Selatosomus impressus**; *Pinus sylvestris, Betula* spp. **Selatosomus nigricornis**; *Salix* spp. in swampy parts of woods. **Elater ferrugineus**; *Fagus sylvatica, Ulmus* spp.

EUCNEMIDAE
Eucnemis capucina; *Fagus sylvatica*. **Dirhagus pygmaeus**; *Fagus sylvatica, Quercus* spp. **Melasis buprestoides**; *Fagus sylvatica*. **Hylis olexai**; *Fagus sylvatica*.

DERODONTIDAE
Laricobius erichsoni; *Abies* spp., *Pseudotsuga menziesii*.

ANOBIIDAE

Dryophilus anobioides; *Cytisus scoparius, Rubus fruticosus*. **D. pusillus**; *Picea* spp., *Abies* spp., *Larix* spp. **Ochina ptinoides**; *Hedera helix*. **Xestobium rufovillosum**; in old trees of *Quercus* and *Salix*, and occasionally other species of tree. **Ernobius angusticollis**; *Picea* spp., *Pinus* spp. **E. mollis**; *Pinus* spp. and other Coniferous trees. **E. nigrinus**; *Pinus* spp. **Gastrallus immarginatus**; *Quercus* spp. **Anobium inexpectatum**; *Hedera helix*. **Hadrobregmus denticollis**; in old *Quercus* and *Crataegus*. **Xyletinus longitarsis**; *Quercus* spp., *Ulmus* spp. **Anitys rubens**; in decayed *Quercus*.

BOSTRICHIDAE

Rhyzopertha dominica; in decayed *Quercus* and other trees.

TROGOSSITIDAE

Nemozoma elongatum; under bark of *Ulmus* spp.

PELTIDAE

Ostoma ferrugineum; *Pinus sylvestris*. **Thymalus limbatus**; *Fagus sylvatica*.

CLERIDAE

Tillus elongatus; *Fagus sylvatica*. **Opilo mollis**; associated with *Crataegus* spp. **Thanasimus formicarius**; *Pinus* spp. and other conifers. **Thanasimus rufipes**; *Pinus* spp. and other conifers.

MELYRIDAE

Hypebaeus flavipes; *Quercus* spp. **Sphinginus lobatus**; *Quercus* spp.

LYMEXYLIDAE

Hylecoetus dermestoides; *Betula* spp., *Fagus sylvatica, Pinus* spp., *Quercus* spp. **Lymexylon navale**; *Quercus* spp.

KATERETIDAE

Brachypterolus linariae; *Linaria* spp. **B. pulicarius**; *Linaria* spp. **B. vestitus**; *Antirrhinum* spp. (*Antirrhinum latifolium* in parks and gardens). **B. antirrhini**; *Antirrhinum* spp. **Brachypterus glaber**; *Urtica* spp. **B. urticae**; *Urtica* spp. **Kateretes rufilabris**; *Juncus* spp., *Carex* spp. **K. pedicularius**; *Carex* spp. **K. pusillus**; *Carex* spp.

NITIDULIDAE

Pria dulcamarae; *Solanum dulcamara, Solanum nigrum*. **M. atramentarius**; *Lamiastrum galeobdolon*. **M. atratus**; *Rosa* spp. **M. bidens**; *Clinopodium vulgare*. **M. bidentatus**; *Genista tinctoria*. **M. brevis**; *Helianthemum canum*. **M. brunnicornis**; *Stachys sylvatica*. **M. coracinus**; yellow Cruciferae. **M. corvinus**; *Campanula trachelium*. **M. difficilis**; *Lamium album*. **M. erichsoni**; *Hippocrepis comosa*. **M. erythropus**; *Lotus corniculatus*. **M. exilis**; *Thymus praecox* sub. sp. *arcticus*. **M. flavimanus**; *Rosa* spp. **M. fulvipes**; *Sinapis arvensis*. **M. gagathinus**; *Mentha arvensis*. **M. haemorrhoidalis**; *Lamiastrum galeobdolon*. **M. incanus**; *Nepeta cataria*. **M. kunzei**; *Lamiastrum galeobdolon*. **M. lugubris**; *Thymus praecox* sub. sp. *arcticus*. **M. morosus**; *Lamium album*. **M. nanus**; *Marrubium vulgare*. **M. nigrescens**; *Trifolium repens*. **M. obscurus**; *Teucrium scorodonia*. **M. ochropus**; *Stachys palustris*. **M. ovatus**; *Glechoma hederacea*. **M. pedicularis**; *Stachys officinalis*. **M. planiusculus**; *Echium vulgare*. **M. rotundicollis**; *Sinapis arvensis, Sisymbium officinale*. **M. ruficornis**; *Ballota nigra*. **M. serripes**; *Galeopsis angustifolia*. **M. solidus**; *Helianthemum nummularium*. **M. subrugosus**; *Jasione montana, Campanula glomerata*. **M. umbrosus**; *Prunella vulgaris, Prunella grandifolia*. **M. viduatus**; *Galeopsis tetrahit*. **M. viridescens**;

Sinapis arvensis, Brassica spp. other yellow *Cruciferae*. **Pityophagus ferrugineus**; *Pinus* spp.

CUCUJIDAE

Laemphloeus monilis; *Fagus sylvatica*, under bark. **Cryptolestes spartii**; *Cytisus scoparius*, in dead branches. **Leptophloeus clematidis**; *Clematis vitalba*.

CRYPTOPHAGIDAE

Telmatophilus brevicollis; *Sparganium erectum*. **T. caricis**; *Sparganium erectum*. **T. sparganii**; *Sparganium erectum*.

BIPHYLLIDAE

Biphyllus lunatus; *Fraxinus excelsior*, in the fungus *Daldinia concentrica* on dead branches. **Diplocoelus fagi**; *Fagus sylvatica*, under bark.

BYTURIDAE

Byturus tomentosus; adults on many flowers particularly those of *Rubus idaeus*.

PHALACRIDAE

(Note. Species of **Phalacrus** are associated with the mentioned plants that are infected by Smut Fungi, *Ustilaginales*).

Phalacrus caricis; *Carex* spp. **P. corruscus**; cereal crops, *Triticum, Hordeum* etc. **P. fimetarius**; *Brachypodium pinnatum*. **P. substriatus**; *Carex* spp., *Narthecium ossifragum*. **Olibrus aeneus**; *Tanacetum vulgare, Matricaria* spp., *Artemisia* spp. **O. affinis**; *Hypochoeris* spp., *Tragopogon pratensis*. **O. corticalis**; *Conyza canadensis, Senecio* spp. **O. flavicornis**; *Leontodon autumnalis*. **O. liquidus**; *Pilosella* spp. **O. millefolii**; *Achillea millefolium*. **O. pygmaeus**; *Crepis* spp., *Filago vulgaris, Leontodon* spp. **Stilbus oblongus**; *Typha latifolia*.

CERYLONIDAE

All species are found under bark of various trees, particularly those that are mentioned.

Cerylon fagi; *Fagus, Quercus* spp. **C. ferrugineum**; *Fagus sylvatica, Betula* spp., *Quercus* spp., *Acer* spp. **C. histeroides**; *Fagus, Quercus* spp., *Pinus* spp., *Ulmus* spp.

COCCINELLIDAE

Clitostethus arcuatus; *Hedera helix*. **Coccidula scutellata**; *Typha* spp. **Scymnus nigrinus**; *Pinus* spp. **S. auritus**; *Quercus* spp. **S. suturalis**; *Pinus* spp. **Nephus quadrimaculatus**; *Pinus sylvestris, Hedera helix*. **Chilocorus renipustulatus**; *Salix* spp. **Exochomus quadripustulatus**; *Pinus* spp., *Picea* spp. **Aphidecta obliterata**; *Pinus* spp., *Picea* spp. **Coccinella hieroglyphica**; *Calluna vulgaris*. **Anatis ocellata**; *Pinus* spp. **Neomysia oblongoguttata**; *Pinus* spp., *Picea* spp.

COLYDIIDAE

Cicones variegata; *Fagus, Carpinus*, under bark. **Aulonium ruficorne**; *Larix* spp., *Pinus* spp., under bark. **Aulonium trisulcum**; *Ulmus* spp. under bark.

TENEBRIONIDAE

Bolitophagus reticulatus; in the fungus *Fomes fomentarius* growing on *Betula* spp. (Scotland only). **Diaperis boleti**; in the fungus *Piptoporus betulinus* on *Betula* spp. **Scaphidema metallicum**; *Ulmus* spp. **Corticeus bicolor**; *Ulmus* spp., *Fraxinus*. **C. fraxini**; *Pinus* spp.

C. linearis; *Picea* spp., *Pinus* spp. *C. unicolor*; *Fagus*, *Quercus* spp., *Betula* spp. **Prionychus ater**; in decaying wood of *Quercus* spp., *Fraxinus*, *Salix* spp., *Ulmus* spp., *Malus* spp. and *Prunus* spp. *P. melanarius*; *Fagus sylvatica*. **Pseudocistela ceramboides**; *Quercus* spp. **Mycetochara humeralis**; *Quercus* spp., *Fagus sylvatica*, *Acer* spp., *Prunus* spp.

SALPINGIDAE
Salpingus castaneus; *Pinus* spp.

PYTHIDAE
Pytho depressus; *Pinus sylvestris* (Scotland only).

PYROCHROIDAE
Pyrochroa coccinea; *Quercus* spp. *P. serraticornis*; *Fagus*, *Quercus* spp., *Ulmus* spp. **Schizotus pectinicornis**; *Betula* spp.

MELANDRYIDAE
Anisoxya fuscula; *Fagus sylvatica*, *Fraxinus excelsior*, *Salix* spp., *Acer campestre*. **Abdera biflexuosa**; *Quercus* spp., *Fraxinus excelsior*. *A. flexuosa*; *Alnus* spp., *Salix* spp. *A. quadrifasciata*; *Carpinus betulus*, *Quercus* spp., *Fagus sylvatica* and occasionally other trees. *A. triguttata*; *Pinus* spp. **Phloiotrya vaudoueri**; *Carpinus betulus*, *Fagus sylvatica*, *Quercus* spp., *Fraxinus excelsior*. **Hypulus quercinus**; *Corylus avellana*, *Quercus* spp. **Zilora ferruginea**; in the fungus *Hirschioporus abietinus* on *Pinus* spp., *Larix* spp. and *Abies* spp. **Melandrya barbata**; *Fagus sylvatica*, *Quercus* spp. *M. caraboides*; *Betula* spp., *Quercus* spp., *Fagus sylvatica*, *Ulmus* spp.

SCRAPTIIDAE
Species of **Anaspis** are attracted to flower blossom particularly that of *Crataegus* spp.

MORDELLIDAE
Mordellistena acuticollis; *Cirsium arvense*. *M. nanuloides*; *Artemisia maritima*. *M. parvula*; *Achillea* spp., *Artemisia* spp. *M. pumila*; *Cirsium* spp.
(Note. Members of this family are attracted to the flower heads of the Umbelliferae particularly to *Heracleum sphondylium*.).

OEDEMERIDAE
Chrysanthia nigricornis; *Pinus* spp. (Scotland only.)

MELOIDAE
Lytta vesicatoria; *Fraxinus excelsior*, *Ligustrum* spp.

ADERIDAE
Aderus oculatus; *Quercus* spp., *Tilia* spp. *A. populneus*; *Quercus* spp., *Tilia* spp.

CERAMBYCIDAE
Larvae of Longhorn Beetles feed upon the woody parts of various trees and shrubs and it is often the nature and condition of this such as thickness of bark, moisture content or state of decay that is more impor-

tant than the actual species of tree on which the eggs are deposited. Adults will be found at the breeding sites or on nearby flower-heads, particularly those of *Rubus fruticosus* and *Heracleum spondylium*. The following list is not exhaustive but just gives the major hosts.

Prionus coriarius; *Quercus* spp., *Fagus sylvatica*, *Pinus* spp., *Abies* spp., *Betula* spp. *Arhopalus rusticus*; *Pinus* spp. *A. tristis*; *Pinus* spp., *Picea* spp. *Asemum striatum*; *Pinus* spp., *Picea* spp., *Larix* spp., *Abies* spp. *Tetropium castaneum*; *Larix* spp., *Picea* spp., *Pinus* spp. *T. gabrieli*; *Larix* spp., *Pinus* spp., *Picea* spp., *Abies* spp. *Rhagium bifasciatum*; *Pinus* spp., *Picea* spp., *Abies* spp., *Betula* spp. *R. inquisitor*; *Pinus* spp., *Picea* spp., *Larix* spp., *Cedrus* spp. *R. mordax*; *Pinus* spp., *Picea* spp., *Quercus* spp., *Betula* spp., *Fraxinus excelsior*. *Stenocorus meridianus*; *Salix* spp., *Prunus* spp., *Fraxinus excelsior*. *Acmaeops collaris*; *Quercus* spp. *Leptura rubra*; *Larix* spp., *Picea* spp., *Abies* spp. *L. sanguinolenta*; *Pinus* spp. *L. scutellata*; *Fagus sylvatica*. *L. sexguttata*; *Fagus sylvatica*. *Judolia cerambyciformis*; *Quercus* spp., *Betula* spp., *Castanea sativa*. *J. sexmaculata*; *Pinus* spp., *Abies* spp. *Trinophylum cribratum*; *Quercus* spp. *Obrium brunneum*; *Picea* spp., *Larix* spp., *Pinus* spp. *O. cantharinum*; *Populus tremula*, *Malus* spp. *Aromia moschata*; *Salix* spp. *Pyrrhidium sanguineum*; *Quercus* spp. *Phymatodes alni*; *Quercus* spp., *Alnus glutinosus*. *P. testaceus*; *Quercus* spp. *Anaglyptus mysticus*; *Crataegus* spp. *Lamia textor*; *Salix* spp. *Mesosa nebulosa*; *Quercus* spp. *Pogonocherus fasciculatus*; *Pinus* spp., *Picea* spp. *P. hispidulus*; *Malus* spp., *Pyrus communis*. *P. hispidus*; *Malus* spp., *Pyrus communis*. *Leiopus nebulosus*; *Quercus* spp. *Acanthocinus aedilis*; *Pinus* spp. *Saperda carcharias*; *Populus* spp., *Salix* spp. *S. populnea*; *Populus* spp., *Salix* spp. *S. scalaris*; *Populus* spp., *Salix* spp. *Oberea oculata*; *Salix* spp. *Stenostola dubia*; *Tilia* spp. *Tetrops praeusta*; *Malus* spp., *Crataegus* spp.

BRUCHIDAE

Bruchus atomarius; *Vicia* spp., *Lathyrus* spp. *B. loti*; *Lotus corniculatus*, *Lathyrus pratensis*. *B. rufimanus*; *Lathyrus* spp., *Vicia* spp. *B. rufipes*; *Lathyrus* spp., *Vicia* spp. *Bruchidius cisti*; *Helianthemum nummularium*, *Lotus corniculatus*. *B. olivaceus*; *Onobrychis viciifolia*. *B. villosus*; *Cytisus scoparius*.

CHRYSOMELIDAE

Macroplea appendiculata; *Myriophyllum* spp., *Potamogeton* spp. *M. mutica*; *Potamogeton pectinatus*, *Ruppia* spp., *Zannichellia palustris*. *Donacia aquatica*; *Glyceria* spp. and other emergent vegetation at the edge of rivers, ponds and lakes where other members of the genus will also be found. *D. bicolora*; *Sparganium erectum*. *D. cinerea*; *Carex* spp., *Sparganium* spp., *Typha* spp., *Phragmites australis*. *D. clavipes*; *Phragmites australis*. *D. crassipes*; *Nymphaea alba*, *Nuphar lutea*. *D. dentata*; *Sagittaria sagittifolia*. *D. impressa*; *Scirpus lacustris*. *D. marginata*; *Sparganium erectum*. *D. obscura*; *Carex* spp. *D. semicuprea*; *Glyceria* spp. *D. simplex*; *Glyceria* spp., *Carex* spp., *Sparganium* spp., *Typha* spp. *D. sparganii*; *Sparganium* spp., *Glyceria* spp. *D. thalassina*; *Scirpus* spp., *Carex* spp. *D. versicolorea*; *Potamogeton natans*. *D. vulgaris*; *Sparganium* spp., *Typha* spp., *Carex* spp., *Scirpus* spp. *Plateumaris affinis*; *Carex* spp. *P. braccata*; *Phragmites australis*. *P. discolor*; *Carex* spp. *P. sericea*; *Carex* spp., *Iris pseudacorus*, *Nuphar* spp., *Nymphaea alba*. *Orsodacne cerasi*; *Crataegus* spp., *Prunus* spp. *O. lineola*; *Crataegus* spp. *Bromius obscurus*; *Epilobium angustifolium*. *Zeugophora flavicollis*; *Populus tremula*. *Z. subspinosa*; *Populus tremula*. *Z. turneri*; *Populus tremula* (Scotland only). *Lema cyanella*; *Cirsium arvense*. *Oulema melanopa*; *Avena sativa* and other Gramineae. *Crioceris asparagi*; *Asparagus officinalis*. *Lilioceris lilii*; cultivated species of *Lilium*. *Labidostomis tridentata*; *Betula* spp. *Gynandrophthalma affinis*; *Corylus avellana*. *Cryptocephalus aureolus*; adults on heads of *Pilosella* spp. and other yellow Compositae. *C. coryli*; *Corylus avellana*. *C. decemmaculatus*; *Salix* spp. *C. exiguus*; *Salix* spp. *C. hypochaeridis*; adults on heads of *Pilosella* spp. and other yellow Compositae. *C. labiatus*; *Quercus* spp., *Betula* spp., *Corylus avellana*. *C. moraei*; *Hypericum* spp. *C. nitidulus*; *Betula* spp., *Corylus avellana*. *C. parvulus*; *Betula* spp. *C. punctiger*; *Betula* spp., *Salix* spp. *C. pusillus*; *Betula* spp. *C. querceti*; *Quercus* spp. *C. sexpunctatus*; *Betula* spp., *Corylus avellana*. *Timarcha goetttingensis*; *Galium* spp. *T. tenebricosa*; *Galium* spp. *Chrysolina banksi*; *Ballota nigra*, *Plantago lanceolata*.

C. brunsvicensis; *Hypericum* spp. *C. cerealis*; *Thymus praecox*. *C. fastuosa*; *Galeopsis tetrahit*. *C. haemoptera*; *Plantago* spp. in sandy coastal areas. *C. hyperici*; *Hypericum* spp. *C. menthastri*; *Mentha aquatica*. *C. oricalcia*; *Anthriscus sylvestris*. *C. polita*; various species of Labiatae. *C. sanguinolenta*; *Linaria* spp. *C. varians*; *Hypericum* spp. **Gastrophysa polygoni**; *Polygonum* spp., *Rumex* spp. **G. viridula**; *Rumex* spp. **Phaedon armoraciae**; *Veronica beccabunga*. **P. cochleariae**; *Brassica* spp., *Rorippa* spp. **P. concinnus**; *Cochlearia anglica*, *Triglochin maritima*. **P. tumidulus**; *Heracleum sphondylium*. **Hydrothassa glabra**; *Ranunculus* spp. **H. hannoveriana**; *Caltha palustris*. **H. marginella**; *Ranunculus* spp., *Caltha palustris*. **Prasocuris junci**; *Veronica beccabunga*. **P. phellandrii**; *Oenanthe aquatica*, *Oenanthe crocata*. **Plagiodera versicolora**; narrow-leaved species of *Salix*. **Chrysomela aenea**; *Alnus glutinosa*. **C. populi**; *Salix* spp., *Populus* spp. **C. tremula**; *Populus tremula*, *Salix* spp. **Phytodecta decemnotata**; *Populus tremula*, *Salix* spp. **P. olivacea**; *Cytisus scoparius*. **P. pallida**; *Corylus avellana*, *Sorbus aucuparia*. **P. viminalis**; *Salix* spp. **Phyllodecta laticollis**; *Populus* spp. **P. polaris**; dwarf mountain species of *Salix*. **P. vitellinae**; *Populus* spp., *Salix* spp. **P. vulgatissima**; *Salix* spp., *Populus* spp. **Galerucella calmariensis**; *Lythrum salicaria*. **G. grisescens**; *Hydrocharis morsus-ranae*, *Lysimachia vulgaris*. **G. lineola**; *Salix* spp. particularly *S. viminalis*. **G. nymphaeae**; *Nymphaea alba*, *Nuphar lutea*. **G. pusilla**; *Lythrum salicaria*. **G. sagittariae**; *Sagittaria sagittifolia*, *Rumex hydrolapathum*. **G. tenella**; *Filipendula ulmaria*. **Pyrrhalta viburni**; *Viburnum lantana* & *V. opulus*. **Lochmaea caprea**; *Salix* spp. particularly *S. caprea*. **L. crataegi**; *Crataegus* spp. **L. suturalis**; *Calluna vulgaris*. **Phyllobrotica quadrimaculata**; *Scutellaria galericulata*. **Luperus flavipes**; *Betula* spp., *Salix* spp., *Alnus glutinosa*, *Corylus avellana*. **L. longicornis**; *Betula* spp., *Salix* spp., *Alnus glutinosa*. **Calomicrus circumfusus**; *Genista tinctoria*, *Cytisus scoparius*, *Ulex* spp. **Agelastica alni**; *Alnus glutinosa*. **Sermylassa halensis**; *Galium* spp. **Phyllotreta aerea**; *Brassica* spp., *Sinapis* spp. **P. atra**; *Brassica* spp. **P. consobrina**; *Brassica* spp., *Sinapis* spp. **P. cruciferae**; *Brassica* spp., *Sinapis* spp. **P. diademata**; *Cardamine* spp., *Cochlearia* spp. **P. nemorum**; *Brassica* spp. **P. nigripes**; *Brassica* spp. **P. nodicornis**; *Reseda lutea*. **P. ochripes**; *Alliaria petiolata*, *Cardamine* spp., *Rorippa* spp. **P. tetrastigma**; *Cardamine* spp., *Nasturtium* spp. **P. undulata**; *Brassica* spp. **P. vittata**; *Raphanus* spp., *Brassica* spp. **Aphthona atrocaerulea**; *Euphorbia* spp. **A. atrovirens**; *Origanum vulgare*. **A. euphorbiae**; *Euphorbia* spp. **A. herbigrada**; *Helianthemum nummularium*. **A. lutescens**; *Lythrum salicaria*. **A. melancholica**; *Euphorbia* spp. **A. nigriceps**; *Geranium pratense*. **A. nonstriata**; *Iris pseudacorus*. **Longitarsus absynthii**; *Artemisia maritima*. **L. aeruginosus**; *Eupatorium cannabinum*. **L. agilis**; *Scrophularia* spp. **L. anchusae**; *Echium vulgare*, *Cynoglossum officinale*. **L. atricillus**; *Trifolium* spp. and other Leguminosae. **L. ballotae**; *Ballota nigra*, *Marrubium vulgare*. **L. brunneus**; *Aster tripolium*, *Thalictrum* spp. **L. curtus**; *Echium vulgare*, *Symphytum* spp., *Pulmonaria* spp. **L. dorsalis**; *Senecio* spp. **L. exoletus**; *Echium vulgare*, *Cynoglossum officinale*. **L. ferrugineus**; *Mentha* spp. **L. flavicornis**; *Senecio* spp. **L. fowleri**; *Dipsacus fullonum*, *Thymus* spp., *Glechoma hederacea*. **L. ganglbaueri**; *Senecio* spp. **L. gracilis**; *Senecio jacobaea* & *S. vulgaris*. **L. holsaticus**; *Pedicularis* spp. **L. jacobaeae**; *Senecio jacobaea*. **L. kutscherea**; *Plantago* spp. **L. lycopi**; *Lycopus europaeus*, *Calamintha* spp, *Mentha* spp., *Nepeta cataria*. **L. melanocephalus**; *Plantago* spp. **L. membranaceus**; *Teucrium scorodonia*. **L. nasturtii**; *Symphytum* spp. and other Boraginaceae. **L. nigerrimus**; *Utricularia* spp. **L. nigrofasciatus**; *Verbascum* spp., *Scrophularia* spp. **L. obliteratus**; *Thymus* spp., *Origanum vulgare*. **L. ochroleucus**; *Senecio* spp. **L. parvulus**; *Linum* spp. **L. pellucidus**; *Convolvulus arvensis*. **L. plantagomaritimus**; *Plantago maritima*. **L. pratensis**; *Plantago* spp. **L. quadriguttatus**; *Cynoglossum officinale*, *Echium vulgare*. **L. reichei**; *Aster tripolium*, *Stachys palustris*. **L. rubiginosus**; *Calystegia sepium*. **L. rutilus**; *Scrophularia auriculata*. **L. succineus**; *Achillea* spp., *Eupatorium cannabinum*, *Leucanthemum vulgare*, *Artemisia* spp. **L. suturalis**; *Lithospermum officinale*. **L. suturellus**; *Senecio* spp. **L. tabidus**; *Verbascum* spp.

(Note. The common **Longitarsus luridus** will be found on many plants particularly various species of Compositae.).

Altica brevicollis; *Corylus avellana*. **A. britteni**; *Calluna vulgaris*, *Erica* spp. **A. ericeti**; *Erica tetralix*. **A. lythri**; *Epilobium hirsutum*, *Lythrum salicaria*. **A. oleracea**; *Calluna vulgaris*, *Erica* spp., *Epilobium* spp., *Oenothera* spp. **A. palustris**; *Lythrum salicaria*, *Epilobium hirsutum* & *E. parviflorum*. **A. pusilla**; *Helianthemum nummularium*, *Sanguisorba minor*. **Hermaeophaga mercurialis**; *Mercurialis perennis*. **Batophila aerata**; *Rubus* spp. **B. rubi**;

Rubus spp. particularly *R. idaeus*. **Lythraria salicariae**; *Lythrum salicaria*. **Ochrosis ventralis**; *Solanum dulcamara, Matricaria* spp. **Crepidodera ferruginea**; *Cirsium* spp., *Carduus* spp. **C. impressa**; *Limonium vulgare*. **C. transversa**; *Cirsium* spp. **Derocrepis rufipes**; *Vicia* spp. **Hippuriphila modeeri**; *Equisetum* spp. particularly *E. palustre*. **Chalcoides aurata**; *Salix* spp., *Populus* spp. **C. aurea**; *Populus* spp., *Salix* spp. **C. fulvicornis**; *Salix* spp., *Populus* spp. **C. nitidula**; *Populus* spp., *Salix* spp. **C. plutus**; *Salix* spp., *Populus* spp. **Epitrix atropae**; *Atropa bella-donna*. **E. pubescens**; *Solanum dulcamara*. **Podagrica fuscicornis**; *Malva sylvestris* and other species of Malvaceae. **P. fuscipes**; *Malva sylvestris* and other species of Malvaceae. **Mantura chrysanthemi**; *Rumex acetosella*. **M. matthewsi**; *Helianthemum nummularium*. **M. obtusata**; *Rumex* spp. **M. rustica**; *Rumex* spp. **Chaetocnema concinna**; *Polygonum* spp., *Rumex* spp. **Sphaeroderma rubidum**; *Cirsium* spp., *Carduus* spp. **S. testaceum**; *Cirsium* spp., *Carduus* spp., *Centaurea* spp. **Dibolia cynoglossi**; *Galeopsis* spp., *Ballota nigra, Stachys* spp. **Psylliodes affinis**; *Solanum dulcamara*. **P. attenuata**; *Humulus lupulus*. **P. chalcomera**; *Carduus* spp. **P. chrysocephala**; *Brassica* spp. and other Cruciferae. **P. cuprea**; *Brassica* spp. and other Cruciferae. **P. dulcamarae**; *Solanum dulcamara*. **P. hyoscyami**; *Hyoscyamus niger*. **P. luridipennis**; *Rhynchosinapis wrightii*. **P. marcida**; *Cakile maritima*. **P. napi**; *Brassica* spp. and other Cruciferae. **P. picina**; *Lythrum salicaria*. **P. weberi**; *Rorippa* spp. **Pilemostoma fastuosa**; *Inula conyza*. **Cassida hemisphaerica**; *Silene vulgaris, S. uniflora*. **C. murraea**; *Pulicaria dysenterica*. **C. nebulosa**; *Beta* spp. and other Chenopodiaceae. **C. rubiginosa**; *Cirsium* spp., *Carduus* spp. **C. sanguinosa**; *Achillea ptarmica*. **C. vibex**; *Cirsium* spp., *Centaurea* spp., *Carduus* spp. **C. viridis**; *Mentha aquatica*. **C. vittata**; *Spergularia* spp.

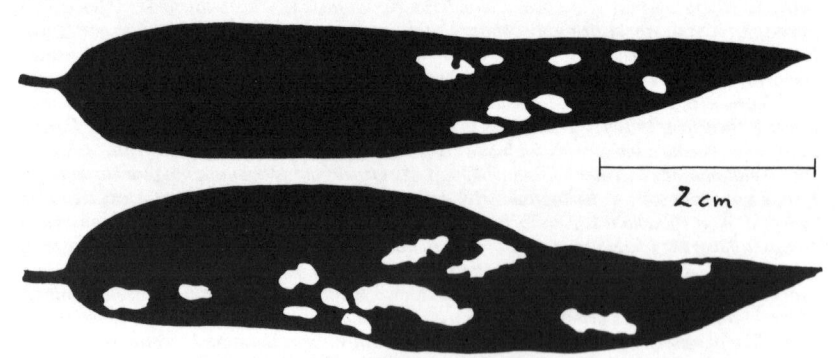

CHRYSOMELIDAE
Galerucella sagittariae: adult feeding on *Polygonum amphibium*

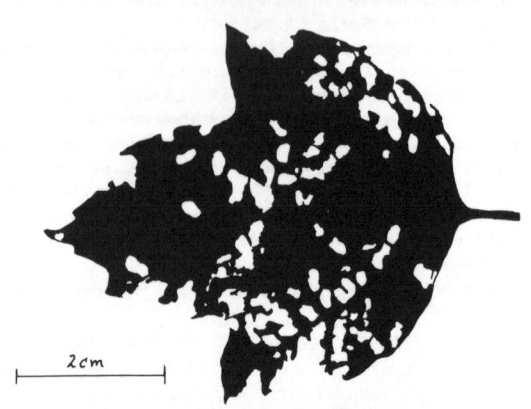

CHRYSOMELIDAE
Pyrrhalta viburni (Pk.): larval feeding on Guelder-rose

NEMONYCHIDAE
Rhinomacer attelaboides; *Pinus sylvestris*.

ANTHRIBIDAE
Anthribus resinosus; in the fungus *Daldinia concentrica* on *Fraxinus excelsior* and occasionally other trees. *Tropideres sepicola*; *Quercus* spp. *T. niveirostris*; *Quercus* spp. **Platystomus albinus**; in decaying *Quercus* spp., *Betula* spp., *Salix* spp. **Brachytarsus fasciatus**; feeds as larva in coccids on *Crataegus* spp. *B. nebulosus*; *Crataegus* spp.

ATTELABIDAE
Attelabus nitens; *Quercus* spp. **Apoderus coryli**; *Corylus avellana*. **Rhynchites caeruleus**; *Crataegus* spp., *Pyrus* spp., *Malus* spp., *Sorbus torminalis*. *R. cupreus*; *Sorbus aucuparia*. *R. cavifrons*; *Quercus* spp. *R. aeneovirens*; *Quercus* spp. *R. aequatus*; *Crataegus* spp. *R. germanicus*; herbaceous Rosaceae. *R. interpunctatus*; *Quercus* spp. *R. longiceps*; *Salix* spp. *R. nanus*; *Betula* spp. *R. pauxillus*; *Crataegus* spp. *R. tomentosus*; *Salix* spp., *Populus tremula*. **Byctiscus betulae**; *Populus* spp., *Salix* spp., *Corylus avellana*, *Betula* spp. *B. populi*; *Populus* spp. **Deporaus mannerheimi**; *Betula* spp. **Deporaus betulae**; *Betula* spp., *Alnus glutinosa*, *Corylus avellana*. **Rhynchites olivaceus**; *Quercus* spp.

APIONIDAE
Apion affine; *Rumex acetosa*. *A. curtirostre*; *Rumex acetosella* & *R. acetosa*. *A. hydrolapathi*; *Rumex obtusifolius* & *R. hydrolapathum*. *A. lemoroi*; *Polygonum aviculare* agg. *A. limonii*; *Limonium* spp. *A. marchicum*; *Rumex acetosella*. *A. sedi*; *Sedum* spp. *A. violaceum*; *Rumex* spp. (not *acetosella*). *A. malvae*; *Malva sylvestris* and other species of Malvaceae. *A. rufirostre*; *Malva sylvestris* and other species of *Malva*. *A. aeneum*; *Malva sylvestris* and other species of Malvaceae. *A. radiolus*; *Malva sylvestris* and other species of Malvaceae. *A. soror*; *Althaea officinalis*. *A. pallipes*; *Mercurialis perennis*. *A. semivittatum*; *Mercurialis annua*. *A. urticarium*; *Urtica* spp. *A. difficile*; *Genista tinctoria*. *A. fuscirostre*; *Cytisus scoparius*. *A. genistae*; *G. anglica*. *A. ulicis*; *Ulex* spp. *A. cruentatum*; *Rumex acetosa*. *A. hydrolapathi*; *Rumex* spp. *A. frumentarium*; *Rumex acetosella*. *A. miniatum*; *Rumex* spp. (Docks). *A. rubens*; *Rumex acetosella*. *A. sanguineum*; *Rumex acetosella*. *A. atomarium*; *Thymus* spp. *A. cineraceum*; *Prunella vulgaris*. *A. flavimanum*; *Origanum vulgare*. *A. vicinum*; *Mentha aquatica*. *A. pubescens*; *Trifolium* spp. *A. seniculus*; *Trifolium* spp. *A. confluens*; *Chamomilla recutita*, *Matricaria perforata* and perhaps other species of mayweed. *A. stolidum*; *Leucanthemum vulgare*. *A. brunnipes*; *Filago* spp., *Gnaphalium* spp. *A. carduorum*; *Cirsium* spp., *Carduus* spp. *A. dentirostre*; *Ulex* spp. *A. lacertense*; *Cirsium* spp., *Carduus* spp. *A. onopordi*; *Carduus* spp., *Centaurea nigra* agg. *Cirsium* spp., *Arctium* spp. *A. dispar*; *Anthemis cotula*. *A. hookeri*; *Matricaria perforata* & *M. maritima*. *A. laevigatum*; *Matricaria* spp. & *Chamomilla* spp. *A. ebeninum*; *Lathyrus pratensis Lotus uliginosus.*, *Onobrychis viciifolia*. *A. immune*; *Cytisus scoparius*. *A. striatum*; *Cytisus scoparius*. *A. aethiops*; *Vicia* spp. *A. astragali*; *Astragalus glycyphyllos*. *A. pisi*; *Medicago* spp. perhaps other Leguminosae. *A. punctigerum*; *Vicia sepium*. *A. curtisi*; *Trifolium* spp. *A. gyllenhali*; *Vicia* spp. particularly *V. cracca*. *A. intermedium*; *Onobrychis viciifolia*. *A. loti*; *Lotus corniculatus*. *A. meliloti*; *Melilotus* spp. *A. minimum*; *Salix* spp. *A. ononis*; *Ononis* spp. *A. afer*; *Lathyrus* spp., *Vicia* spp. *A. reflexum*; *Onobrychis viciifolia*. *A. scutellare*; *Ulex* spp., esp. *Ulex minor*. *A. sicardi*; *Lotus uliginosus*. *A. simile*; *Betula* spp. *A. spencii*; *Vicia cracca*. *A. tenue*; *Medicago* spp., *Melilotus* spp. *A. viciae*; *Vicia cracca*. *A. virens*; *Trifolium repens*. *A. vorax*; *Vicia* spp. particularly *V. cracca* & *V. sepium*. *A. waltoni*; *Hippocrepis comosa*. *A. cerdo*; *Vicia cracca*. *A. craccae*; *Vicia cracca* and occasionally other *Vicia* & *Lathyrus* spp. *A. pomonae*; *Vicia sativa* agg. *A. subulatum*; *Lathyrus pratensis* and occasionally *Vicia* spp. *A. apricans*; *Trifolium* spp. particularly *T. pratense*. *A. assimile*; *Trifolium* spp. esp. *T. pratense*. *A. dichroum*; *Trifolium* spp. particularly *T. repens*. *A. difforme*; *Trifolium* spp. and occasionally on *Polygonum* spp. *A. dissimile*; *Trifolium arvense*. *A. filirostre*; *Medicago* spp. *A. laevicolle*; *Trifolium* spp. esp. *T. pratense*. *A. nigritarse*; *Trifolium* spp. particularly *T. campestre* & *T. dubium*. *A. ononicola*; *Ononis* spp. *A. schoenherri*; *Trifolium* spp. *A. trifolii*; *Trifolium*

spp. *A. varipes*; *Trifolium* spp. *Nanophyes gracilis*; *Lythrum portula*. *N. marmoratus*; *Lythrum salicaria*.

CURCULIONIDAE

All the weevils feed upon plant material in some form or other, and many species are found upon a wide variety of plants. Only those species with specific host plants are listed below.

Otiorhynchus ligustici; *Anthyllis vulneraria*. **Phyllobius pomaceus**; *Urtica dioica*. **Polydrusus confluens**; *Cytisus scoparius*, *Ulex* spp. **Strophosomus curvipes**; *Calluna vulgaris*. **S. melanogrammus**; many trees particularly *Corylus avellana*. **S. nebulosus**; *Calluna vulgaris*, *Erica* spp. **S. sus**; *Calluna vulgaris*, *Erica* spp. **Liophloeus tessulatus**; found on *Hedera helix*, feeds on Umbelliferae. **Sitona ambiguus**; *Vicia sepium*. **S. cambricus**; *Lotus uliginosus*. **S. cylindricollis**; *Melilotus* spp. **S. gemellatus**; *Ononis repens*, *Medicago* spp. **S. griseus**; Leguminosae in sandy areas. **S. hispidulus**; *Trifolium* spp., *Medicago* spp. **S. humeralis**; *Medicago* spp., *Trifolium* spp. **S. lepidus**; *Trifolium* spp. **S. lineatus**; *Trifolium* spp., *Vicia* spp., *Lathyrus* spp., *Pisum sativum* and almost all other Leguminosae. **S. lineellus**; *Lotus corniculatus*, *Trifolium repens*. **S. macularius**; *Onobrychis viciifolia*. **S. ononidis**; *Ononis repens*, *O. spinosa*. **S. puncticollis**; *Trifolium* spp., *Lotus* spp., *Vicia* spp. **S. regensteinensis**; *Cytisus scoparius*, *Ulex* spp. **S. striatellus**; *Ulex* spp., *Cytisus scoparius*. **S. sulcifrons**; *Trifolium* spp., *Medicago* spp., *Vicia* spp. **S. suturalis**; *Vicia* spp. **S. waterhousei**; *Lotus corniculatus*. **Coniocleonus nebulosus**; *Calluna vulgaris*, *Erica* spp. **Chromoderus affinis**; *Chenopodium* spp., *Atriplex* spp. **Cleonus piger**; *Cirsium arvense*, *Carduus* spp. **Lixus algirus**; *Cirsium* spp., *Centaurea* spp., *Malva* spp. **L. scabricollis**; *Beta* spp. **L. paraplecticus**; *Oenanthe crocata*, *Sium latifolium*. **L. vilis**; *Erodium cicutarium*. **Larinus planus**; Thistles, esp. *Cirsium vulgare*. **Rhinocyllus conicus**; Thistles, esp. *Cirsium vulgare*. **Hypera adspersa**; *Apium nodiflorum;* *Oenanthe* spp. **H. dauci**; *Erodium cicutarium*. **H. fuscocinerea**; *Medicago* spp. **H. meles**; *Medicago sativa*, *Trifolium pratense*. **H. nigrirostris**; *Ononis repens*, *O. spinosa*. **H. ononidis**; *Ononis* spp. **H. pastinacae**; *Daucus carota*. **H. plantaginis**; *Lotus corniculatus*. **H. postica**; *Medicago* spp., *Trifolium* spp. **H. punctata**; *Trifolium* spp. **H. rumicis**; *Rumex* spp. **H. suspiciosa**; *Lotus* spp., *Lathyrus* spp. **H. venusta**; *Ulex* spp. esp. *U. minor*. **Limobius borealis**; *Geranium* spp. particularly *G. pratense*. **L. mixtus**; *Erodium cicutarium*. **Cionus alauda**; *Scrophularia* spp., *Verbascum* spp. **C. hortulanus**; *Scrophularia* spp., occasionally *Verbascum* spp. & *Buddleja* spp. **C. longicollis**; *Verbascum thapsus*. **C. nigritarsis**; *Verbascum nigrum*. **C. scrophulariae**; *Scrophularia* spp., and occasionally *Verbascum* spp. & *Buddleja* spp. **C. tuberculosus**; *Scrophularia* spp. particularly *S. auriculata*. **Cleopus pulchellus**; *Scrophularia* spp., *Verbascum* spp. **Hylobius abietis**; *Pinus* spp., *Picea* spp. **H. transversovittatus**; *Lythrum salicaria*. **Liparus coronatus**; *Anthriscus sylvestris*. **L. germanus**; *Heracleum sphondylium*. **Leiosoma deflexum**; *Ranunculus repens*, *Anemone nemorosa*, *Caltha palustris*. **L. oblongulum**; *Ranunculus* spp. **Syagrius intrudens**; *Pteridium aquilinum*. **Gronops inaequalis**; *Atriplex* spp. **G. lunatus**; *Spergularia* spp. **Pissodes castaneus**; *Pinus* spp. **P. pini**; *Pinus* spp. **P. validirostris**; *Pinus* spp. **Magdalis armigera**; *Ulmus* spp. **M. barbicornis**; *Malus* spp., *Crataegus* spp., *Prunus* spp. **M. carbonaria**; *Betula* spp. **M. cerasi**; *Crataegus* spp., *Malus* spp., *Pyrus* spp., *Quercus* spp. **M. duplicata**; *Pinus* spp. **M. memnonia**; *Pinus* spp. **M. phlegmatica**; *Pinus* spp. **M. ruficornis**; *Prunus* spp. **Anoplus plantaris**; *Betula* spp. *A. roboris*; *Alnus glutinosa*. **Tanysphyrus lemnae**; *Lemna* spp. **Eremotes ater**; *Pinus* spp. **Rhyncolus gracilis**; *Fagus sylvatica*. **R. truncorum**; *Acer* spp. **Trachodes hispidus**; old twigs of *Fagus sylvatica* & *Quercus* spp. **Cryptorhynchus lapathi**; *Salix* spp. **Stenopelmus rufinasus**; *Azolla filiculoides*. **Bagous petro**; *Utricularia* spp., *Ceratophyllum* spp. **Bagous cylindrus**; *Glyceria* spp. **B. binodulus**; *Stratiotes aloides*. **B. brevis**; *Ranunculus flammula*. **B. collignensis**; *Equiseteum fluviatile*. **B. limosus**; *Potamogeton natans*, *P. lucens*. **B. longitarsis**; *Myriophyllum* spp. **B. lutulosus**; *Juncus* spp. particularly *J. subnodulosus*. **B. nodulosus**; *Butomus umbellatus*. **B. subcarinatus**; *Ceratophyllum submersum*. **B. tempestivus**; *Potamogeton* spp. **B. glabrirostris**; *Stratiotes aloides*. **B. lutosus**; *Sparganium erectum*. **B. lutulentus**; *Equiseteum fluviatile*. **B. puncticollis**; *Hydrocharis morsus-ranae*, *Elodea* spp., *Stratiotes aloides*. **Hydronomus alismatis**; *Alisma plantago-aquatica*. **Dorytomus affinis**; *Populus tremula*. **D. dejeani**; *Populus tremula*. **D. filirostris**; *Populus* spp. ***D. hirtipennis***; *Salix* spp. ***D. longimanus***; *Populus* spp. ***D. majalis***; *Salix* spp. ***D. melanophthalmus***; *Salix* spp. ***D. rufatus***; *Salix* spp. ***D. salicinus***; *Salix cinerea, S. caprea, S. aurita*. ***D. salicis***; *Salix repens*. ***D. taeniatus***; *Salix* spp., *Populus* spp. ***D. tortrix***; *Populus tremula*. ***D. tremulae***; *Populus alba, P. tremula*. ***D. validirostris (ictor)***; *Populus* spp. ***Notaris*** spp. and ***Thryogenes*** spp. are found on various rushes, sedges and other waterside plants but their true associations are as yet not fully worked out. ***Grypus equiseti***; *Equisetum* spp. particularly *E. palustre*. ***Pachytychius haematocephalus***; *Lotus corniculatus*. ***Pseudostyphlus pillumus***; *Matricaria perforata, Chamomilla recutita*. ***Smicronyx coecus***; *Cuscuta epithymum*. ***S. jungermanniae***; *Cuscuta epithymum*. ***S. reichi***; *Centaurium erythraea*. ***Mononychus punctumalbum***; *Iris foetidissima, Iris pseudacorus*. ***Rutidosoma globulus***; *Populus tremula*. ***Coeliodes dryados***; *Quercus* spp. ***C. erythroleucos***; *Quercus* spp. ***C. nigritarsis***; *Betula* spp. ***C. ruber***; *Quercus* spp. ***C. rubicundus***; *Betula* spp. ***Zacladus exiguus***; small-flowered *Geranium* spp. ***Z. geranii***; *Geranium pratense, G. sylvaticum, G. sanguineum*. ***Stenocarus umbrinus***; *Papaver rhoeas, P. somniferum*. ***Micrelus ericae***; *Calluna vulgaris, Erica* spp. ***Cidnorhinus quadrimaculatus***; *Urtica dioica*. ***Ceuthorhynchidius barnevillei***; *Achillea millefolium*. ***C. dawsoni***; *Plantago coronopus*. ***C. horridus***; *Cirsium* spp., *Carduus* spp. ***C. rufulus***; *Plantago lanceolata, P. maritima*. ***C. thalhammeri***; *Plantago maritima*. ***C. troglodytes***; *Plantago lanceolata*. ***Ceutorhynchus alliariae***; *Alliaria petiolata*. ***C. angulosus***; *Stachys palustris, Galeopsis tetrahit*. ***C. arquatus***; *Lycopus europaeus*. ***C. asperifoliarum***; *Cynoglossum officinale, Echium vulgare* and other species of Boraginaceae. ***C. assimilis***; *Brassica* spp., *Cardaria draba, Sinapis* spp., *Sisymbrium* spp. and other species of Cruciferae. ***C. atomus***; *Arabidopsis thaliana*. ***C. campestris***; *Leucanthemum vulgare*. ***C. cochleariae***; *Cardamine pratensis, Cardamine* spp. ***C. constrictus***; *Alliaria petiolata*. ***C. contractus***; *Sisymbrium* spp. and other Cruciferae. ***C. depressicollis***; *Fumaria* spp. ***C. erysimi***; *Capsella bursa-pastoris*. ***Ceutorhynchus euphorbiae***; *Myosotis* spp. ***C. floralis***; Cruciferae in general. ***C. geographicus***; *Echium vulgare*. ***C. hepaticus***; *Sinapis* spp., *Brassica* spp. and other similar species of Cruciferae. ***C. hirtulus***; *Erophila verna*. ***C. litura***; *Cirsium* spp. ***C. marginatus***; *Hypochoeris* spp. and perhaps other closely related Compositae. ***C. melanostictus***; *Lycopus europaeus, Mentha aquatica*. ***C. mixtus***; *Fumaria* spp., *Corydalis claviculata*. ***C. moelleri***; *Hieracium* spp. *Leontodon* spp. ***C. parvulus***; *Lepidium* spp. esp. *Lepidium heterophyllum*. ***C. pectoralis***; *Cardamine* spp. *Rorippa* spp. ***C. pervicax***; *Cardamine pratensis*. ***C. pictitarsis***; *Sisymbrium officinale* and occasionally other species of Cruciferae. ***C. pilosellus***; *Taraxacum* spp. ***C. pleurostigma***; *Brassica* spp. and at times other species of Cruciferae. ***C. pollinarius***; *Urtica dioica*. ***C. pulvinatus***; *Descurainia sophia*. ***C. pumilio***; *Teesdalia nudicaulis*. ***C. punctiger***; *Taraxacum* spp. ***C. pyrrhorhynchus***; *Sisymbrium officinale*. ***C. quadridens***; *Sisymbrium officinale* and many other species of Cruciferae. ***C. querceti***; *Rorippa palustris*. ***C. quercicola***; *Fumaria* spp. ***C. rapae***; *Descurainia sophia, Sisymbrium officinale* and other species of Cruciferae. ***C. resedae***; *Reseda luteola*. ***C. rugulosus***; *Chamomilla recutita* and other species of mayweed. ***C. sulcicollis***; *Sisymbrium officinale* and other species of Cruciferae. ***C. terminatus***; *Daucus carota*. ***C. thomsoni***; *Alliaria petiolata*. ***C. timidus***; *Sisymbrium officinale*. ***C. triangulum***; *Achillea millefolium*. ***C. trimaculatus***; *Carduus nutans*. ***C. turbatus***; *Cardaria draba*. ***C. unguicularis***; *Arabis hirsuta*. ***C. verrucatus***; *Glaucium flavum*. ***C. viduatus***; *Stachys palustris*. ***Eubrychius velutus***; *Myriophyllum* spp. ***Litodactylus leucogaster***; *Myriophyllum* spp., *Potamogeton* spp. ***Rhinoncus albicinctus***; *Polygonum amphibium*. ***R. bruchoides***; *Polygonum persicaria*. ***R. castor***; *Rumex acetosella*. ***R. inconspectus***; *Polygonum amphibium*. ***R. pericarpius***; *Rumex* spp. ***R. perpendicularis***; *Polygonum* spp. ***Phytobius canaliculatus***; *Myriophyllum* spp. ***P. comari***; *Potentilla palustris, Lythrum salicaria*. ***P. olssoni***; *Lythrum portula*. ***P. quadricornis***; *Polygonum amphibium, P. lapathifolium*. ***P. quadrinodosus***; *Polygonum amphibium*. ***P. quadrituberculatus***; *Polygonum* spp. ***P. waltoni***; *Polygonum hydropiper*. ***Amalus scortillum***; *Polygonum aviculare*. ***Amalorrhynchus melanarius***; *Nasturtium officinale, N. microphyllum*. ***Drupenatus nasturtii***; *Nasturtium officinale, N. microphyllum*. ***Tapinotus sellatus***; *Lysimachia vulgaris*. ***Poophagus sisymbrii***; *Nasturtium officinale, N. microphyllum, Rorippa amphibia*. ***Orobitis cyaneus***; *Viola* spp. ***Baris analis***; *Pulicaria dysenterica*. ***B. laticollis***; *Sisymbrium officinale, Brassica oleracea*. ***B. lepidii***; *Nasturtium* spp. *Rorippa* spp. and other Cruciferae in marshy areas. ***B. picicornis***; *Reseda lutea*. ***B. scolopacea***; *Halimione portulacoides*. ***Limnobaris pilistriata***; *Carex* spp. ***L. t-album***; *Carex* spp. ***Anthonomus bituberculatus***; *Crataegus* spp. ***A. brunnipennis***; *Potentilla erecta*. ***A. chevrolati***; *Crataegus* spp. ***A. conspersus***; *Sorbus aucuparia*. ***A. humeralis***; *Malus* spp. ***A. pedicularius***;

Crataegus spp. ***A. piri***; *Malus* spp. ***A. pomorum***; *Malus* spp. ***A. rubi***; *Rosa* spp., *Rubus* spp., *Fragaria* spp. and at times other species of Rosaceae. ***A. rufus***; *Prunus spinosa*. ***A. ulmi***; *Ulmus* spp. ***A. varians***; *Pinus sylvestris*. **Furcipus rectirostris**; *Prunus padus*. **Brachonyx pineti**; *Pinus sylvestris*. **Curculio betulae**; *Betula* spp. ***C. glandium***; *Quercus* spp. ***C. nucum***; *Corylus avellana*. ***C. pyrrhoceras***; *Quercus* spp. ***C. rubidus***; *Betula* spp. ***C. salicivorus***; *Salix* spp. ***C. venosus***; *Quercus* spp. ***C. villosus***; *Quercus* spp. **Tychius crassirostris**; *Melilotus alba*. ***T. flavicollis***; *Lotus corniculatus*. ***T. junceus***; *Medicago* spp. and other clover-like Leguminosae. ***T. lineatulus***; *Trifolium* spp. ***T. meliloti***; *Melilotus* spp. ***T. parallelus***; *Cytisus scoparius*. ***T. polylineatus***; *Trifolium* spp. ***T. pusillus***; *Trifolium* spp. ***T. quinquepunctatus***; *Lathyrus* spp., *Vicia* spp. ***T. schneideri***; *Anthyllis vulneraria*. ***T. stephensi***; *Trifolium* spp. ***T. tibialis***; *Trifolium* spp. **Miccotrogus picirostris**; *Trifolium* spp. **Sibinia arenariae**; *Spergularia* spp. ***S. potentillae***; *Spergula arvensis*. ***S. primitus***; *Spergularia* spp., *Sagina* spp. ***S. sodalis***; *Armeria maritima*. **Ellescus bipunctatus**; *Salix* spp. **Acalyptus carpini**; *Salix* spp. **Miarus campanulae**; *Campanula* spp. ***M. degorsi***; *Campanula* spp. ***M. graminis***; *Campanula* spp. ***M. plantarum***; *Campanula* spp. **Mecinus circulatus**; *Plantago* spp. particularly *P. lanceolata*. ***M. collaris***; *Plantago maritima*, *P. coronopus*. ***M. janthinus***; *Linaria vulgaris*. ***M. pyraster***; *Plantago* spp. particularly *P. lanceolata*. **Gymnetron antirrhini**; *Linaria vulgaris*. ***G. beccabungae***; *Veronica beccabunga*. ***G. collinum***; *Linaria vulgaris*. ***G. labile***; *Plantago lanceolata*. ***G. linariae***; *Linaria vulgaris*. ***G. melanarium***; *Veronica chamaedrys*. ***G. pascuorum***; *Plantago lanceolata*. ***G. rostellum***; *Veronica chamaedrys*. ***G. villosulum***; *Veronica anagallis-aquatica*, *V. catenata*. **Rhynchaenus alni**; *Ulmus* spp. ***R. avellanea***; *Corylus avellana*, *Quercus* spp. ***R. decoratus***; *Salix purpurea*. ***R. fagi***; *Fagus sylvatica*. ***R. foliorum***; *Salix* spp. ***R. iota***; *Myrica gale*. ***R. pilosus***; *Quercus* spp. ***R. populi***; *Salix* spp., *Populus* spp. ***R. pratensis***; *Centaurea nigra* agg. ***R. quercus***; *Quercus* spp. ***R. rusci***; *Betula* spp. ***R. salicis***; *Salix* spp. ***R. stigma***; *Salix* spp., *Alnus glutinosa* and at times other catkin-bearing trees. ***R. testaceus***; *Alnus glutinosa*. **Ramphus oxyacanthae**; *Crataegus* spp. ***R. pulicarius***; *Salix* spp.

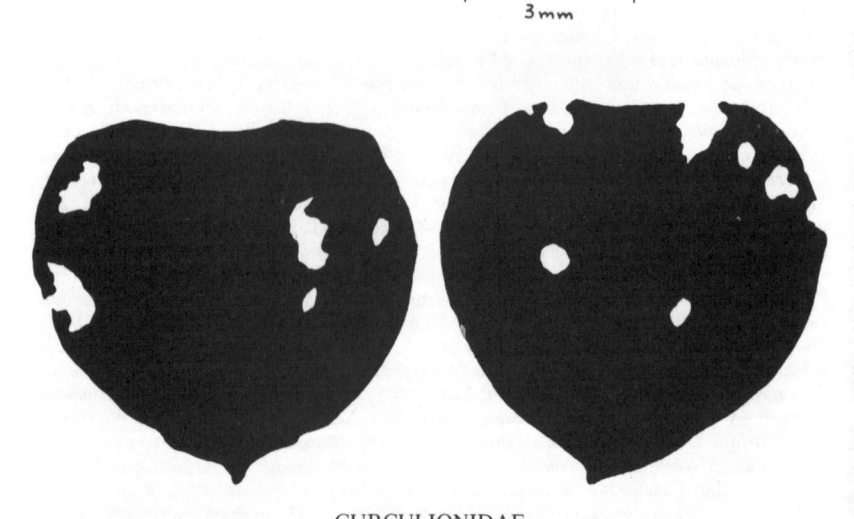

CURCULIONIDAE
Anthonomus brunnipennis (Curtis): adult feeding on flower petals of *Potentilla erecta*

CURCULIONIDAE
Otiorhynchus porcatus (Hb.): adult feeding on cultivated Primula

CURCULIONIDAE
Barynotus obscurus (F.): adult feeding on African Violet

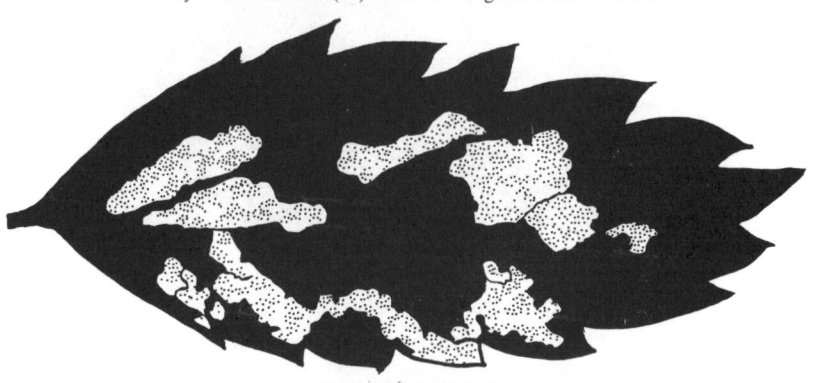

CURCULIONIDAE
Phytobius comari (Hb.): larval feeding on *Potentilla palustris*.

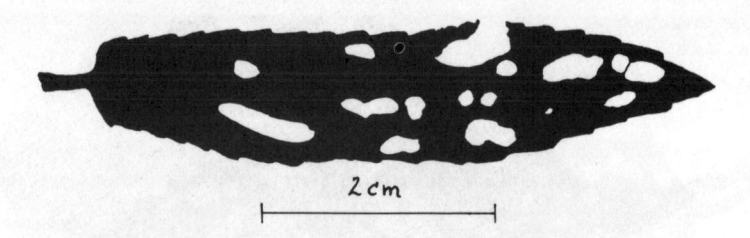

CURCULIONIDAE
Ceutorhynchus viduatus (Gyll.): adult feeding on *Stachys palustris*

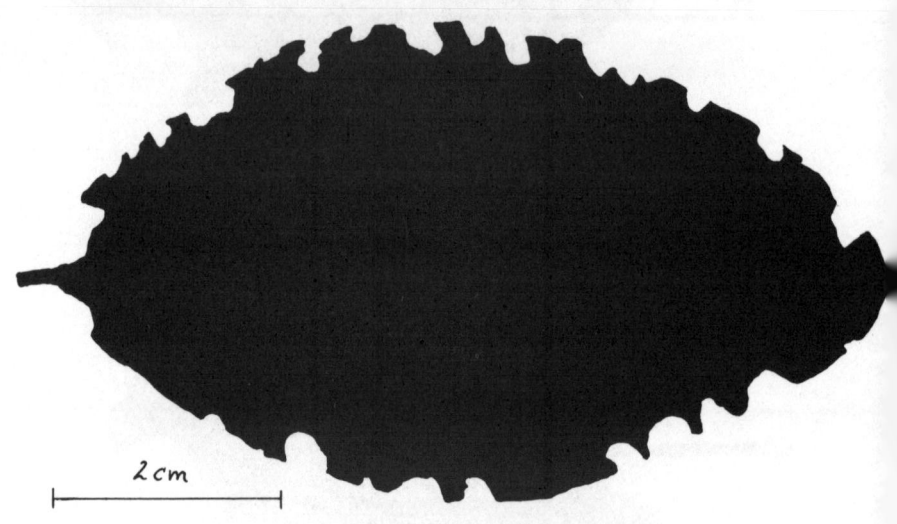

CURCULIONIDAE
Sitona lineatus (L.): adult feeding on Laburnum

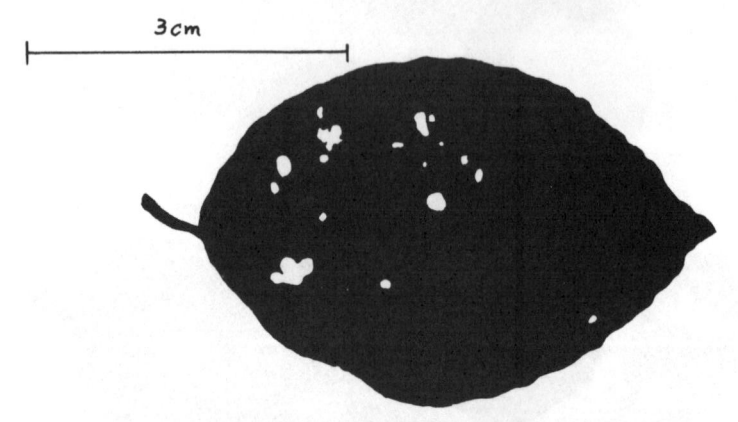

CURCULIONIDAE
Rhynchaenus fagi (L.): adult feeding on Beech

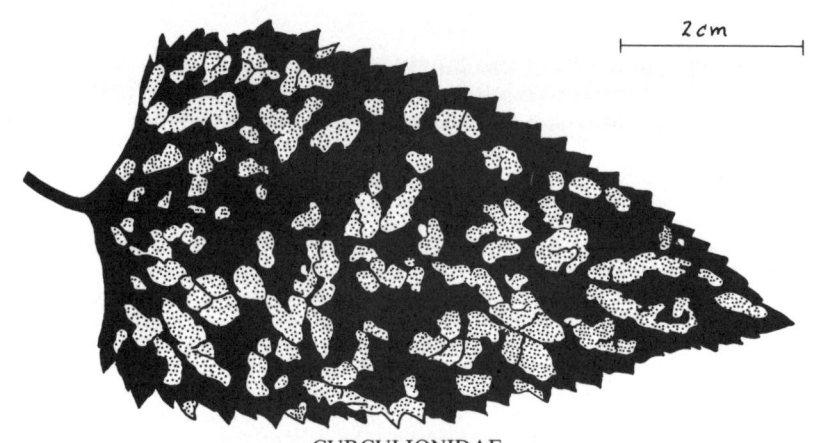

CURCULIONIDAE
Cionus scrophulariae (L.): larval feeding on *Scrophularia nodosa*

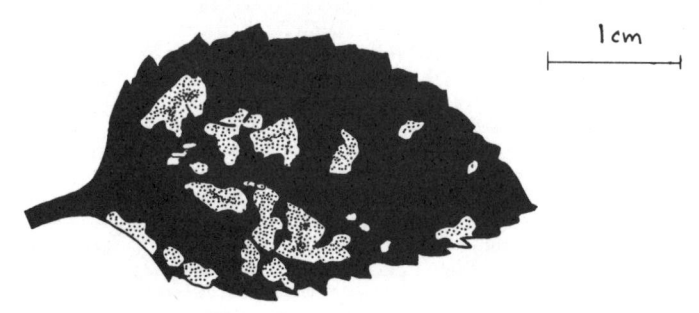

CURCULIONIDAE
Cleopus pulchellus (Hb.): larval feeding on *Scrophularia nodosa*.

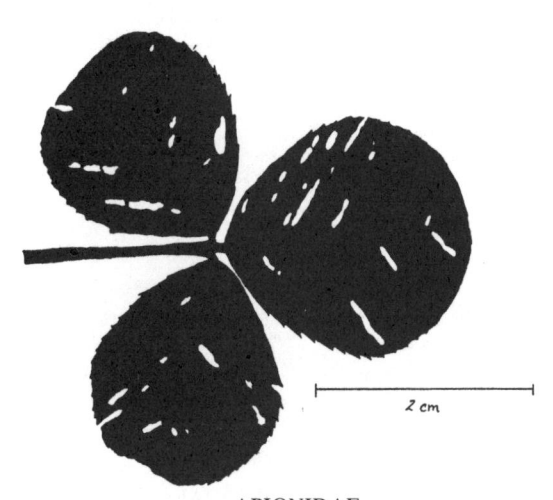

APIONIDAE
Apion dichroum Bedel: adult feeding on White Clover

SCOLYTIDAE

The larvae and adults of this family are to be found beneath the bark, and for some species well into the sapwood, of various trees and shrubs. Only the major hosts are listed here but most species will also occur in a range of similar or related hosts.

Scolytus intricatus; *Quercus* spp. and at times *Castanea sativa*, *Ulmus* spp., *Fagus sylvatica*, *Populus* spp., *S. laevis*; *Ulmus* spp., *S. mali*; *Pyrus* spp., *Prunus* spp., *Crataegus* spp., *Ulmus* spp., *S. multistriatus*; *Ulmus* spp. *S. ratzeburgi*; *Betula* spp. *S. rugulosus*; *Prunus* spp., *Malus* spp., *Pyrus* spp. *S. scolytus*; *Ulmus* spp. **Hylesinus crenatus**; *Fraxinus excelsior*. **H. oleiperda**; *Fraxinus excelsior*. **Leperisinus varius**; *Fraxinus excelsior*. **L. orni**; *Fraxinus excelsior*. **Acrantus vittatus**; *Ulmus* spp., *Fraxinus excelsior*. **Xylechinus pilosus**; *Abies* spp., *Picea* spp., *Larix* spp. **Kissophagus hederae**; *Hedera helix*. **Hylastinus obscurus**; *Ulex* spp., *Cytisus scoparius*. **Phloeophthorus rhododactylus**; *Ulex* spp., *Cytisus scoparius*. **Phloeosinus thujae**; *Chamaecyparis* spp., *Cupressus* spp., *Thuja* spp. **Hylurgops palliatus**; *Abies* spp., *Pinus* spp. *Picea* spp., *Larix* spp. **Hylastes angustatus**; *Pinus* spp., *Picea* spp. **H. ater**; *Pinus* spp. **H. attenuatus**; *Pinus* spp. **H. brunneus**; *Pinus* spp. **H. canicularius**; *Picea* spp. **H. opacus**; *Pinus* spp. occasionally in *Fraxinus excelsior* and *Ulmus* spp. **Tomicus minor**; *Pinus* spp., occasionally in *Abies* spp. and *Picea* spp. **T. piniperda**; *Pinus* spp., occasionally in *Picea* spp., *Abies* spp. and Larix spp. **Polygraphus poligraphus**; *Picea* spp., *Pinus* spp. **Dryocoetinus alni**; *Alnus glutinosa*, *Fagus sylvatica* and *Corylus avellana*. **D. villosus**; *Quercus* spp., *Fagus sylvatica* and *Castanea sativa*. **Dryocoetes autographus**; *Picea* spp., *Pinus* spp., *Larix* spp., *Abies* spp. **Lymantor coryli**; *Corylus avellana* and at times in other broadleafed trees. **Xylocleptes bispinus**; *Clematis vitalba*. **Taphrorychus bicolor**; *Fagus sylvatica*, *Quercus* spp. and many other broadleafed trees. **Xyloterus domesticum**; *Quercus* spp., *Fagus sylvatica*. **X. lineatum**; *Abies* spp., *Picea* spp., *Pinus* spp., *Larix* spp. **X. signatum**; *Quercus* spp., *Fagus sylvatica*. **Cryphalus abietis**; *Abies* spp., *Picea* spp. **C. piceae**; *Picea* spp., *Abies* spp. **Ernoporus caucasicus**; *Tilia* spp. **E. fagi**; *Fagus sylvatica*, and at times *Quercus* spp. and *Betula* spp. **E. tiliae**; *Tilia cordata*. **Trypophloeus asperatus**; *Populus* spp., *Salix* spp. **T. granulatus**; *Populus* spp. **Xyleborus dispar**; *Castanea sativa*, *Quercus* spp., *Fagus sylvatica* and many other broadleafed trees. **X. dryographus**; *Quercus* spp., *Castanea sativa* and at times *Fagus sylvatica* and *Ulmus* spp. **X. saxeseni**; *Quercus* spp., *Fagus sylvatica* and *Castanea sativa*. **Pityophthorus lichtensteini**; *Pinus* spp. **P. pubescens**; *Pinus* spp., *Picea* spp. **Pityogenes bidentatus**; *Pinus* spp., *Picea* spp. **P. chalcographus**; *Pinus* spp., *Picea* spp., *Abies* spp., *Larix* spp. **P. quadridens**; *Abies* spp., *Picea* spp. **P. trepanatus**; *Pinus* spp. **Ips acuminatus**; *Pinus* spp., *Larix* spp. **I. cembrae**; *Larix* spp., *Pinus* spp., *Picea* spp. **I. sexdentatus**; *Pinus* spp. **I. typographus**; *Picea* spp., *Abies* spp., *Pinus* spp. **Orthotomicus erosus**; *Pinus* spp. **O. laricis**; *Pinus* spp., *Larix* spp. **O. suturalis**; *Pinus* spp.

PLATPODIDAE

Platypus cylindrus; *Fagus sylvaticus*, *Castanea sativa*, *Fraxinus excelsior*.

Reference

Grey-Wilson, C. & Blamey, M. 1989. The Illustrated Flora of Britain and Northern Europe. Hodder & Stoughton.

BEETLES ASSOCIATED WITH STORED FOOD PRODUCTS

Dr. John Muggleton

(Prepared with permission of the Ministry of Agriculture, Fisheries and Food, Central Science Laboratory, Slough. Crown Copyright Reserved)

1. INTRODUCTION

Not unreasonably, the average coleopterist usually confines his activities to searching for, and studying, those beetles found outdoors. In doing so he ignores the 500 or so species, worldwide, that have been found associated with stored food products or with other manufactured goods of plant or animal origin. I hope that this account will stimulate interest in these beetles by drawing attention to their existence, indicating where they may be found and how they may be bred. These beetles are of special interest as many are important cosmopolitan pests, destroying and spoiling food throughout the world, often in those countries where it can be spared the least. Although a number of species have been the subject of detailed laboratory investigations, there remains much to be learnt from studies of their distribution, habits and life histories outside the laboratory.

The term "stored food products" is used to include any item of plant or animal origin which is stored in man-made premises. It includes unprocessed items such as cereals and other grains, for example rapeseed and legumes, together with dried fruit, herbs and nuts. It also includes processed items such as manufactured foods, animal feedstuffs, tobacco, animal skins, wool and textiles. These commodities may be stored for long periods in warehouses, mills, grain stores, barns, maltings, food factories, shops and, of course, the coleopterist's home. They are also temporarily held in the ships (often in containers) and vehicles used to transport them between and within the countries of the world.

Stored product beetles have been associated with man probably for as long as he has been storing food; the first sowing of cereals has been traced back to Syria in *c.* 9000 BC. Certainly the association has been in existence for several thousand years; *Tribolium* spp. have been recorded from Egyptian tombs dating from *c.* 2500 BC and the tomb of Tutankhamun (*c.* 1340 BC) contained at least six species all associated with stored food products today, including *Oryzaephilus surinamensis*, at present the most important beetle pest of stored grain in the United Kingdom. With such a long association with man it is perhaps not surprising that many of these beetles are rarely found away from stored commodities. Some of them are confined to one product or to a group of products. Bruchids, for example, occur in the seeds of legumes, while *Sitophilus* and *Rhyzopertha* show a marked preference for whole grain. Others, such as *Oryzaephilus*, *Lasioderma* and Ptinids, are more general feeders.

Some of the beetles found with stored food may also occur in other situations, such as haystacks, bird's nests (a well known reservoir for infestations), bee and wasp nests and cellars. Further, a few species sometimes found in store-places are probably breeding in, or are attracted to, decaying or mouldy foods (e.g. *Ahasverus advena, Typhaea stercorea*). timber in the fabric of the buildings, the droppings of rats, mice and birds or accumulations of damp debris and floor sweepings (e.g. *Cryptophagus* spp.). Also, species which can only survive and breed on living plants may be brought in from the field with freshly harvested grain and legumes (e.g. *Sitona* spp.). These latter species are usually of secondary importance as pests of stored products, but some are of interest as they could be at a transitional stage to becoming pests in stores.

2. COLLECTING METHODS

I have on one occasion found a single specimen of *Oryzaephilus surinamensis* crawling up a grass stem in a deep Cotswold valley far from any grain store. On another occasion an evening meal outside a hotel restaurant in the heart of the French countryside was interrupted by a succession of stored product beetles landing on the table, such that a surreptitious visit to my bedroom for a collecting tube was necessary. Investigation after the meal revealed that the source was a nearby grain silo. Such events are the exception and in general the coleopterist wanting to study stored product beetles will have to search the stored goods, the buildings in which they are stored and the vehicles in which they are transported, if he is to have any success in finding them. The notes below refer exclusively to collecting in these habitats.

Permission: Permission must always be sought from the owner or manager before entering any premises to collect insects.

Safety: Great care must be taken when visiting grain stores, as heaps of grain are inherently unstable. Keep well clear of the edges of such heaps and the walling supporting them, never enter bins or silos storing grain. Some grain is stored admixed with insecticide or may have a surface treatment of insecticide dust, so you must be aware of the hazards of contact with organophosphorus and other insecticides. Grain handling and processing machinery can also be dangerous, so avoid searching in places where such machinery is installed, even if it does not appear to be working at the time. When an infestation with rats is observed or suspected there is a possibility of infection with Weil's disease, so gloves should be worn, and cuts and abrasions should be covered.

Searching: In any premises the best form of pest control is hygiene, so it follows that the best collecting grounds are those mills or warehouses that are in constant use and are not kept very clean. In such places accumulations of food are allowed to remain in odd corners and cavities for long periods and infestations are common. The best places to find specimens are among small heaps of debris and floor-sweepings, and especially in damp and dark corners; behind the boarding around walls; at points where sacks or other containers of food are in contact with one another, or with the walls and floor; in cracks and crevices; and in the vicinity of accumulations of rodent and bird droppings or nest material. Many dead specimens can be found in cobwebs and on window-ledges. Finally, of course, one should search in and around the stored products themselves. Some species are active when it is dark and can be more easily obtained them.

Trapping: Various baits or traps can be used and some are now marketed commercially for use in grain stores, but only two types, pitfall traps and bait bags, are likely to be of use to the amateur who will be able to make them for himself. Pitfall traps can be used in grain and other commodities. A plastic container such as a pint-size plastic beer "glass" is ideal and should be inserted in the grain with the rim level with the surface of the grain. Surface dwelling beetles will fall in, but as many species can climb shiny surfaces it is necessary to put some water with a few drops of washing-up liquid into the bottom of the container to trap the beetles. The pitfall traps must be placed on a level area or else they will rapidly fill with grain. Pitfall traps made of glass should not be used because of the potential hazard of broken glass being left in the grain. Baiting with food is another method that has been exploited commercially, and "bait bags" can be purchased from agricultural suppliers. These bags measure 100 × 160mm and contain a mixture of broken carob (locust bean), wheat grain and peanuts, enclosed in a nylon mesh of sufficient size to allow the entry of *Sitophilus* spp. The bait bags should be placed around the wall or floor margins, in large crevices, on ledges or on the surface of the stored commodity. They are examined by shaking the bag and its contents over a white tray or sheet, and any beetles falling out can be collected up using a pooter. The bags will attract a wide range of species and can be used in grain stores, mills and any other premises where the presence of stored product beetles is suspected. Obviously one could experiment with the content of the bait bags to find that most effective for certain species, or for the widest range of species in a particular situation. Trapping will often reveal the presence of species not found by searching or sieving because of their low population density. Both pitfall traps and bait bags should be left in place for seven to ten days to be most effective. In commercial stores and farm premises care should be taken to ensure that they are put somewhere where they will not be disturbed. When placing traps on grain it is useful to tie them to a one metre marker cane in case the grain moves and buries the trap, and care should be taken to ensure that all the traps are removed from the stored commodity at the end of the visit. The bait bags are also very attractive to rats and mice and so should not be put in areas where there is obvious rodent activity. Moisture will also attract stored product beetles and pieces of wet sacking or small piles of moistened food placed on the floor will act as attractants for a variety of species.

Sieving: Beetles can be separated from various stored products, floor sweepings and debris, by sieving. A "nest" of sieves is best used for this as it will allow the material to be passed through a range of meshes in a single operation. In practice the combination of a sieve with a 2mm aperture with another of 0.7mm aperture will separate most beetles from grain and smaller debris, with most beetles passing through the 2mm mesh and being retained by the 0.7mm mesh. The larger beetles, grain or other coarse material will be retained by the 2mm mesh. The contents of the sieve can be examined by emptying them onto a white tray or sheet and any beetles present can be removed using a pooter.

Identification: The species most frequently found in the British Isles can be identified using the standard identification works. Two specialist publications, one by Halstead (1986), and the other by Hinton and Corbet (1980) are worth noting and details can be found in the bibliography. The former work provides keys to the families of stored product beetles on a worldwide basis, while the latter keys the common pests to species. There is also the detailed but much older work by Hinton (1945) dealing with many of the important pest genera. Unfortunately only the first part of this study was completed. Although many of the species are cosmopolitan, with the result that even the most exotic foods may be infested by commonplace species, from time to time unusual species, that cannot be keyed out in the standard British or European works, will be found. These can be sent to the M.A.F.F. Central Science Laboratory, London Road, Slough, Berks., SL3 7HJ, which specialises in stored product insects, or to the Department of Entomology, British Museum (Natural History), Cromwell Road, London, SW7 5BD, for identification; a charge, however, is usually made for this service, so you should enquire first.

3. Breeding

The Study of Life Histories: The discovery of larvae or pupae of stored product beetles will be followed by the need to breed them through to the adult stage in order to find out to which species they belong, but the study of life histories is also, of course, an activity in its own right and one that appeals to many coleopterists; it is to these to whom the following notes are especially addressed. The situations in which stored product beetles are found are usually man-made and thus breeding and life history studies are easily carried out under conditions very close to those in which the beetles are normally found. This is a unique advantage of studying this group of beetles. Rearing storage beetles is particular easy provided that a few simple rules are followed in relation to the temperature and moisture of the food, its nature and the hygiene of the cultures.

Temperature: Up to a certain point the warmer the environment the faster an insect will develop. There is, however, an optimum temperature at which development is quickest and mortality lowest. In laboratory studies the optimum temperature for development generally lies between 25°C and 30°C (77°F to 86°F). However, stored product beetles will develop quite well at room temperature, and it is interesting to speculate just how often temperatures of 25°–30°C will be found in the unheated premises in which food and grain is often stored in Great Britain. Even species like *Oryzaephilus surinamensis* will survive several days in a domestic refrigerator. The optimum temperature for one species will not necessarily be that for another. If the temperature rises above the optimum, the usual results are increasing sterility of the adults with rise in temperature, and the death of all stages at the lethal temperature zone. If the temperature is lower than the optimum, the developmen of the insect is longer, provided all other conditions are the same, until a temperature is reached at which development can barely take place and at which no eggs are laid. This is the threshold of development, and again differs for different species. As laboratory workers have tended to work with beetle strains which have become acclimatised to laboratory conditions, the amateur can make useful contributions to our knowledge of stored-product beetles by using populations freshly collected from the field to investigate the highest and lowest temperatures at which breeding will occur.

A maximum–minimum thermometer hung near the cultures and read daily is the easiest means of recording the temperature of the surrounding air. It must be remembered, however, that the temperature inside the culture jar may be rather different and a thermometer inserted into the culture medium will give a better idea of the actual temperature of the culture.

Moisture: The next most important factor is moisture, not so much visible water, as the amount in the air and the food medium. That in the air may be measured as the relative humidity (R.H.) and that in the food as the moisture content; both these measurements are expressed as percentages. The humidity of the air can be found by means of a whirling hygrometer, wet and dry bulb thermometer, paper hygrometers and other apparatus. The moisture content of the food is difficult to determine accurately without access to a special drying oven.

Many species will breed poorly or not at all if the R.H. is below 50%, and an R.H. of 60–70% seems to be the optimum for most species. Breeding will occur at higher humidities but in such situations the food rapidly becomes mouldy and unsuitable for many beetles. For a sealed culture kept at 70% R.H. and 25°C mould should not become a problem for 10 weeks or more. If mould growth appears earlier this is an indication that the conditions in the culture jar are too humid, or that the moisture content of the food was too high to start with. If the food goes mouldy rapidly it is best discarded, and the culture started afresh with new food. As mentioned earlier, for some species (e.g. *Typhaea stercorea*) mould may be an important part of their diet and, for such species, the best results are obtained when the food supports a vigorous growth of mould. This can be achieved by adding a dampened pad of paper or cotton wool to the food when the culture is set up.

Without access to a laboratory, achieving the correct moisture content of the food is difficult. A simple way would be to store the food in a place where the relative humidity of the air was known and was close to that which you wished to achieve. After one or two weeks the food would equilibrate to the R.H. of the surrounding air. This method has the disadvantage of exposing the food to infestation by other organisms. Alternatively the food can be left for one to two weeks in an airtight container in which is placed some water saturated with an appropriate chemical salt, but care must be taken to ensure that the solution does not come into contact with the food. The principle of this method is that the food will absorb moisture until it is in equilibrium with the air in the container, and when the air is enclosed over certain salt solutions it has a fairly constant vapour pressure, depending on the strength of the solution. Thus at 20°C and over a saturated solution of potassium carbonate the R.H. of the air would be about 45%; over calcium or magnesium nitrate it would be about 55%; over sodium nitrite or ammonium nitrate about 65%; over sodium chloride about 78%, and over distilled water about 98%. At lower temperatures the R.H.s will mostly be a little higher. The purest salts obtainable should be used and the food should be spread in a thin layer. These are the simplest solutions to use; however when accurate work involving a range of humidities is contemplated, it would be necessary to use the appropriate strength potassium hydroxide solutions which cannot be easily produced by the amateur coleopterist.

Many of the stored product beetles will drink water eagerly and, although not essential, its availability may prolong their life and increase productivity. Among the beetles that might benefit from having water available are *Carpophilus hemipterus, Tenebrio molitor* and many of the Ptinids. Water can be provided in a corked glass tube with a blotting-paper wick. The water is put into the tube together with a strip of blotting-paper which should reach to the bottom of the tube and pass between the cork and the side of the tube before projecting for about 25mm from the top of the tube. As long as there is water in the tube and the cork is not too tight a fit, the exposed portion of the blotting-paper will remain damp. The water tube should be pushed down into the food in such a way that the blotting-paper does not touch the surface of the food as, if it does, the food will go mouldy.

Food: As with any other group of beetles the food requirements of stored product beetles will differ from species to species and this will apply to artificial, as well as natural, diets. The food must also be in an appropriate form to allow development to take place. Thus, *Sitophilus* spp. require whole grains of wheat for successful breeding as the larvae develop inside the wheat grain; other species, with free-living larvae, can be given ground wheat. Some species (e.g. *Dermestes*) require food of animal origin, but for most species an entirely vegetarian diet is acceptable.

There are two approaches to feeding, one is to provide the beetles with whatever food they were found on, the other is to provide a standard culture medium. The former approach will be rewarding as there is often little information on the range of natural foods on which development can be completed, and this could be a study in itself. The disadvantage is that such foods collected with the beetles, or directly from their habitat, may be infested with other organisms, including mites, bacteria, protozoans and moulds, detrimental to the beetles. The nature of the food may also make it difficult to handle; sticky dates or Turkish Delight infested with *Oryzaephilus* can be rather messy! If, however, one's principal concern is not with the range of natural foodstuffs, but with convenience, hygiene or rearing large numbers then a standard culture medium can be employed. Table 1 gives examples of some culture media and the species that will feed on them. The list is neither exhaustive nor exclusive and some species may be reared on more than one medium. As some grain is treated in store with insecticides, care needs to be exercised in choosing a supplier of grain; if possible find a supply that is said to be insecticide free. The sudden death, or greatly reduced productivity, of a whole culture on transfer to a fresh supply of food can often be due to contamination with insecticides. Several of the media require the use of whole ground wheat and this can be obtained by putting wheat grains through a coffee grinder, using a fairly fine setting. Dried yeast seems to be a useful addition to most media, especially to those with a low nutritive value, such as plain flour.

A General Culturing Method: The most convenient containers for cultures of stored product beetles are glass jars. The 1lb (454g) and 2lb (908g) sizes sold for storing jams and other preserves are the most useful, but good results can be obtained with most sizes and shapes of jars, although tall, narrow jars where the surface area of the food will be small in relation to its depth, are best avoided. Sufficient food should be put into the jar to occupy one third to one half of its volume. Different species have different requirements; *O. surinamensis* rarely penetrates more than the first 25mm of the standard culture medium, whereas the superficially (in sizes and shape) similar *C. ferrugineus* will burrow right to the bottom of the food. A piece of crumpled kitchen paper should be placed on top of the food. This paper provides a larger surface area for the adults and larvae and thus helps to reduce contact between individuals, and lessens the chances of cannibalism and reduces other effects of overcrowding. The paper also provides additional egg-laying and pupation sites for some species. Where necessary a pad of damp paper or a water tube can be added to the food at this stage (see above).

Although some species will pupate in the food itself, others require special pupation sites which often serve to protect them against the cannibalistic tendencies of their own adults. Members of the genera *Tenebrio, Tenebriodes* and *Dermestes*, in particular, will attack their own uncovered pupae, especially when they are kept under too dry conditions, but many members of other genera will also exhibit cannibalism if the cultures become overcrowded. To overcome such problems it is advisable to supply large corks (e.g. for *Dermestes* and *Tenebriodes*), lumps of cotton wool (e.g. for *Gnatocerus cornutus* and *Tribolium destructor*) or small rolls of sacking or corrugated cardboard. These should be placed on top of the food in the jars, where they will provide adequate pupation sites. Only experience will tell if such measures are necessary in the particular conditions in which you are rearing your beetles.

Once the culture jars are prepared the adult beetles (or larvae) can be introduced. The number put into each jar will depend on the size of the adults and the numbers available. For most genera (e.g. *Sitophilus, Tribolium* and *Oryzaephilus*) up to 100 adults in a 2lb jar is acceptable, but for the larger beetles (e.g. *Tenebrio*) there should be perhaps no more than 10 in a 2lb jar. These are maximum numbers to avoid overcrowding; it is, of course, perfectly possible to start cultures with far fewer adults; a single female *Oryzaephilus surinamensis* will produce several hundred offspring in a couple of generations. The jars then need to be sealed both to keep out unwelcome guests and to keep in the inhabitants, as some stored product beetles can fly and others can climb glass. Glass-climbing ability can differ between members of the same genus, thus the *Oryzaephilus surinamensis* strains kept at the Slough Laboratory can climb glass whereas the *O. mercator* strains cannot. This is an important point to bear in mind when handling the beetles, as those that cannot climb glass are much easier to contain. The most effective method of closing the jars is to seal a filter-paper or blotting-paper circle to the rim of the jar with melted wax, and then as an added insurance to cover the top of the jar with a piece of cloth held in place by a rubber band (two are safer!). This method contains the insects but allows gaseous exchange through the paper. If wax and filter- or blotting-paper are not available then cloth held in place by rubber bands would be an alternative, but the bands must be very tight as many of the beetles can pass through very small holes. Similarly the normal screw-top of the jar could be used, but this would need to be punctured with holes that would allow air, but not the occupants, to pass through. I have known *Oryzaephilus surinamensis* to escape by walking along the screw thread of a closed screw-topped bottle. If the cultures are overcrowded some species (e.g. *Sitophilus* spp.) will chew their way out through the covering material.

Once set up, the culture jars should be placed in a warm and fairly moist place (60–70%; R.H.). This level of humidity may be difficult to obtain in a modern centrally-heated house and should be regarded as an ideal. Some breeding will take place with cooler conditions/lower humidities, but the productivity of the beetles may be greatly reduced. It is also worth remembering that a local microclimate will be set up in the jar itself and it is unlikely that this will be same as that in the room in which the beetles are being kept. Thus one's notes should state that the *culture* has been kept at, say 25°C and 70% R.H., rather than that the *insects* have been reared under these conditions. Jars should not be stored in direct sunlight, but may, otherwise, be put anywhere else. Cultures will breed perfectly well if kept in total darkness, and, indeed, many of their natural habitats will be totally dark. The response of these beetles to photoperiod is an area that would repay study.

Culture hygiene: The cultures are a rich source of food to other organisms and are susceptible to infestation by moulds, mites and booklice, as well as by other species of beetle. All of these unwanted guests will compete for the food and may make conditions in the culture jar unsuitable for its intended occupants. Mites of the genera *Tyrophagus* and *Suidasia* will outcompete many of the stored product beetles, and *Caloglyphus* species may also eat the beetle eggs, as will the predatory mite species. Except when rearing species such as *Typhaea stercorea* and *Ahasverus advena*, if the food is too damp mould growth will soon make it unsuitable for consumption. Booklice can build up to large population densities in culture jars, but appear to have only nuisance value. Different species of beetle will compete with each other for the food and even eat each other's eggs, larvae and pupae, so it is important to keep single species cultures. The problems of overcrowding have already been mentioned as this may lead to cannibalism and the spread of disease. It is worth noting here that some genera, in particular *Tribolium* and *Cryptolestes*, appear vulnerable to sporozoan infections.

Heat sterilisation of the culture medium at 70°–90°C is often advocated as a method of preventing the infestation of cultures and the spread of disease. This does, of course, mean that the food will become very dry and need moistening, and there is the possibility that some ingredients will be denatured by the heat. The writer prefers storage for at least two weeks in a domestic deep freeze (at $-18°C$) as an alternative. (N.B. The freezing compartment of a refrigerator is not sufficiently cold for this purpose.) This method has been used for many years without problems and may even be a more effective way of killing sporozoan spores. It also has the advantage that the food can be prepared beforehand and stored for long periods until required. The food must be brought to room temperature before the beetles are introduced into it. Mould growth can be inhibited by ensuring that the jars are completely dry before use and that the food is not too moist. The danger from mite infestation is considerable, but can be overcome by scrupulous hygiene. As well as being in the food, mites may also be on the working surfaces and floating in the air, so culture jars should be stored upside down and rinsed thoroughly with hot water before use. Once set up, culture jars should be left open for as short a time as possible. Food should be stored in a closed container, preferably in a freezer. The prevention of mite infestation is another good reason for sealing the tops of the jars with filter or blotting paper, rather than using some other covering with gaps through which mites and booklice can creep.

When the food supply is exhausted or the culture becomes overcrowded it will be necessary to move some adults to fresh food. When this is done, it is important to transfer only the beetles and not the old culture medium which may contain fungal spores and frass.

Further studies: Although many of these beetles have been studied in the laboratory, there are still plenty of opportunities for the amateur to make novel observations while rearing them. Studies can be made on the range of temperatures at which the species will breed and on the length of the life cycle at various temperatures. They will be even more useful if it is possible to measure or control the humidity. The effects that differing photoperiods may have on life cycles and productivity may well produce interesting observations. The acceptability of various foods and the extent to which the life cycle can be completed on them is another area worthy of investigation. Whatever investigations are made it will be necessary to keep notes and it is particularly important to record the temperature and humidity at which the cultures are kept, the date on which the culture was set up and the number of adults, or other stage, used to start the culture. The other information required will depend on the nature of the investigation. Much of this information can be recorded on a label attached to the culture jar, but a permanent record must also be made and, if all this is done, it should be possible to produce useful contributions to our knowledge of this diverse group of beetles. One final item that should be remarked on is that the rearing methods described above were designed for use in the laboratory. The amateur coleopterist may want to consider how these methods might be adapted and improved to better suit his needs.

4. Bibliography

Work on stored product beetles is generally published in those scientific journals concerned with applied entomology or pest control, in particular in the *Journal of Stored Products Research* published by Pergamon Press. Occasionally papers may appear in more general journals such as the *Entomologist's Monthly Magazine*. I have listed below a small selection of publications that may be useful to someone tackling this grouping of beetles for the first time. I have already mentioned the identification works; that by Hinton and Corbet (1980) should enable the most frequently found species to be identified. The book by Sokoloff (1974), while dealing only with the genus *Tribolium*, shows what might be studied in less well-known species and genera in relation to ecology, biology and nutrition. The paper by Solomon (1951) gives details of the solutions needed to condition the culture media to the required humidity. The books by Hickin (1974) and Mourier and Winding (1975) can be recommended to the more general reader.

References

Halstead, D.G.H., 1986. Keys for the identification of beetles associated with stored products. I – Introduction and key to families. *J. stored Prod. Res.*, **22**, pp. 163–203.

Hickin, N.E., 1974. *Household Insect Pests*. Associated Business Programmes, London.

Hinton, H.E., 1945. *A Monograph of the Beetles Associated with Stored Products*, Vol. 1. British Museum (Natural History), London.

Hinton, H.E. & Corbet, A.S., 1980. *Common Insect Pests of Stored Food Products, a guide to their Identification*. Economic Series No. 15. 6th edn. British Museum (Natural History), London.

Mourier, H. & Winding, O., 1975. *Collins Guide to Wild Life in House and Home*. Collins, London.

Sokoloff, A., 1974. *The Biology of Tribolium*, Vol. 2. Clarendon Press, Oxford.

Solomon, M.E., 1951. The control of humidity with potassium hydroxide, sulphuric acid, and other solutions. *Bull. ent. Res.*, **42**, pp. 543–554.

Table 1. Some useful culture media and the species that feed on them.

Constituents	Ratio of constituents by weight	Species
Whole wheat	1:0	*Rhyzopertha dominica* *Sitophilus* spp.
Whole ground wheat, rolled oats and yeast	5:5:1	*Oryzaephilus* spp. *Cryptolestes* spp. *Ahasverus advena* *Tenebrio* spp.
Fishmeal, yeast and bacon ends	16:1:4	*Dermestes* spp.
Fishmeal, yeast (and a piece of felt)	16:1	*Anthrenus verbasci* *Attagenus pellio*
Whole ground wheat, fishmeal and yeast	8:4:1	*Ptinus* spp. *Alphitobius diaperinus* *Gnatocerus cornutus* *Tribolium destructor*
Whole ground wheat and yeast	10:1	*Lasioderma serricorne* *Stegobium paniceum* *Latheticus oryzae* *Palorus ratzeburgii*
Wholewheat flour and yeast	12:1	*Tribolium castaneum* *Tribolium confusum*

BEETLES ASSOCIATED WITH ANTS AND ANT NESTS

BY G.B. WALSH (REVISED BY J. COOTER)

The late H.St.J.K. Donisthorpe years ago published our standard work on collecting myrmecophilous coleoptera and from time to time copies may be had in secondhand bookshops or from specialist dealers.

The notes that follow are taken from the first edition of this "Handbook" modified only very slightly and with nomenclature brought up to present day (1990) standards.

Spring and autumn are the best times for working ants nests and the preferred time is the morning as after the sun has been shining and heating up the outer part of the nest the beetles tend to seek the cooler interior.

The three best species of ants to work and the ones with which most beetles are found are:
Formica rufa L.
Lasius fuliginosus (Latr.)
Lasius flavus (F.)

However, beetles are to be found with other species too, notably
Lasius niger (L.)
Formica fusca L.
Formica sanguinea Latr.
Myrmica ruginodis Nyl.
Myrmica rubra (L.)
Myrmica scabrinodis Nyl.
Tapinotus erraticum (Latr.)

Some ants'-nest beetles are exceedingly constant in keeping to their own hosts, others seem a little more cosmopolitan.

Formica rufa L. – Generally several nests can be found in a restricted area and it is good practice to find a group which can be easily worked. Early in the spring one or two bricks or flattish rough stones and a piece of wood or thick branch can be placed on the upper sides and top of the nest mound. A bunch of long grass, twisted, can be pushed into a hole in the side of the nest. After setting these "traps" the nest must be inspected regularly, say, once a fortnight, but in warmer weather more regularly as the ants' constant re-working of the nest material will result in the "traps" soon becoming buried and lost to the coleopterist.

Inspection begins by setting out the necessary equipment, and some people tie their cuffs or put cycle clips around their ankles to prevent too many ants getting under one's clothing. A sheet can be laid close to the mound, or a cloth bag with wide mouth open held above the mound. The bricks, stones and wood are then quickly but carefully picked up and dropped in the bag or put on the sheet. The tunnels and passageways under the trap are inspected for beetles and any that are seen can be collected (with luck and great speed) by the pooter. Next the "traps" are given a sharp tap to dislodge loose material from their undersurfaces. This material is inspected as is the undersurface of the "trap" – again any beetles are collected at once with the aid of the pooter. Some inhabitants are very slow to move, so a long and careful look through a seething mass of ants is necessary. It is good practice, once the bulk of the ants have left the collecting sheet, to sift the debris over a new sheet or tray and examine the siftings for small-fry and sluggish movers (e.g. Histerids).

Next the grass twist is bodily removed from the nest and unwound on the sheet and the debris similarly inspected.

The paths and runs used by the ants around the nest should be inspected, litter and moss sifted and a search made under any nearby logs or stones – such places being the haunt of, for example, *Zyras humeralis* (Grav.)

I (J.C.) have used these methods personally in the Wyre Forest and found it a very good way to get *Quedius brevis* Er. – a species that, in my limited experience, frequented the underside of wood or log "traps" placed on the nest mound. *Clytra quadripunctata* (L.) is to be found by searching or sweeping/beating any overhanging vegetation – the females drop their eggs onto the nest mound, but take care to keep off the ant-infested mound themselves lest they become food for the nest.

A second method which I have employed in the New Forest and in Glen Tanar, always with a friend, is also described by Donisthorpe. It involves the wholesale removal of handfuls of nest material for sifting and inspection over a sheet. More ants are antagonised by such methods, so there is perhaps a greater need to guard against these active little creatures running up one's sleeves and trouser-legs or down one's neck. Thus doing this work with a friend has advantages as each worker can divide time between looking on the sheet and observing the progress of the ants on one's companion. The double-handfuls are dropped into the sieve and rapidly sifted. Ants can then be removed from one's person, the residue placed near the nest and a little time allowed for the bulk of the ants to clear the sheet. The debris is inspected, again keeping a sharp look out for small and sluggish species, and of course those beetles that mimic ant locomotion. The debris is then resifted and re-inspected. Residue can be taken home for inspection, but knowing what good results friends have had purely by inspecting the nest siftings in the field, it is perhaps unnecessary to risk frayed tempers resulting from the release of large ants indoors. The siftings can be tipped back onto or near the nest and the ants will soon make good any damage. In the Highlands I have found nearly full grown larvae of *Cetonia cuprea* in nest mounds, and have successfully reared adults. The method is simple, the largest larvae only are taken and placed in a stout box or tin with a quantity of nest material. The whole is taken home and left for a few weeks. Careful inspection will reveal the progress, and after a while the adults can be removed.

Donisthorpe gives the following myrmecophilous coleoptera associated with *F. rufa* –

Myrmetes piceus (Pk.)

Dendrophilus punctatus (Hb.)
D. pygmaeus (L.)
Hetaerius ferrugineus (Ol.)
Ptenidium formicetorum Kraatz
Ptilium myrmecophilum (Allibert)
Acrotrichis montandoni (Allibert)
Stenichnus bicolor (Denny)
S. godarti (Latr.)
Othius myrmecophilus Kies.
Leptacinus formicetorum Märkel
Gyrohypnus atratus (Heer)
Platydracus latebricola (Grav.)
P. stercorarius (Ol.)
Heterothops spp. (? *niger* Kraatz)
Quedius brevis Er.
Amischa analis (Grav.)
Amidobia talpa (Heer)
Notothecta confusa (Märkel)
N. flavipes (Grav.)
Lyprocorrhe anceps (Er.)
Atheta sodalis (Er.)
Drusila canaliculata (F.)
Zyras humeralis (Grav.)
Dinarda maerkeli Kies.
Oxypoda haemorrhoa (Mannh.)
O. formiceticola Märkel
O. recondita Kraatz
Thiasophila angulata (Er.)
Haploglossa pulla (Gyll.)
Aleochara ruficornis Grav.
Batrisodes venustus (Reichenbach)
Cetonia aurata (L.) – larvae and pupae
C. cuprea F. – larvae and pupae
Monotoma angusticollis Gyll.
M. conicicollis Aube
Coccinella distincta Faldermann
Clytra quadripunctata (L.) – larvae and pupae

Donisthorpe lists special habitats for the following species:

Aleochara ruficornis – a rare species, taken in moss and by sweeping near the nest. *Dinarda maerkeli* is almost always found by means of the clump of wood placed on the nest (though Walsh has found it by sifting the nest contents).

Lasius fuliginosus (Latr.) – This is a tree-nesting ant, generally utilising oak or beech, often dead trees. Locating a nest can be difficult because the ants use set runs which extend some distance from the nest entrance. Having discovered the nest, one can observe the ants entering the interior of the tree by means of holes at the roots or in the trunk. All such cavities should be "plugged" with grass twists, and if the tree is hollow, a large twist can be placed inside. A piece of old damp wood loosely wrapped in paper or an old bone placed in the nest entrance or cavity is often productive. After some days, the grass and any other "traps" can be carefully removed and shaken over a sheet. Chinks in the tree should be carefully inspected and any loose bark carefully lifted and placed on the sheet.

The runs used by the ants will also repay examination and any stones or logs, litter and leaves near them ought to be examined or sifted. Tussocks growing near the tree can be cut and shaken over the tray and it is often profitable to sift or otherwise inspect any "saw-dust" ejected from the tree by the ants.

The following beetles are found with *Lasius fuliginosus* –
Dendrophilus punctatus (Hb.)
Ptenidium formicetorum Kraatz
P. gresneri Er.
Othius myrmecophilus Kies.
Gyrohypnus atratus (Heer)
Heterothops spp. (? *praevius* Er.)
Quedius brevis Er.
Amischa analis (Grav.)
Notothecta confusa (Märkel)
Plataraea brunnea (F.)
Liogluta nitudula (Kraatz)
Atheta hepatica (Er.)
Atheta consanguinea (Eppelsheim)
Drusila canaliculata (F.)
Zyras cognatus (Märkel)
Z. funestus (Grav.)
Z. laticollis (Märkel)
Z. limbatus (Pk.)
Z. lugens (Grav.)
Z. haworthi Stph.
Z. humeralis (Grav.)
Ilyobates bennetti Donisthorpe
Amarochara bonnairei (Fauv.)
Oxypoda haemorrhoa (Mannh.)
O. vittata Märkel
Homoeusa acuminata (Märkel)
Haploglossa gentilis (Märkel)
H. pulla (Gyll.)
Thiasophila angulata (Er.)
T. inquilina (Märkel)
Aleochara ruficornis Grav.
Amauronyx maerkeli (Aube)
Batrisodes venustus (Reichenbach)
Amphotis marginata (F.)

The species of *Zyras* have the habit of rolling up when disturbed and may remain motionless for some time, so may be easily overlooked. *Amphotis marginata* is often taken under loose bark but has been taken by sifting dead leaves in the ants' runs. *Leptinus testaceus* Müll is sometimes found with this ant.

Lasius flavus (F.) – nests either under a stone, generally a flint, or in a mound of earth. With earth-mound nests little can be done and in general this type of nest seems quite unproductive. Those built under a stone are more productive. The stone should be quickly and carefully lifted and placed on the sheet or in a cloth bag. The tunnels and passageways under the stone should then be inspected. Often the ants sieze beetles and take them down into the nest as they do with their own larvae and pupae. The underside of the stone can then be inspected and afterwards carefully replaced over the nest.

The following species occur with *Lasius flavus* –
Hetaerius ferrugineus (Ol.)
Sunius bicolor (Ol.)
Othius myrmecophilus Kies.
Platydracus stercorarius (Ol.)
Lamprinodes saginatus (Grav.)
Amischa analis (Grav.)
Drusilla canaliculata (F.)
Zyras limbatus (Pk.)
Homoeusa acuminata (Märkel)
Amauronyx maerkeli (Aube)
Claviger testaceus Preyssler
C. longicornis Müller

Formica fusca L. nests either under stones or in old posts and stumps; the former should be treated in the same manner as *L. flavus* and in the latter the stump or post can be broken up over a sheet and sifted. The following beetles have been taken with *Formica fusca* –
Hetaerius ferrugineus (Ol.)
Neuraphes carinatus (Muls.)
Lamprinoides saginatus (Grav.)
Drusilla canaliculata (F.)
Zyras limbatus (Pk.)
Lomechusa emarginata (Pk.)
L. paradoxa Grav.
Dinarda pygmaea Wasm.
Homoeusa acuminata (Märkel)
Aleochara ruficornis Grav.
Amauronyx maerkeli (Aube)
Batrisodes venustus (Reichenbach)
Opatrum sabulosum (L.)

These associations between ants and beetles have been noted:

ANT	BEETLE
Ponera coarctata (Latr.)	*Drusilla canalculata* (F.)
	Lamprinodes saginatus (Grav.)
	Tychobythinus glabratus (Rye)
Myrmica rubra (L.)	*Lomechusa emarginata* (Pk.)
	L. paradoxa (Grav.)
	Drusilla canaliculata (F.)
	Zyras collaris (Pk.)
	Platydracus stercorarius (Ol.)
Myrmica ruginodis Nyl.	*Lomechusa emarginata* (Pk.)
	Drusilla canaliculata (F.)
	Lamprinodes saginatus (Grav.)
	Platydracus stercorarius (Ol.)
	P. latebricola (Grav.)
Myrmica sulcinodis Nyl.	*Lomechusa emarginata* (Pk.)
	Drusilla canaliculata (F.)
Myrmica scabrinodis Nyl.	*Lomechusa emarginata* (Pk.)
	Zyras limbatus (Pk.)
	Drusilla canalculata (F.)
	Amischa analis (Grav.)
	Platydracus stercorarius (Ol.)

	Othius myrmecophilus Kies. *Batrisodes venustus* (Reich.)
Leptothorax acervorum (F.)	*Drusilla canaliculata* (F.)
Tetramorium caespitum (L.)	*Drusilla canaliculata* (F.) *Platydracus stercorarius* (Ol.)
Lasius niger (L.)	*Homoeusa acuminata* (Maerk.) *Zyras limbatus* (Pk.) *Drusilla canaliculata* (F.) *Claviger testaceus* Preyssler *C. longicornis* Müll. *Opatrum sabulosum* (L.)
Lasius alienus (Foerst.)	*Drusilla canaliculata* (F.) *Claviger testaceus* Preyssler
Lasius brunneus (Latr.)	*Dendrophilus punctatus* (Hb.) *Ptenidium formicetorum* Kr. *P. turgidum* Thom. *Acrotrichis montandoni* (All.) *Euconnus pragensis* (Machulka) *Stenichus bicolor* (Denny) *S. godarti* (Latr.) *Eutheia formicetorum* Reitt. *Aleochara sanguinea* (L.) *Haploglossa gentilis* (Maerk.) *H. pulla* (Gyll.) *Oxypoda recondita* Kr. *Ilyobates propinquus* (Aube) *Tachyusida gracilis* (Er.) *Zyras limbatus* (Pk.) *Drusilla canaliculata* (F.) *Atheta nitidula* (Kr.) *Atheta sodalis* (Er.) *Amischa analis* (Grav.) *Euryusa optabilis* Heer *E. sinuata* Er. *Quedius scitus* (Grav.) *Xantholinus angularis* Ganglb. *Leptacinus formicetorum* Maerk. *Othius myremcophilus* Kies. *Batrisodes venustus* (Reich.) *B. laportei* (Aube) *B. adnexus* (Hampe) *Symbiotes latus* Redt. *Ptinus subpilosus* Sturm *Dryophthorus corticalis* (Pk.)
Lasius umbratus (Nyl.)	*Acrotrichis montandoni* (All.) *Atheta consanguinea* (Epp.) *Zyras humeralis* (Grav.) *Claviger longicornis* Müll.
Lasius mixtus (Nyl.)	*Homoeusa acuminata* (Maerk.) *Claviger longicornis* Müll.

Formica pratensis Retz.   *Ptenidium formicetorum* Kr.
  Acrotrichis montandoni (All.)
  Leptacinus formicetorum Maerk.
  Oxypoda formiceticola Maerk.
  O. haemorrhoa (Mann.)
  Monotoma angusticollis Gyll.
  Cetonia cuprea (F.)

Formica exsecta Nyl.   *Dendrophilus punctatus* (Hb.)
  Othius myrmecophilus (Kies.)
  Amischa analis (Grav.)
  Amidobia anceps (Er.)
  Notothecta flavipes (Grav.)
  Drusilla canaliculata (F.)
  Zyras limbatus (Pk.)
  Dinarda hagensi Wasm.
  Oxypoda haemorrhoa (Mann.)

Formica sanguinea Latri.   *Hetaerius ferrugineus* (Ol.)
  Othius myrmecophilus (Kies.)
  Quedius brevis Er.
  Lamprinodes saginatus (Grav.)
  Notothecta flavipes (Grav.)
  Drusilla canaliculata (F.)
  Zyras limbatus (Pk.)
  Lomechusoides sturmosa (F.)
  Oxypoda haemorrhoa (Mann.)
  O. recondita Kr.

References (Identification of Ants)

Bolton, B. & Collingwood, C.A. 1975. Hymenoptera, Formicidac. *Handbk. Ident. Br. Insects.* **6(3c)**: 34 pp. Royal Ent. Soc. London.

Part IV – Beetle Larvae

By M.L. Luff

1. INTRODUCTION

It is obvious to any serious entomologist, although perhaps not thought about by many collectors, that every adult beetle that they seek must first have passed through developmental stages of egg, larva and pupa. Unlike some other insects, many British beetles spend much of their total life (and in particular the winter) as adults. There are, however, common exceptions to this. Some beetles such as the larger wood-boring Cerambycidae, and soil inhabiting Elateridae may spend a year or more as larvae. Others may be more readily found as larvae than as adults. Larvae and adults of the same species frequently share the same habitat, perhaps because beetles, more than any other group of endopterygote insects, are adapted as adults to live in confined spaces and sheltered situations, where larvae also occur. One must be careful, however, not to assume that just because larvae and adults are found together, that they are necessarily of the same species; rearing out of the larvae may be needed to confirm this.

The study of beetle larvae has, however, been neglected by the majority of collectors: a few have reared larvae, but only to obtain perfect adults for their collections. The number who deliberately rear them in order to study their ecology, bionomics, physiology or taxonomy must be very small.

The aim of this chapter is not to direct coleopterists into a specialised channel, but merely to make them "larva conscious". Comparatively few life-histories of our British beetles have been worked out and many of those that have been are but superficially dealt with. Much still remains to be done in larval systematics, for although nearly all the family characters are known, there still remain many genera and species yet to be described and keyed; while many of the "known" species need to be verified by breeding, because systematists are so frequently forced by lack of material to rely on questionable specimens and inadequate descriptions.

It must be emphasised that over-zealous searching for larvae, even more than for adults, may lead to habitat destruction and loss of rare species. The J.C.C.B.I. Code for Insect Collecting applies as much or even more to collecting larvae (which are the future breeding stock of the species) than to adults which may already have bred. In particular, long-term habitats such as old timber should not be intensively worked, and as much material as possible should be left intact.

Fig. 16

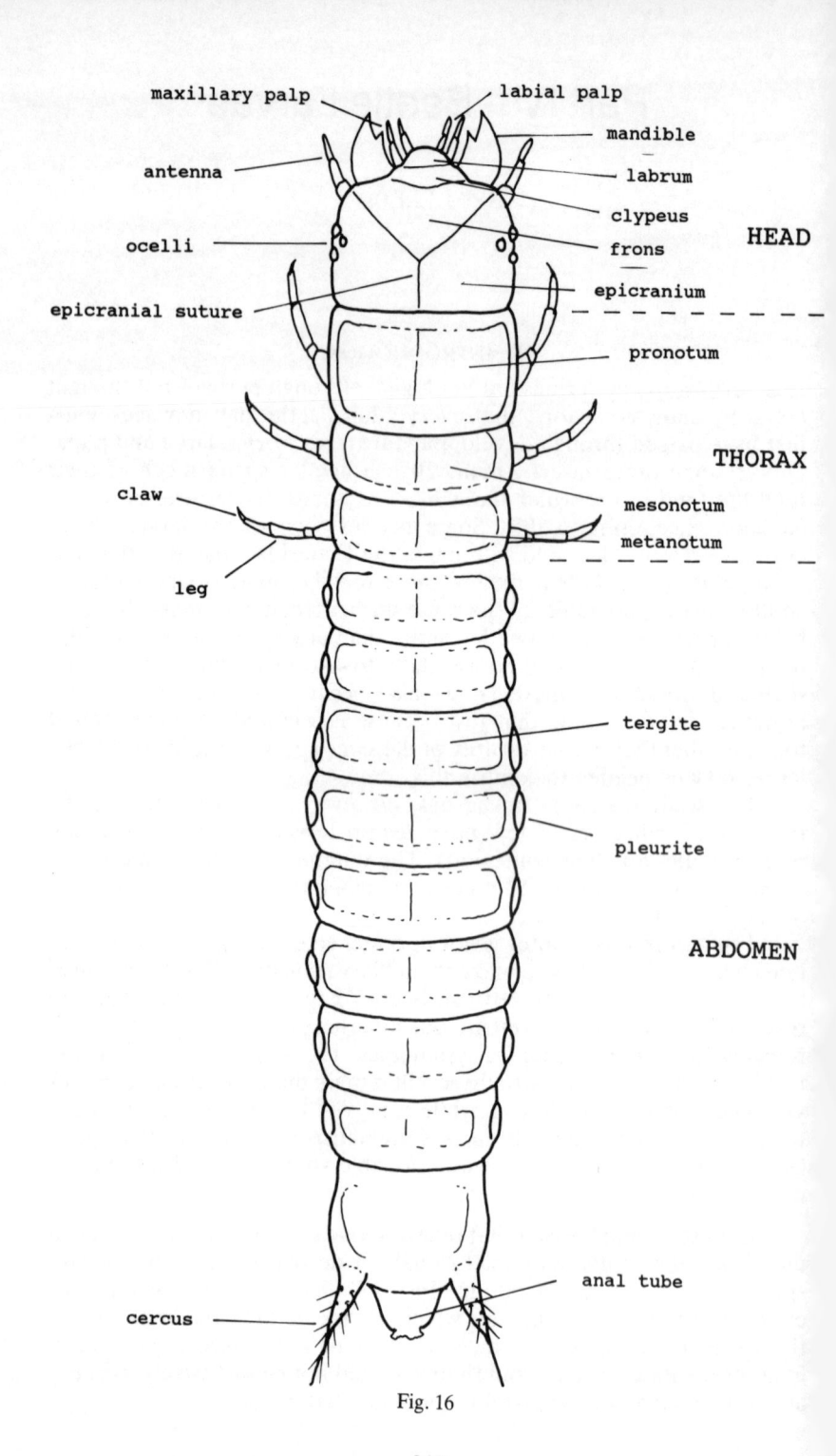

2. Morphology

As the structure of beetle larvae is both more variable and less familiar to the general collector than that of the adults, it is appropriate to outline the general features of beetle larvae. (Fig. 16.) This will both enable them to be recognised as such and serve as a starting point for more detailed identification, if this is to be attempted. It must be remembered that as beetle larvae grow by moulting several times during their life, the size of any species can vary considerably. Generally a fully-grown beetle larva is about 1.5 times the length of the corresponding adult. Most species have between 3 and 5 larval stages or instars, with a moult between each instar. In a few cases their morphology changes markedly between instars; this is called "hypermetamorphosis".

Beetle larvae are seldom brightly coloured; those living in the open tend to be dark brown or black, whereas the majority living in sheltered habitats are pale brown, cream or white, sometimes with a darker head. They generally have a distinct head capsule, unlike many fly larvae, with curved, opposable mandibles, and two sets of palpi (maxillary and labial) as in the adults. Up to six simple eyes or ocelli may be present on each side of the head. The antennae vary from minute two-segmented appendages to long filaments, but are usually three or four-segmented. The thoracic legs range from 5-segmented (plus one or two claws) to completely absent. The abdomen usually has from 8–10 obvious segments, and is without prolegs (in contrast to those found in Lepidoptera) but there may be smaller swellings (*ampullae*) or hooks on particular segments. The apical abdominal segments may bear paired cerci (technically known as *urogomphi*) which can be short and hook-like, or long and slender.

This general structure is usually modified into one of four main morphological types:
1. Campodeiform, as in Carabidae and Staphylinidae. These larvae are active, elongated specimens with long legs (and often long cerci). They hold the mouthparts forwards at the front of the head and are usually predatory.
2. Scarabaeiform, as in Lucanidae, Scarabaeidae and Byrrhidae. These are fat, usually C-shaped grubs, tending to lie on their sides within their larval habitat such as dung or soil. They have well-developed thoracic legs, but their movements are limited by the bulk of their bodies.
3. Eruciform, as in many Chrysomelidae. These are more "caterpillar-like" larvae, with a small head capsule bearing ventrally-directed mouthparts, and short but functional thoracic legs, so that they can crawl over their substrate.
4. Apodous, as in Curculionidae. These are legless grubs, rather like the scarabaeiform type in their C-shaped body, but unable to move other than by bodily contractions; many are found within their food medium such as in plant tissue.

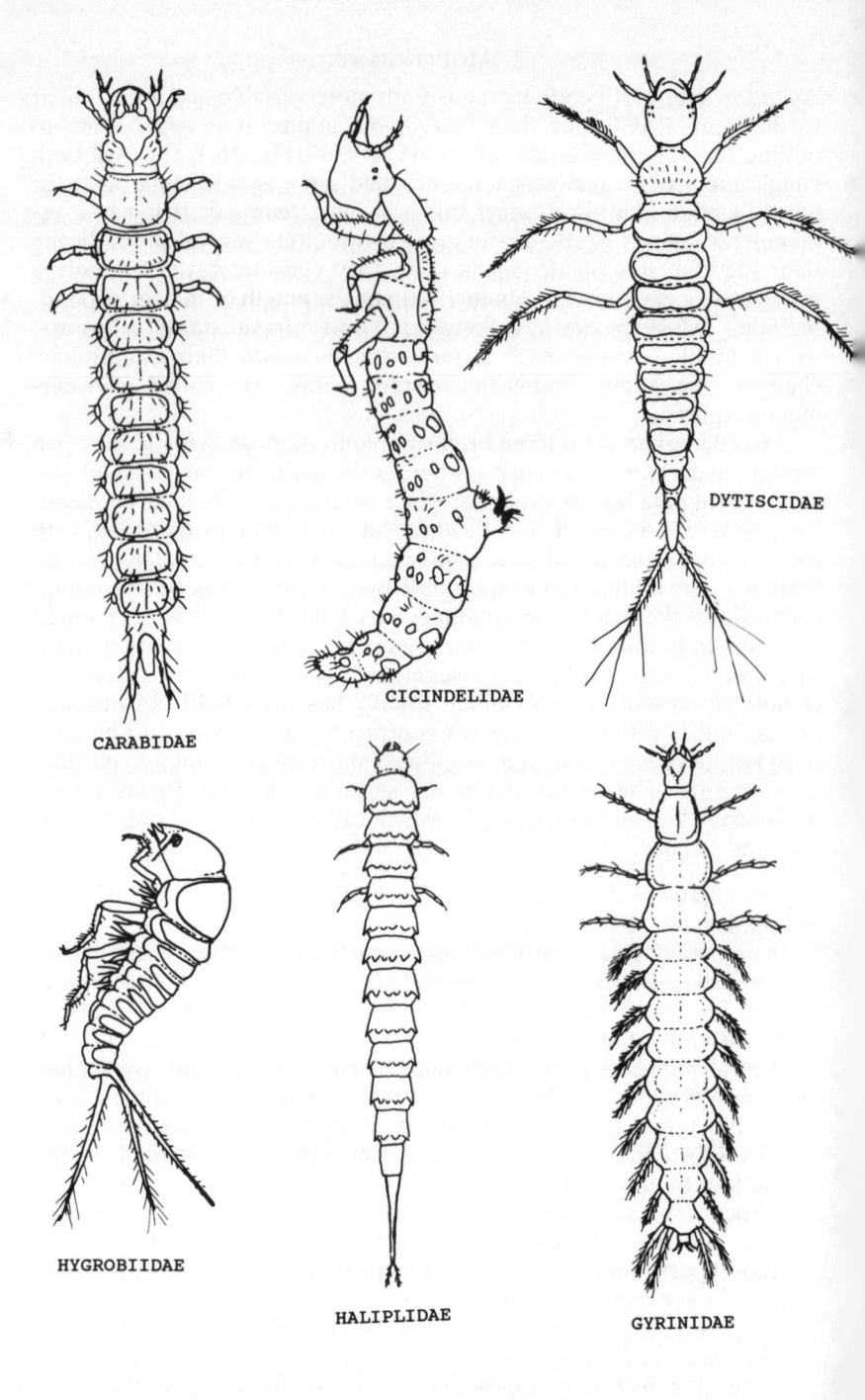

Fig. 17

Intermediate forms occur between these main morphological types, but any one family usually tends to have a characteristic type of larva.

3. IDENTIFICATION

There is currently no complete and easily usable key for the identification, even to family level, of British beetle larvae. The "classical" work is that of van Emden (1942, *Entomologist's Monthly Magazine*, **78**: pp. 206–272 – "Larvae of British Beetles III. Key to families".) There are other papers in the same series on particular families, but all are now difficult to obtain. There are comprehensive works on larvae of the whole Order by Boving & Craighead (1931, reprinted 1953), Ghilarov (1964, in Russian!) and, more recently, a comprehensive volume, with many keys to genera, edited by Klausnitzer (1978 – in German). None of these are readily available to the amateur, however. At the time of writing, there is a volume on the identification of beetle larvae being prepared in the Royal Entomological Society's "Handbooks for the identification of British Insects" series; hopefully this will fill the gap. It would be nice, however, if a beetle larvae key was available in the A.I.D.G.A.P. series sponsored by the Field Studies Council, similar to that already available for the familes of adult beetles.

Keys are available for a few particular families; the best is that to larvae of Coccinellidae, in the excellent little book "Ladybirds" by Majerus & Kearns (1989) in the "Naturalists" Handbooks' series. Some volumes in the "Fauna Entomologica Scandinavica" series (written in English) include keys to larvae, e.g. the Buprestidae. Details of other families are given in the earlier family sections of this Handbook.

4. FAMILY CHARACTERISTICS

The following brief descriptions outline some of the larval characters and habitats of selected families of British beetles. The larval characters of the various families mentioned are not a conclusive guide to their identification and, to facilitate identification in the field, preference has been given whenever possible to the more conspicuous characters. Such a procedure must inevitably leave uncovered many exceptions, but these can be verified by employing more specialised keys, after the beginner has grasped the fundamentals.

SUB-ORDER ADEPHAGA

Larvae usually campodeiform; legs 5-segmented, with one or two movable claws; antennae usually 4-segmented.

CARABIDAE (Fig. 17.) including CICINDELIDAE (Fig. 17.)

Abdomen 10-segmented (including anal tube), form elongate, slender, with paired sub-apical cerci; tarsi with one or two claws. Cicindelid larvae with prominent dorsal hooks on the 5th abdominal segment.

Habitat: On or in the soil. Larvae of *Nebria*, *Notiophilus* and fully-grown *Carabus* and *Pterostichus* are surface-active. Those of some *Har-*

Fig. 18

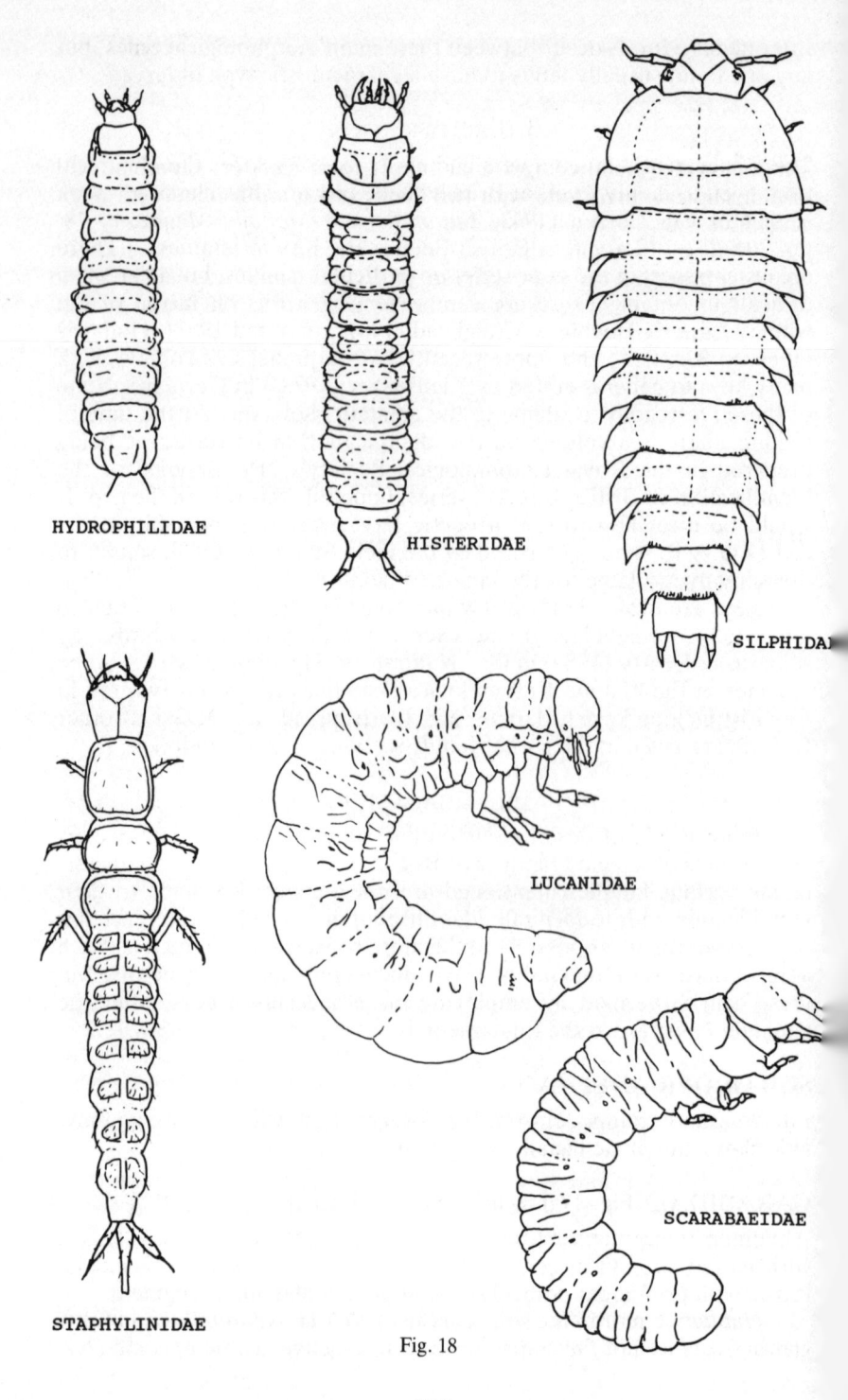

palus store seeds in vertical burrows. *Cicindela* are predatory in vertical burrows.

DYTISCIDAE (Fig. 17.)

Abdomen 8-segmented (no anal tube), form slender, tapering, with slender apical cerci; tarsi with two claws which are fringed with hairs; mandibles large and sickle-shaped.

Habitat: In pools, ditches, ponds, streams, etc. Predaceous on small soft-bodied animals. Larvae of *Ilybius fenestratus* (Fabr.) and other Dytiscids may be dug up from the banks of ponds about March, prior to pupation.

HYGROBIIDAE (Fig. 17.)

Abdomen 8-segmented, the last segment with a long median process and slender cerci; form tapering with very large head and thorax; tufts of ventral gills on thorax and first three abdominal segments.

Habitat: The larva of *Hygrobia hermanni* (Fabr.) is confined to muddy ponds where it feeds on midge larvae in the mud.

HALIPLIDAE (Fig. 17.)

Long, slender-larvae with 9 or 10-segmented abdomen; cerci apical if present; tarsi with a single claw; no gills.

Habitat: In mud or on vegetation in streams, where they feed on filamentous algae such as *Spirogyra*.

GYRINIDAE (Fig. 17.)

Elongated larvae with small head and prominent pairs of tracheal gills on each of the abdominal segments except the last, which has 4 small hooks.

Habitat: On the substrate or aquatic vegetation in the same habitat as the adults, where they are predatory.

SUB-ORDER POLYPHAGA

Legs with at most 4-segments, sometimes absent. Antennae at most 3-segmented. The body form can be any of the four main types.

HYDROPHILIDAE (Fig. 18.)

Variable in appearance, typically elongate, fleshy, with small head and legs (which are vestigal in *Cercyon*), and eight abdominal segments dorsally (the 9th and 10th segments are small and retracted into the 8th); cerci usually small, longer and segmented in *Helophorus* (which superficially resemble many carabid larvae), absent altogether in *Spercheus*.

Habitat: Aquatic species in ponds, etc., particularly where there is an abundance of floating weed; carnivorous. Terrestrial species in similar habitats to those of the adults, including dung and damp soil.

HISTERIDAE (Fig. 18.)

Parallel-sided, somewhat flattened and rather fleshy larvae, with mouthparts directed forwards; labrum absent, mandibles with small retinaculum but no mola; cerci short, two-segmented.

Habitat: In dung, decaying animal and vegetable matter, nests, etc. The predaceous larvae of *Hister 4-maculatus* Linn. and *H. unicolor* Linn. feed on dipterous larvae in dung. Birds' nests, especially those in hollow trees, are worth searching for Histerid larvae; these should be examined as soon as the fledglings have flown. Larvae of *Dendrophilus punctatus* (Herbst) have been taken from a woodpeckers' nest.

SILPHIDAE (Fig. 18.)

Elongated, often flattened and onisciform, with broad thorax and small head. Mouthparts directed downwards; cerci short, two-segmented.

Habitat: Chiefly in soil beneath carcases and other decaying animal matter. *Silpha* larvae usually feed underneath the carcase, but those of *Necrophorus* live in the soil beneath on portions of flesh carried down by adults. Larvae of *Xylodrepa 4-punctata* (Linn.) are predaceous on oak-feeding caterpillars in the spring, and may be beaten from oaks at night.

STAPHYLINDAE (Fig. 18.)

Elongated, campodeiform larvae, often superficially resembling those of Carabidae (but separable by the 4-segmented legs, 3-segmented antennae). Mandibles usually slender, without mola. Cerci usually two-segmented, often long.

Habitat: In soil, dung, fungi, moss, decaying animal and vegetable matter, under bark, etc. Some of these larvae live in dung (*Ontholestes murinus* (Linn.), *Philonthus* spp.), others in saturated moss near brooks (*Lesteva*), in carrion (*Creophilus maxillosus* (Linn.)) and under bark (*Siagonium quadricorne* Kirby), which are predaceous on Scolytid larvae). Many of the smaller species are found in soil-living fungi. The nests of birds harbour the minute larvae of *Microglotta* spp. There are a number of parasitic species, too; for example, larvae of *Velleius dilatatus* (Fabr.) are predatory on hornet larvae; dipterous puparia (especially those in decaying vegetable matter) should be kept as they may prove to contain the parasitic *Aleochara* spp. Ants' nests are particularly rich in Staphylinid larvae and much investigation remains to be done. Mammals' nests also harbour a number of species. Some of the larger Staphyline larvae are active on the soil surface, and may be caught in pitfall traps.

LUCANIDAE (Fig. 18.)

Typical scarabaeiform larvae, fleshy and curved ventrally, abdomen fattest towards its apex; legs small and slender, with stridulatory organs on the mid coxae and hind trochanters.

Habitat: In rotten trees and stumps. The bases of old stumps particularly oak, house the large, fleshy larvae of *Lucanus*. Those of *Dorcus*

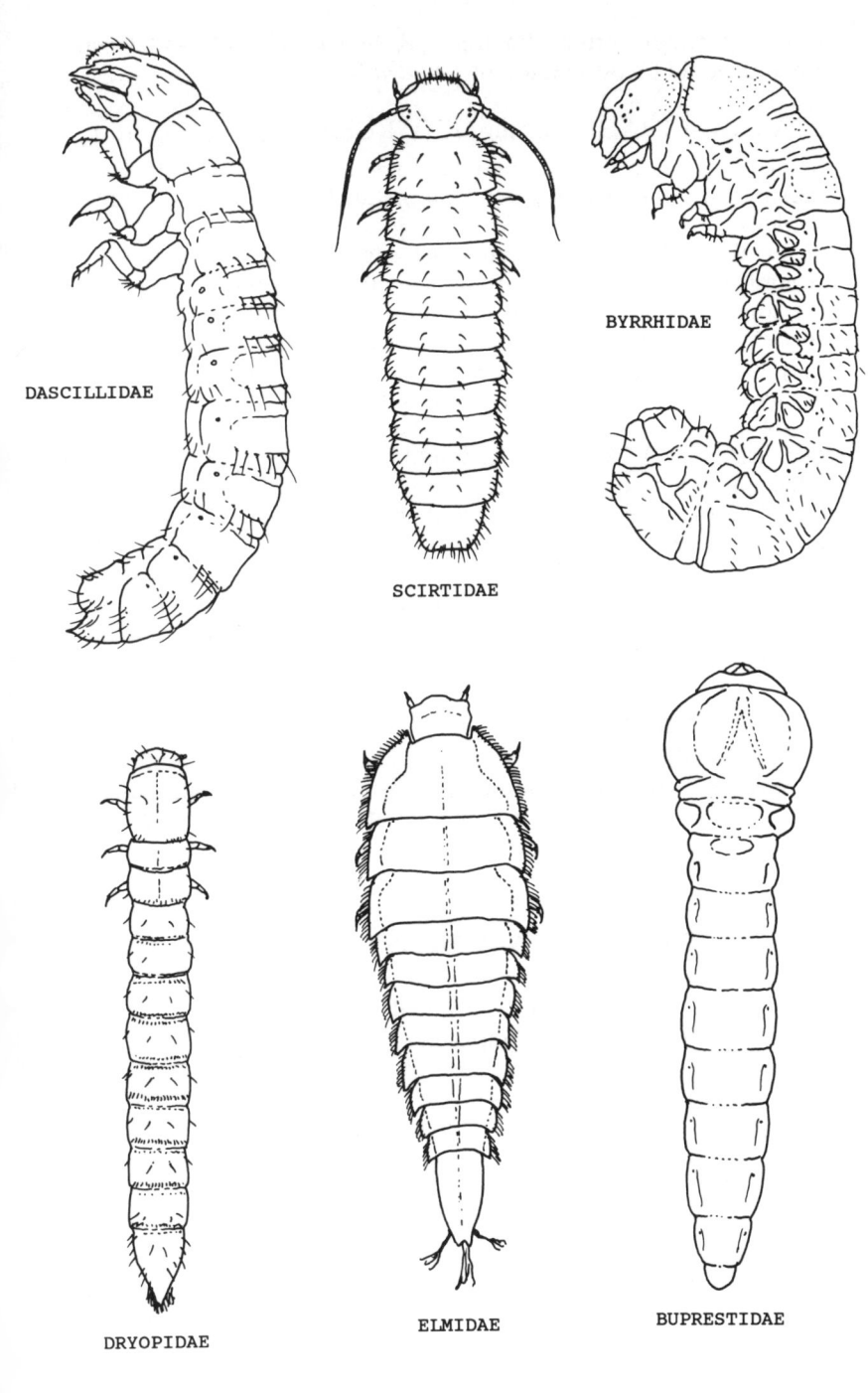

Fig. 19

generally prefer rotten elm trees. Rotten beech, ash and hawthorn stumps should be broken up for *Sinodendron* larvae.

SCARABAEIDAE (including GEOTRUPIDAE, etc.) (Fig. 18.)

Similar to Lucanidae but abdominal segments with extra dorsal folds, making the number of segments not distinct. No stridulatory organ on the legs.

Habitat: In dung or at roots; those feeding in the latter mostly take several years to attain maturity. The majority of species (e.g. *Aphodius*) are coprophagous and it must be borne in mind that the adults of many species carry pellets of dung several inches down into the earth where they lay their eggs. Hence it is even important to examine the earth immediately beneath the dung. Deep holes, roughly ½" in diameter, near rabbit dung in sandy areas indicate the presence of adults of *Typhaeus typhoeus* (Linn.) Larvae will be found in those that appear to be freshly filled up. Deer dung is attractive to some species. The round plump larvae of *Melolontha* and several others of the "Chafer" family are root feeders; these may be obtained by digging or by pulling up old stumps. Larvae of *Cetonia aurata* (Linn.) congregate in colonies beneath rotten elm stumps. *Gnorimus* larvae live in the soft wood mould in hollow oaks and fruit trees. *Trox* and *Saprosites* are exceptional in their habitat, the former living in dried carcases and the latter under bark.

DASCILLIDAE (Fig. 19.)

Elongate, rather cylindrical, fleshy larvae, slightly curved ventrally, with prominent V-shaped dorsal suture on head. Mandibles with mola.

Habitat: In soil, where they feed at roots. Larvae of *D. cervinus* (Linn.) may easily be reared in pots of root fibre and damp sandy or loamy soil.

SCIRTIDAE (HELODIDAE) (Fig. 19.)

Small, flattened, tapering larvae, rather onisciform; antennae long with a multi-articulate thread-like terminal flagellum (which is apparently unique among all holometabolous insect larvae).

Habitat: Aquatic, in ditches, rot-holes in trees, etc. There are several not uncommon species in this group: larvae of some species may be found crawling on rotten submerged branches and leaves in rather shady, foul pools; those of others prefer water weeds and leaves in more open pools, or occur in rain-water in roots or hollow boles of beeches. Larvae of *Helodes minuta* (Linn.) live among mossy stones in streams.

BYRRHIDAE (Fig. 19.)

Superficially rather like Dascillid larvae, but apex of abdomen is more abruptly curved ventrally; the mandibles have no mola, and the dorsal suture of the head is Y-shaped.

Habitat: In soil under turf, mossy tree stumps, roots, etc.

Fig. 20

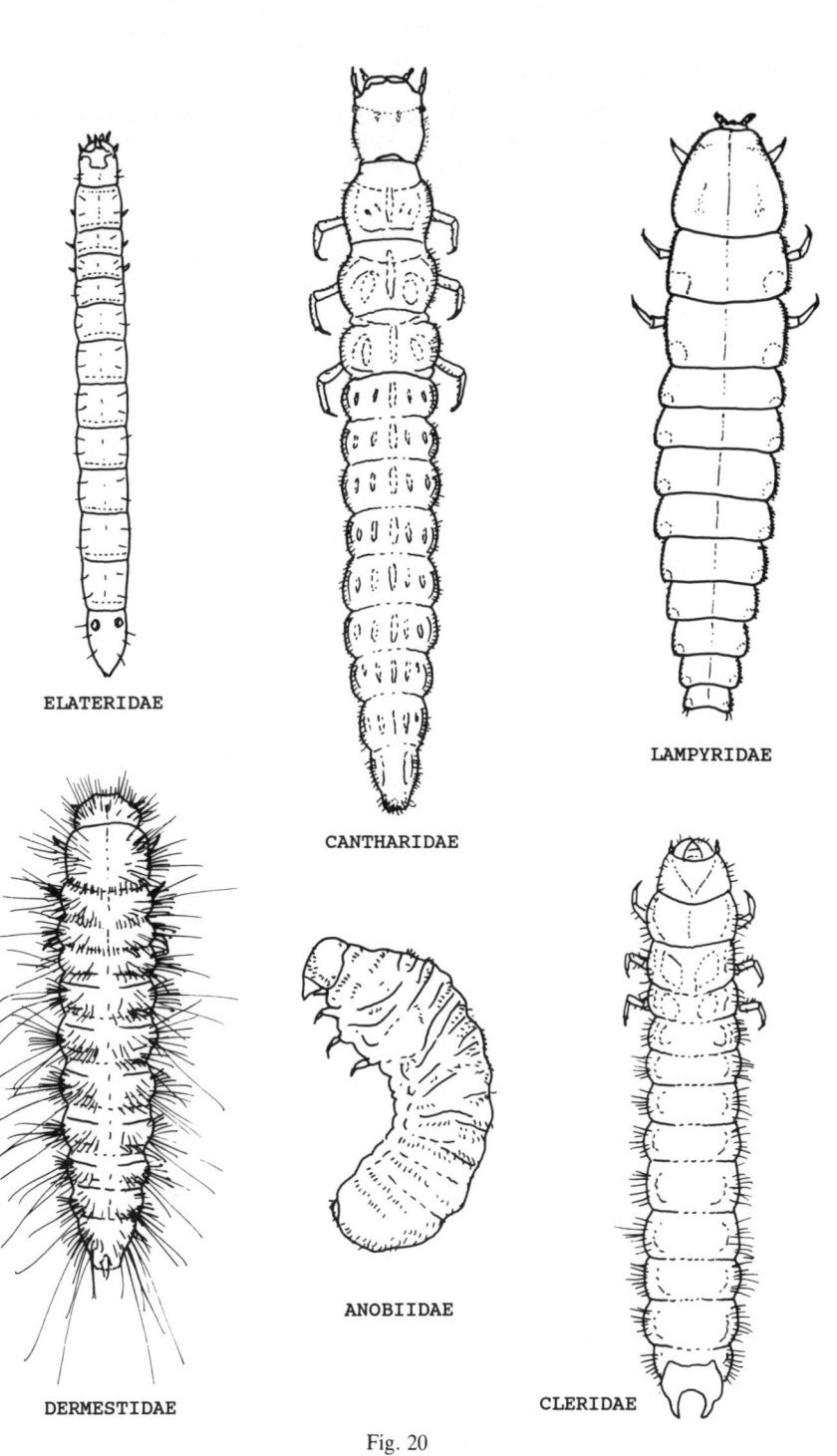

DRYOPIDAE (Fig. 19.)

Elongated, cylindrical, tapering to a point apically, where the ninth abdominal segment has a closeable operculum which covers the retractable 10th segment with its tracheal gills (these absent in some terrestrial *Dryops*).

Habitat: In similar aquatic habitats to the adults; some species are abundant in turf subject to waterlogging for extensive periods annually, as in the moist areas of fields.

ELMIDAE (Fig. 19.)

Similar to Dryopidae, but more flattened, sometimes onisciform, larvae, with three tufts of gills usually protruding from the apical operculum.

Habitat: Crawling beneath stones, etc., in small but swiftly-flowing streams, occasionally in ponds. They are best collected by placing a pond net in a stream and stirring up the gravel or stones or, if roots are present, by shaking them violently so that the larvae are carried out by the current into the net.

BUPRESTIDAE (Fig. 19.)

Cylindrical or flattened, tapering larvae, either head elongate and deeply retracted into prothorax (*Trachys*) or with thorax enlarged; legs absent or minute and labial palps fleshy and unjointed.

Habitat: In leaves and under bark, etc. Larvae of *Trachys* are leaf miners; the brown mines of one species are conspicuous in oak leaves in late summer. Another species is to be found in bramble leaves in late autumn, while larvae of *Trachys troglodytes* Schönh. mine leaves of the Devil's bit Scabious (*Scabiosa succisa* Linn.). *Agrilus* larvae live subcortically in young damaged branches of oaks, etc. *Aphanisticus* larvae occur low down in *Juncus* stems.

ELATERIDAE (Fig. 20.)

Elongated, slender and cylindrical larvae, usually well-sclerotised and orange-brown in colour, giving them their popular name of "wireworms". Labrum fused to head capsule and toothed on its front margin. Apex of abdomen either subconical, or forming a flattened plate with two apical teeth or prongs.

Habitat: Many species feed at roots of plants and some are serious pests. A few, however, live in damaged or rotten trees where they are mainly predatory on other larvae, particularly those of the Cerambycidae. Larvae of *Denticollis linearis* (Linn.) occur in rotten stumps, under bark and in peat. Larvae of *Cardiophorus* are of an unusual appearance, being soft, flaccid and elongate like an extremely slender centipede. Full-grown larvae of *C. asellus* Er. may be dug up in early spring in sandy soil at the roots of heather and pine (usually 4–5" deep). Larvae of *Limonius minutus* (Linn.) occur in the dry, mycelium-filled soil beneath "fairy rings". The sturdy blackish-brown larvae of *Athous villosus* (Geoffr.) live under the bark of oak boughs and are more commonly seen than the nocturnal adults. The orange-brown larvae of *Melanotus rufipes* (Herbst) are abundant under the bark of pine stumps where they feed on *Rhagium* larvae. Another common species to be found in pine stumps is *Elater balteatus* Linn.

CANTHARIDAE (Fig. 20.)

Straight, elongated larvae, campodeiform with ventrally-grooved mandibles. Head with sutures indistinct. Body usually covered with velvety pubescence; no cerci.

Habitat: In loose soil, under logs, or crawling on pathways (*Cantharis*, etc.), where they are frequently seen in the winter even when snow has fallen; hence their popular name "snow-worms". Predaceous on worms, larvae, etc.

LAMPYRIDAE (Fig. 20.)

Similar to cantharids, but flattened and leathery; head covered dorsally by enlarged pronotum, with dorsal sutures distinct, and sides not meeting each other ventrally. Apex of abdomen with a brush-like tuft of hairs.

Habitat: On grass, pathways, etc., where they feed nocturnally on snails. The slightly luminous larva of *Lampyris noctiluca* (Linn.) closely resembles the adult female in general form and colour.

DERMESTIDAE (Fig. 20.)

Short larvae with a rather humped pronotum; body densely covered with usually long (and sometimes branched or barbed) hairs. Cerci small (*Dermestes*) or absent altogether.

Habitat: In dried carcases, skins, woollens, stored products, etc. Larvae of *Dermestes* species mostly feed in dried carcases, except *D. lardarius* Linn., which, as its name implies, is associated with foodstuffs. Old blankets are frequently reported to be infested with larvae of *Attagenus pellio* (Linn.). Larvae of *Megatoma* and *Ctesias* both occur under bark, the latter feeding on particles of insects ensnared in old webs and cocoons of spiders. The brown, hairy larvae of *Anthrenus museorum* (Linn.) are unfortunately too well known to the collector who becomes careless in the maintenance of his collection.

ANOBIIDAE (Fig. 20.)

Ventrally curved, C–shaped larvae, covered with minute spines. With a small head capsule bearing ventral mouthparts; thoracic legs present (unlike superficially-similar weevil larvae). Ocelli absent.

Habitat: In the hard wood of trees, old posts, etc. For example *Anobium punctatum* De G. in rotten oak boughs (as well as in the timber structure of buildings, where it and *Xestobium rufavillosum* De G. do serious damage); *Grynobius excavatus* (Kug.) in holly, birch,

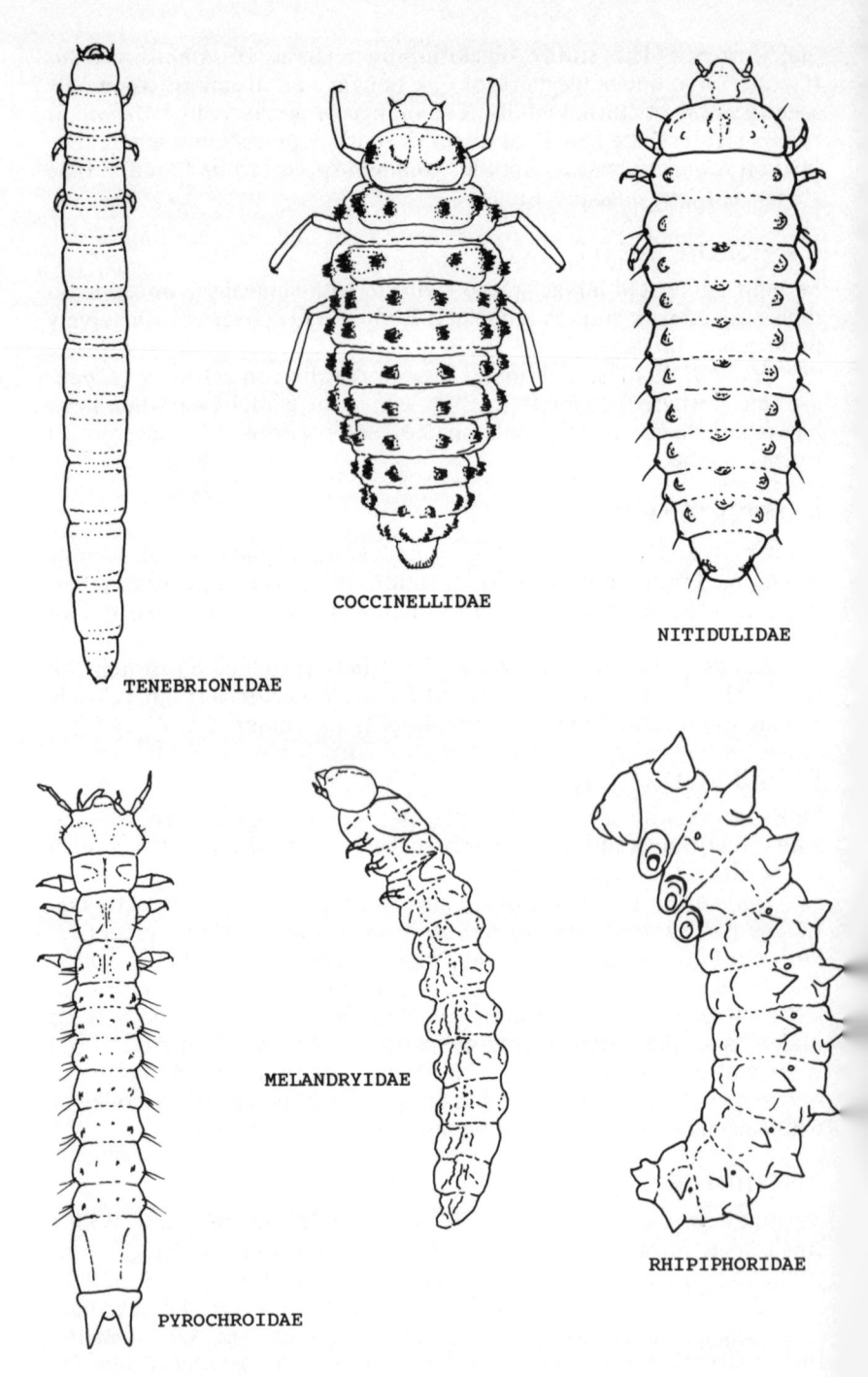

Fig. 21

hawthorn; *Ptilinus pectinicornis* (Linn.) in oak boles; *Xestobium rufovillosum* (De G.) in rotten oak planks and posts in fields. Larvae of *Caenocara bovistae* (Hoffm.) feed in the spongy bases of puffballs.

CLERIDAE (Fig. 20.)

Superficially resembling cantharid larvae, covered with rather longer, finer pubescence and with apical segment of abdomen forming a flat plate which bears two short horn-like cerci.

Habitat: Under bark, in carcases, etc. The pink larvae of *Thanasimus* are predaceous on Scolytid larvae in the bark of pines. *Necrobia* larvae are to be found in old, dry carcases.

NITIDULIDAE (Fig. 21.)

Moderately elongate, somewhat depressed larvae, sometimes with thoracic or abdominal tubercles; cerci short or absent.

Habitat: Under bark (at sap), on flower-heads, in carcases, etc. Larvae of *Nitidula* and *Omosita* often occur in old dry carcases. Larvae of *Rhizophagus* and *Pityophagus* inhabit old borings of Scolytids. Those of *Glischrochilus* feed subcortically in freshly felled oaks. *Cossus* borings often contain numbers of *Omosita* and *Soronia* larvae. Larvae of *Cychramus luteus* (Fair.) live in the spongy bases of puffballs. Those of *Epuraea aestiva* (Linn.) are to be found in nests of *Bombus*. Larvae of *Meligethes aeneus* (F.) and *M. viridescens* can commonly be found in the buds and flowers of Cruciferae, and are pests of oilseed rape.

COCCINELLIDAE (Fig. 21.)

Short but active larvae; body surface usually bearing dorsal setiferous tubercles or spines; often brown or bluish-black, with yellow markings. Legs long, with small apical claw: no cerci.

Habitat: On plants and trees, *e.g.*, heather, thistles, fir, sallow, etc.; mostly predaceous on aphids but a few feeding on mildews, etc. Many are easily reared, but must be provided with ample food if cannibalism is to be avoided.

TENEBRIONIDAE (Fig. 21.)

Elongate, cylindrical larvae, often yellowish-brown and resembling larvae of Elateridae; but the labrum is free (not fused to clypeus) and not toothed along its front margin. Apex of abdomen either rounded or with short prong-like cerci.

Habitat: In damp cellars (*Blaps*) where they feed on rodent droppings, etc.; beneath bark of elm, fir, etc., where they are predaceous on other larvae; at roots of grass and heather (*Cylindronotus*); in granaries (*Palorus, Tribolium, Tenebrio*, etc.); in fungi (*Eledona*, etc.).

PYROCHROIDAE (Fig. 21.)

Parallel-sided and flattened larvae, well-sclerotised; labrum distinct,

MELOIDAE

CERAMBYCIDAE

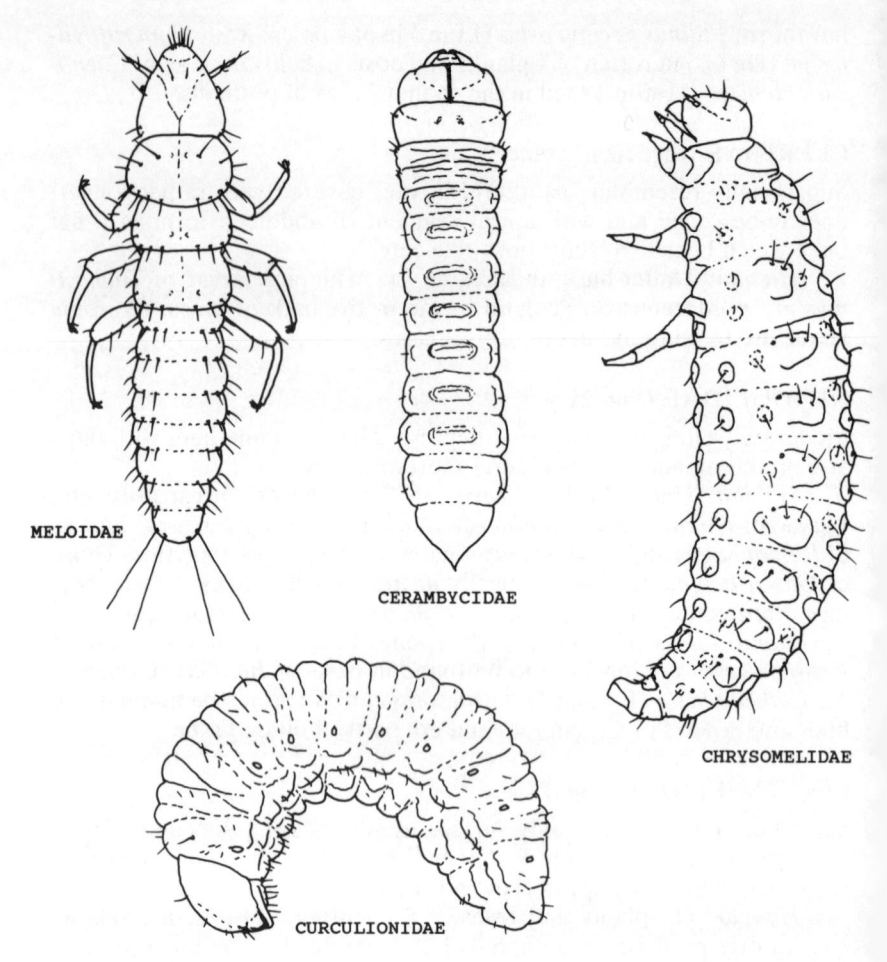

CHRYSOMELIDAE

CURCULIONIDAE

Fig. 22

The figures of whole larvae (Figs. 17–22.) have been drawn specially from preserved specimens wherever possible. In a number of Families, however, larvae were not available, and some existing figures have therefore been re-drawn for this Handbook. These have been based mainly on the classical works of Boving & Craighead (1931, reprinted 1953) "An illustrated synopsis of the principal larval forms of the Order Coleoptera"; Ghilarov (ed. (1964) "Keys for the identification of soil-inhabiting insect larvae" (in Russian); Klausnitzer (ed.) (1978) "Order Coleoptera (Larvae)" (in German). The figure of a weevil larvae is modified from Crowson (1981) "The biology of the Coleoptera", and that of a buprestid from Bily (1982) "The Buprestidae of Fennoscandia and Denmark". The drawings do not always represent any particular species, but have been modified so as to show "typical" features of the family, sometimes based on more than one species.

mouthparts directed forwards. End of abdomen elongated, with a pair of parallel backwardly-directed prongs.

Habitat: Under bark. Larvae of *Pyrochroa coccinea* (Linn.) feed in rotten oaks and those of *P. serraticornis* (Scop.) are to be found in various trees, particularly elm.

MELANDRYIDAE (Fig. 21.)

Elongate but rather fat and fleshy larvae, with large head and thorax; similar to Cerambycidae, but with well-developed legs and more deeply retracted mouthparts.

Habitat: In rotten wood and woody fungi. Larvae of *Conopalpus testaceus* (Ol.) and *Orchesia undulata* Kr. live in rotten, fallen oak boughs, but the pink larvae of *O. micans* Panz. infest the woody bracket fungus of oak and ash. Another fungus-feeding larva is the rare *Hallomenus binotatus* Quens. in the "beef-steak" fungus of oak. Larvae of *Melandrya caraboides* often occur quite commonly in stumps of oak and birch.

RHIPIPHORIDAE (Fig. 21.)

Soft, fleshy larvae, with reduced or vestigial appendages; dorsal surface with conical projections; legs with a single sucker-like claw. Undergoes hypermetamorphosis; the first stage (triangulin) larvae is campodeiform, with long legs, bristle-like antennae and cerci, and 4 or more ocelli on each side of the head.

Habitat: In cells of wasps' nests, ectoparasitic on larvae of *Vespula*, *Metoecus paradoxus* Linn. is the only British representative of this family. Wasps' nest should be examined in late summer for the curious parasitic larvae; the capped, less opaque cells are the most likely to contain the full-grown larvae. To "take" a nest, one should stuff the entrance as deeply as possible with rags and saturate them with chloroform. This should be done at dusk when most of the wasps are inside, care being taken to drive off any late-comers. The nest should then be dug out early the following morning, and any wasps that show signs of activity killed.

MELOIDAE (Fig. 22.)

Similar to Rhipiphoridae, but without dorsal projections on the body; triangulin stage with at most two ocelli on each side.

Habitat: In nests of Apidae where they feed on eggs, food masses, etc. The minute first-instar larvae, which have three claws on each side of their tarsi, rest on flower-heads until the latter are visited by bees, etc., whereupon they attach themselves to them and are carried away; naturally many fail to be successfully transported to the appropriate nests.

CERAMBYCIDAE (including PRIONIDAE, LAMIIDAE) (Fig. 22.)
Fleshy, sub-cylindrical larvae; prothorax broader than remainder of the body; head retractable into the thorax; legs very short (not protruding beyond sides of body) or absent; sometimes with protruberances on the dorsal surface which assist them in moving along their galleries.

Habitat: In living or decayed wood such as the boles of trees, rotten stumps, fallen boughs, roots, posts, palings and twigs; some are confined to the stems of Umbelliferae, etc. The following list shows the great diversity in their lignicolous habitat: *Prionus coriarius* (Linn.) in roots (occasionally stumps and logs) of oak, birch, beech, pine, etc.; suspected roots should be exposed and carefully split open; *Aromia moschata* (Linn.) in main stems of young sallows; *Criocephalus ferus* Muls. in stumps and roots of pines; the oval emergence holes are much larger than those of the common *Asemum striatum* (Linn.); *Tetropium gabrieli* Weise under bark of logs and stumps of larch and pine; *Callidium violaceum* (Linn.) under bark of pine palings; *Phymatodes testaceus* (Linn.) under bark of felled oak logs; *P. alni* (Linn.) in dead oak twigs; *Clytus arietis* (Linn.) under bark of rotten boughs of birch, oak, etc.; *Molorchus minor* (Linn.) under bark of injured or dead branches of Scots pine and spruce; *M. umbellatarum* (von Schreb.) in dead twigs of crab-apple and dead stems of dog rose; *Rhagium bifasciatum* Fabr. in rotten stumps and logs, especially pine; *R. mordax* (De G.) under bark of oak stumps; *Strangalia maculata* (Poda) in rotten sallow and birch branches; *S. 4-fasciata* (Linn.) in oak and alder stumps; *Leiopus nebulosus* (Linn.) under bark of injured oak branches; *Pogonocherus hispidus* (Linn.) in dead twigs of apple, laurel, etc.; *Mesosa nebulosa* (Fabr.) under bark of oak boughs; *Agapanthia villosoviridescens* (De G.) in stems of thistles and *Heracleum sphondylium* in spring and autumn; *Saperda populnea* (Linn.) forms globular swellings (galls) on stems of sallow, aspen and birch saplings; *Phytoecia cylindrica* (Linn.) in stems of Umbelliferae in early spring.

As the majority of longicorn larvae take two or three years to attain maturity, it is possible to find them at any time of the year. Much useful work may be accomplished during winter months in discovering larval colonies either with a view to collecting larvae for rearing or by taking one or two for possible determination by larval keys, and making a sketch map of the exact situation so that it may be visited in the spring for the collection of full-grown larvae or pupae. It may even be convenient to sleeve part of the host plant to ensure that the adults do not escape. The advantages of this method are, firstly, there is more spare time to make a thorough search as most other aspects of field work will have ceased, and, secondly, the discovery of the characteristic emergence holes is greatly facilitated by the absence of undergrowth and foliage. Colonies of several rare longicorns have been found in this way.

CHRYSOMELIDAE (Fig. 22.)

Fleshy larvae, body either slightly curved ventrally or straight; legs usually short, 3- or 4-segmented (or absent in leaf-mining species); mandibles without mola; cerci absent. They are the most "caterpillar-like" of beetle larvae, some resembling lepidopterous or sawfly larvae, but they do not have ventral abdominal prolegs like these other Orders.

Habitat: Mostly on leaves, often large colonies. The lush semi-aquatic flora provides good hunting-grounds, particularly the water dock for *Galerucella* spp., etc. Leaves of guelder rose are often reduced to a network through the ravages of larvae of *Pyrrhalta viburni* (Payk.). Poplar, sallow, hazel, aspen, broom and bedstraws are favourite hosts. The quaint *Cassida* larvae, which are short, oval, slightly depressed and distinctly spiny with a lateral fringe of processes, construct for themselves an "umbrella" of frass and exuvia. They feed externally on leaves of thistles, goosefoot, water mint, etc. The aquatic *Donacia* larvae, which extract oxygen from submerged plants, are most easily procured in the autumn by pulling up these plants and examining their roots for the plump white larvae which make clusters of brown cocoons. Larvae of *Clytra* may be dug out (in their cases) from nests of *Formica rufa*. Larvae of many flea beetles (Halticinae) are miners in leaves, stems or roots of plants.

CURCULIONIDAE (including APIONIDAE, etc.) and SCOLYTIDAE (Fig. 22.)

Rounded and ventrally-curved apodous larvae with a slightly 4-lobed anal segment. The absence of legs distinguishes them from Anobiidae, etc. No characters are known which distinguish all Curculionidae from Scolytidae.

Habitat: In wood, seeds, roots, stored products, etc. The green larvae of *Phytonomus* are to be swept from *Ononis*, *Trifolium*, *Lotus* and *Rumex*. Leaves of *Scrophularia* and *Verbascum* should be examined for the slug-like larvae of *Cionus* which pupate in fibrous cocoons. Certain species are stem feeders such as *Mecinus* in *Plantago maritima*. Larvae of *Cleonus* should be searched for in thistles during August when brown withered stems will indicate their presence; the roots should be examined for swellings. Hazel nuts are often rendered inedible by the presence of *Balaninus* larvae. Many *Apion* larvae infest seed pods of Leguminosae. Beetles of the genera *Attelabus*, *Apoderus*, and *Rhynchites* roll up the leaves of oak, hazel and birch respectively and deposit their eggs therein; these leaves should be gathered in the autumn before they start to fall. Many species cause considerable damage to the roots of plants and trees; for example, the plump larvae of *Hylobius abietis* (Linn.) in roots of fir, *Sitona* larvae in clover and vetch, and larvae of *Phyllobius calcaratus* (Fabr.) in roots of *Ranunculus*. Larvae of *Cryptorhynchidius lapathi* (Linn.) make large galls in willow stems. Many species feed subcortically in stumps and boles. The uncommon larvae of *Magdalis carbonaria* (Linn.) feed under the bark of birch stumps. Larvae of *Platypus cylindrus* (Fabr.) occur locally in old oak stumps where they make vertical galleries in the heartwood. Scolytid larvae form conspicuous and often intricate patterns beneath the bark of elm (*Scolytus scolytus* (Fabr.)), oak (*S. intricatus* (Ratz.)) and of various fruit trees (*S. mali* (Bech.)), *S. rugulosus* (Ratz.)). Sallow catkins should be gathered in April; from these may be reared larvae of *Dorytomus*, etc.

5. Preservation

Beetle larvae, as compared with those of Lepidoptera, are not particularly colourful and this, together with the fact that the majority are so small, renders them unsuitable for general exhibition purposes. Certain species, however, such as the large Lucanid and chafer larvae and the more sclerotised Elaterid and Pyrochroid larvae, if well preserved, make useful and attractive additions to Coleoptera collections and add to their general interest; moreover, if pupae and cocoons are included, life-history series can be prepared illustrating the great diversity of various groups.

The larger, soft, fleshy larvae can be prepared by the method generally adopted for "blowing" lepidopterous larvae, i.e., by enlarging the anus, pressing out the abdominal contents and inflating with hot air until rigid. If, in addition to this, they are then filled with suitably tinted melted wax and allowed to "set", they present a very life-like appearance. Incidentally, cerambycid larvae, especially *Prionus*, cannot be "blown" satisfactorily, because they turn black in the drying process. The wax-like, semi-transparent quality of the integument is best preserved by adopting the following procedure: Place the larva in a test tube containing 30% alcohol (to which one or two drops of acetic acid have been added) and heat at just below boiling point for twenty to thirty minutes; then store in 80% alcohol.

Another satisfactory method, most suited to the smaller, more sclerotised larvae, is as follows: Immerse the larvae for one week (or two, if very large) in 95% alcohol. They should then be dehydrated by immersion for one to three weeks (according to size) in absolute alcohol which should be changed weekly. This should be followed by immersion successively for twelve hours in mixtures of one part xylol to two parts absolute alcohol, two parts xylol to one part absolute alcohol and pure xylol respectively. Finally they should be allowed to dry off and then be mounted in the same way as adults.

The foregoing methods are, of course, unsuitable if one wishes to form an extensive collection of larvae with a view to studying their taxonomy. For this each batch of larvae should be killed, preferably by immersion in a standard fixative such as Carnoy's fluid, made up from:

Glacial acetic acid	10 parts
95% alcohol (industrial)	60 parts
Chloroform	30 parts

Details of this and many other useful preservation techniques are given in the book by Wagstaffe & Fidler (1955, reprinted 1961), "The preservation of natural history specimens, I, Invertebrates". If a fixative is not available, kill larvae in 70% alcohol, to which a few drops of ethyl acetate have been added. They should then be stored in 70% alcohol; the addition of 5% glycerol prevents the liquid from drying out, and keeps the larvae softer.

Larvae should be stored in individual small glass or polythene tubes, corked with plastic stoppers, or plugged with cotton wool. In the latter case they should then be kept inverted in a well-sealed bottle (such as a Kilner jar) which is itself filled with 70% alcohol, and checked regularly to prevent its drying out. If alcohol-preserved larvae do dry out and shrivel up, they can often be restored by immersion for up to 24 hours in cold 5% potassium hydroxide solution (followed by washing in water prior to re-storage in 70% alcohol).

Data, biological notes and any observations made during collecting larvae should be carefully recorded in a field notebook. An example might be:—

Jar 2   *Ilybius fenestratus* (Fabr.)   DYTISCIDAE
Tube 18   Surrey: Mytchett. 22.iii.1989
           Dug up under grass, near lake; 2 specimens preserved.
Tube 274   Surrey: Ash Vale. 27.ii.1990
           Dug up from banks of canal; 2 specimens and exuvia preserved; 2 being reared (T.O.).

And on the other side of the page, observations thus:

Beaker 274   Larvae kept in damp soil in beaker; food, small worms. 20.iv, two larvae became inert and slightly curled having pressed out pupal cells. 26.iv, both pupated; exuvia preserved; soil moistened every third day. 11–12.v, adults emerged; took two days to become pigmented. Larvae apparently pupate several yards away from water. (And so forth.)

All larvae should be labelled, using labels written either in indian ink, or with one of the readily available disposable micropoint nylon-tiped drawing pens. But be sure to use one that is waterproof, and test the ink for alcohol-fastness before using it for labels. Each tube should have a label giving date as in the notebook referred to above, together with a unique number which relates the tube (and specimens in it) to that book.

For the serious study of larvae, their identification requires examination of cleared and mounted specimens under the high magnification of a compound microscope, using transmitted light. There are several standard methods available of making such slide preparations of larvae. One of the quickest is as follows:

Heat larvae gently on a hot plate in 5 or 10% potassium hydroxide, until the internal tissues have dissolved. Wash them in water, and then warm them (on a cooler hotplate) in chloralphenol (a solution of chloral hydrate in phenol). This renders them transparent; they can then either be examined on a slide in the same liquid, or mounted in Berlese's mountant, made from–

Distilled water	20ml
Chloral hydrate	150g
Gum arabic	15g
Glucose syrup	10ml
Acetic acid	5ml

This and similar mounting media are also available from commercial entomological or biological suppliers. It has the advantage of being reversible, that is the cover slip and larvae can be soaked off the slide by immersion in water, if needed at a later date.

REARING LARVAE

BY DR. M.L. LUFF

Introduction

Whereas a Lepidopterist may spend as much time searching for the immature as for the mature stages of butterflies and moths, the collector of beetles, rarely carries out similar methods of collecting, for the following reasons:
1. The feeding habits of beetle larvae are very varied, some being carnivorous, actively seeking living prey; some feed upon carrion or upon dung; many burrow in the soil or in the solid wood of trees, and others are parasites or inquilines in the nests of ants and other creatures.
2. Many beetles require several years in which to complete their metamorphoses.
3. In general, the larvae of beetles are unattractive aesthetically, devoid of the beauty of form and coloration shown so conspicuously by the caterpillars and, frequently, by the pupae of Lepidoptera.
4. Much more care, attention and elaboration of apparatus (and, consequently much more time) is, in general, required to rear beetles successfully.

Nevertheless, there is a greater feeling of self-satisfaction in bringing successfully to maturity some interesting beetle than in the rearing butterflies and moths. It happens, too, not infrequently, that the coleopterist in search of a definite species finds on searching its known habitat that he is too early, the season perhaps being late and that it is represented by fully, or almost fully, grown larvae. It is pity to neglect these, for from them it may be possible with due care to rear the beetles. Many Coleoptera have phytophagous larvae, to rear which, even from the egg-stage, is almost as simple as rearing caterpillars. While the careful and detailed working-out of the life histories of beetle species is rather the province of the specialist or research student and requires incessant application and much time, the conscientious working coleopterist may still add not a little to the knowledge of the bionomics of beetles.

The whole essence of rearing larvae is to reproduce as accurately as possible the actual conditions under which they normally exist. The chief drawback to rearing them is the duration of the larval period, for the majority take at least a year to attain maturity, while some take three or more years.

Such annual or extended life-cycles are usually synchronised with the time of year by an obligatory resting phase or "diapause" at some stage in the development of the beetle. This may be in the egg (rarely), larva, pupa or (often) immature adult stage of the life cycle. As the timing of this phase is often linked to natural changes in outdoor environmental features such as day length and temperature, it is important to subject larvae being reared to natural conditions wherever possible, rather than keeping them indoors "in the warm" to hasten their development.

A further very important point to observe is the exact nature of the food. This is not always at first apparent; for instance, a larva might be found under bark, but this does not necessarily mean that it is lignivorous; it may be predaceous on other larvae or feed on moulds, hyphae of fungi, dead insects, or even spider webs. Hence, until the beginner has "become better acquainted with their habits", it is advisable to collect a reasonably large supply of the wood, fungus, or whatever the larvae are found in, so that their exact feeding habits may be observed in captivity. The more experienced the collector becomes, the less this precaution will prove necessary. Another equally important point to observe is the actual *condition* of the food, especially with regard to its degree of moisture and decay. Generally speaking, it is better to have the food too moist rather than too dry; moreover, it is fatal to give larvae dry rotten wood when they have been living in sound wood. Breeding experiments have proved that the successful development of lignivorous larvae depends to a greater extent on the nature and condition of the wood (i.e., the thickness of bark, presence of sap, degree of moisture, stage of deterioration, etc.) than on the species of tree from which they were obtained, since the majority of these larvae are undoubtedly polyphagous.

Two fatal conditions in rearing are drought and mould, and continuous inspection is necessary if these are to be avoided. Food should be moistened and examined for signs of mould periodically. The maintenance of good ventilation is the surest way of inhibiting mildews; non-porous rearing containers should never be exposed to sunlight as this causes excessive condensation. The growth of mould in soil may be inhibited by occasionally sprinkling the surface with water containing 5 parts to the hundred of common salt.

No special breeding cages or other apparatus are required for general rearing purposes, because they can all be improvised from jam-jars, flower-pots, etc.; but a few dozen glass-topped tins roughly 3" in diameter and $1\frac{1}{2}$" deep are extremely useful: the bottoms should be covered with pieces of damp blotting paper to maintain a humid atmosphere or to absorb excess moisture as the case may be.

One can also use plastic pots or beakers, covered by cling-film which is perforated in a few places to aid ventilation. As an alternative to blotting paper, the base can be filled to a depth of 1cm with plaster of Paris, which is then moistened with water. The remainder of this Chapter will outline rearing techniques which can be tried for larvae living in a range of habitats. Remember, however, that there is no sure guide to successful rearing of any beetle species. Be prepared to experiment, and adjust the basic methods to each individual species. The satisfaction of rearing a difficult species will be all the greater if you have had to devise your own scheme in order to be succeed. If rearing of a rare species is achieved, consider returning the adult(s) to their natural habitat, so that the species is conserved. But be wary about releasing reared species away from their normal environments, as outlined in the "Code for Insect Introductions", published by the Joint Committee for Conservation of British Insects.

During the course of rearing, a number of larvae may be found to be parasitised: do not throw them away. Preserve some of the parasite larvae and place the host and its contents in a separate tin. The resulting imagines, together with their larvae and full data, should be sent to an authority on the group. Such material is badly needed.

Finally every rearing cage or container should bear a label with the number of the species as recorded in the catalogue; self-adhesive labels are suitable for this purpose.

Terrestrial predatory larvae

This description applies to many species of Carabidae and Staphylinidae, as well as Cantharidae and Lampyridae and most Cocinellidae. Larvae may be obtained either by collecting in the field, or by extraction from the substrate in a Tullgren funnel or similar device. If gravid female beetles are found (recognisable by their distended abdomens), they may lay eggs if kept on a moist medium such as moist sieved soil or peat (as supplied by garden centres for use as a seedling compost). When any larvae hatch, they can be treated similarly to those collected in the field.

The problem in rearing such larvae is that they must be kept individually in order to avoid cannibalism. Each can be maintained in a small container with a moistened absorbent base (blotting paper, plaster of Paris, soil or peat). The addition of a further layer or piece of moss provides convenient shelter for the larva. Food, either live prey such as Collembola, small dipterous larvae, or chopped-up pieces of larger food such as mealworm larvae, earthworms, etc. should be added regularly. Be careful, however, to remove any dead, uneaten food before it decomposes. Some larvae will eat unnatural food such as small pieces of raw or cooked meat. It is always preferable, however, to find an apparently "natural" food if possible. Cool and moist conditions must be ensured, and the food changed daily. Lampyrid (glow-worm) larvae feed on snails such as *Helix* and *Limnaea* species, and will display their light if gently handled. They pupate, like many terrestrial larvae, among moss and leaf litter on, or near the surface of the soil. Cocinellids may also be reared as above, fed with leaves bearing colonies of aphids. Detailed advice on rearing ladybirds is given in "Ladybirds" (Naturalists' Handbooks, No. 10) by Majerus & Kearns, 1989.

Larvae of tiger beetles (*Cicindela*) can be found in vertical burrows on dry, exposed sandy or peaty soil. In order to rear the larvae, blocks of soil about 25cm deep and 15cm square, with a burrow in the centre should be dug up, and held together by wrapping in polythene bags. Place the soil in a south facing, sunny position, and feed the larvae with small caterpillars or similar food, held in forceps. Keep the soil surface slightly dampened. When the larvae pupate in the burrows, cover the surface with polythene or glass to prevent drying out.

Aquatic predatory larvae

This includes mainly larvae of Dytiscidae, but also other Adephaga such as Gyrinidae and Hygrobiidae, and the aquatic species of Hydrophilidae.

Dytiscid larvae are best obtained in the late summer and reared by placing them individually – as they are cannibalistic – in a shallow glass or earthenware dish almost filled with water. It is advisable to use a quantity of the water from which they were originally taken as this would contain plenty of minute crustacea to support them during their early stages. Some small aquatic plants should be thrown in to provide anchorage for the larva and to enable it to crawl out on to *terra firma* (a piece of sheet cork is suitable for the latter purpose). The dish and its contents should then be lowered into a larger tank (preferably of glass so that the metamorphoses may be followed more closely), the bottom of which has been covered with about 1" of moist loam and has a steep bank of loam at one end (Fig. 23A). Some Dytiscid larvae, particularly those of *Dytiscus*, pupate more readily when confronted by a vertical "wall". Larvae will feed on freshwater crustacea, worms and even on small tadpoles. Always procure and preserve the larval exuvium as soon as the pupa is revealed. When it is desired to rear several aquatic species for general interest and observation a shallow pool should be constructed (see Macan, 1982, *The Amateur Entomologist* No. 17). Straight-sided glass beakers have proved quite suitable for rearing species up to the size of *Ilybius ater* De Geer. These are provided with a basal depth of soil, fine silt, gravel or mud from the bed of a pond. Do not let this mud be much deeper than 2cm and plant in it a sprig of some small aquatic plant such as *Ranunculus fluitans, Luronium* (*Alisma*) *natans* or *Elodea*, and also a small fragment of some aquatic moss. Care must be taken to avoid too much vegetation or it will be very difficult to keep the life history under observation. After the plants have been introduced, the beaker is filled with water from a pond up to within a half-inch of the top and is covered with a square of glass or piece of cling film held down with a rubber band. The beaker is then set aside for at least a week for the silt to settle and for the plant to take root. Pondwater should always be used since it introduces an abundance of small Crustacea and other suitable food for the young larvae.

It is possible, by varying the plant introduced into the beaker and by the type of pond from which the water is taken, to vary the conditions from acid (Sphagnum bog) to alkaline (chalk pond), consequently widening the range of species it is possible to rear.

With these small aquaria two separate courses are open. Either they may be used exclusively for breeding *ex ovo*; or larvae taken in the field may be reared. In the former case adult beetles, one species only to each beaker, are introduced and experience has shown that it is best to

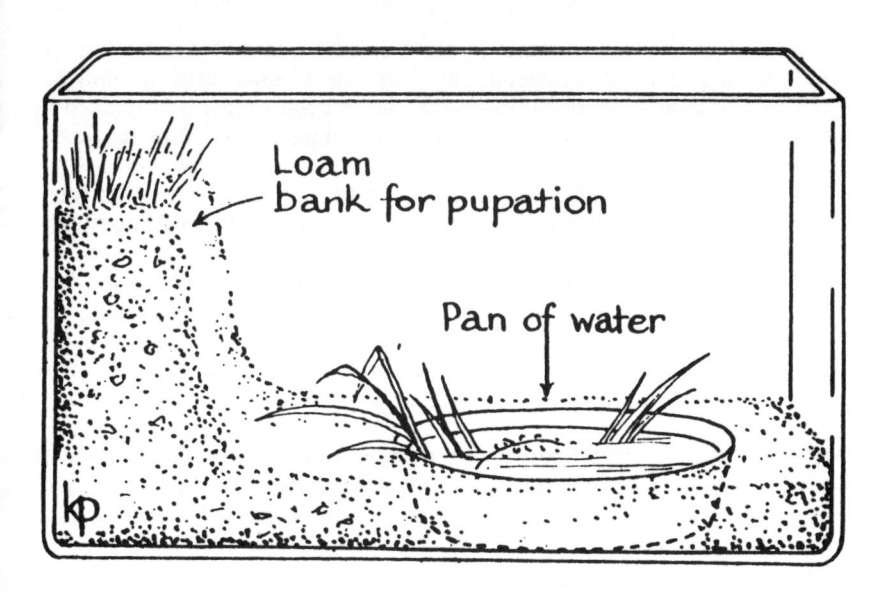

Fig. 23A. An Aquarium Terrarium

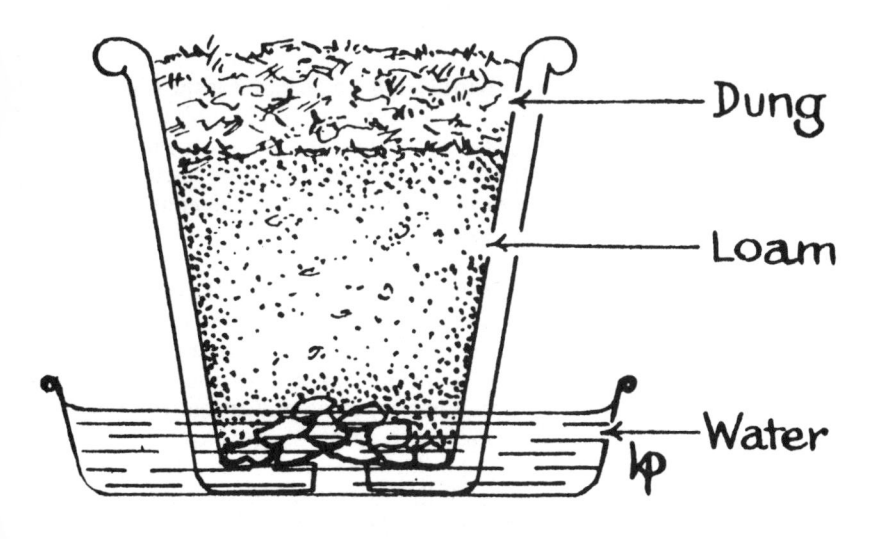

Fig. 23B. A pot for dung beetle larvae

use two females to each male to allow selection of mate. It is advisable to note the number of specimens introduced, since after a sufficient number of eggs have been observed, the imagines must be removed or they will devour either the eggs or the young larvae. The particular season at which a breeding aquarium is begun is of no importance, since it is possible to keep the imagines alive for up to two years under these conditions and they will breed in their due season. All that is required is that the adults be fed at intervals: the larvae of *Chironomus* spp. are excellent for this purpose. For a beaker with three specimens of an *Agabus* fifteen to twenty *Chironomus* larvae should be an adequate supply for perhaps ten days. Very little is known of the egg-laying habits of the carnivorous water beetles and it is always necessary to make regular observations to ascertain the periods and sites of oviposition. The eggs are laid either at random on the plants, frequently in the axils of the leaves, or in the mud at the bottom, either just on the surface or even about half an inch below it. Once oviposition has been observed, constant watch must be kept until sufficient eggs are judged to have been laid: the adults are then removed. Emergence usually takes place in five to ten days, but it has been recorded that eggs of *Agabus chalconatus* form *melanocornis* Zimm. have overwintered, emergence taking place more then 240 days after laying. The young larvae will usually be satisfied with Entomostraca and Cladocera and this will suffice to rear nearly all the Hydroporines to full size, more food being introduced from cultures as may be required. Cannibalism among the larvae is of very varying prevalence and only experience will show the optimum number of larvae that can safely be allowed. It is worth while to allow for more eggs being laid than this optimum number as cannibalism will make use of the excess as valuable food supply. With Agabines and Colymbetines feeding with *Chironomus* larvae becomes necessary when the larvae increase in size. Up to a limit, the more food offered the more rapid is growth.

All aquatic Adephaga pass through three larval instars and pupation takes place out of the water. With the technique outlined above, it is necessary to remove the larvae from the aquaria to terraria for pupation. When they are ready for this transfer they become restless and will be noticed trying to escape from the water. They are then plump and the fat bodies may be seen to be well developed.

For a terrarium one can use either a smaller beaker or fairly shallow glass bowl. Ordinary soil is fairly tightly pressed into it. Divide the bowls into about four compartments, either with glass plates or stiff board, and place one larva in each compartment. The dividing plates must reach right up to the covering glass or the larvae will wander around and this may lead to confusion and mistakes in identification when dealing with field-caught material, unless care is taken to ensure only pure cultures in both aquaria and pupation terraria. It is best to have a few bits of dead grass or moss lying on the surface of the soil in the terraria, for some species seem to prefer to build their pupal cells on the surface, using the dead grass as rafters in the mud roof. Others will excavate a fairly deep burrow before enlarging the end to form a pupal cell.

It is worthwhile to mention that certain species appear normally to breed during the winter months, so that it is possible to maintain the interest during the whole year.

With species of a larger size, such as *Hydaticus, Acilius* and *Dytiscus*, breeding from the egg requires larger containers to serve as aquaria. Rather stronger plants will be required, since these genera oviposit within the stem tissues.

Rearing larvae captured in the field may all be done in beaker aquaria. even with the largest *Dytiscus*. As when rearing *ex ovo*, care must be taken that the individuals placed in tumblers are a pure "culture" or otherwise serious mistakes in identification of larvae will be inevitable. For this reason this method is not so satisfactory in the results achieved, since without a lot of experience it is very difficult to compare and identify an active living larva with preserved material. In genera such as *Hydroporus* the specific characters are so fine that it is often very difficult to be sure of the specific identity or of the difference between two preserved specimens, let alone a living and a dead specimen.

Adult Gyrinidae are more difficult to keep alive in aquaria for any lenth of time, but gravid females may be kept long enough for them to lay their eggs. Larvae can then be reared as with Dytiscidae; when fully grown the third instar becomes dark brown, and needs to climb out of the water onto a vertical surface on which to pupate. Either stones at the edge of the aquarium, or stems such as those of *Phragmites*, sticking vertically out of the water, will suffice. The larvae pupate in cocoons formed from debris in the water, and adults will emerge after a few weeks.

Adult Hydrophilidae such as *Hydrochara* and *Hydrophilus* require the same technique as is used for the larger Dytiscids; but they are very much more cannibalistic. *Helochares, Enochrus, Cymbiodyta* and *Hydnobius* may be fairly satisfactorily reared in the beaker aquaria; but it has been found that there is a high rate of cannibalism among them and that their restlessness brings them rapidly into dangerous contact with one another.

Many of the remaining genera of the Palpicornia have the larvae either terrestrial or semi-aquatic; only a few are truly aquatic. For such genera as *Limnebius, Ochthebius, Helophorus, Hydrochus, Anacaena, Paracymus*, aquaria-terraria provide the best means of rearing them successfully. Bowls may be filled with soil from the edge of a pond, shaped in such a way as to leave a depression occupying one-third or half of the bowl. The depression is filled with water to a varying depth, depending on the genus intended to be reared. It may be from one or two millimetres up to two or three centimetres. The "bank" portion is provided with small plants and a sprig of moss or another small aquatic plant and a young plant of *Glyceria* is planted in the "pool". As always, the aquarium-terrarium must be allowed to settle down for a few days. Adult beetles may then be introduced. The position in which the egg-masses are deposited varies enormously in this family. Some place the cases direct in the water, others partly in the water and others in or on the drier portions of the bank. Both *Limnebius* and *Ochthebius* are, like *Hydraena*, feeders on Confervae and the larvae are as easily drowned as are most *Hydraena* larvae. The aquaria-terraria intended for them are therefore nearly without water – wet mud suits them very well. Most of the other genera mentioned above prefer a wetter situation. Feeding is occasionally required and very small *Lumbricus*, freshly killed, are quite suitable. Cannibalism is advantageous in that it allows one to avoid some of the tedious time spend "feeding" the aquaria-terraria.

According to Angus (1973, *Trans. R. Ent. Soc. Lond.*, **125**: pp. 1–26), larvae of *Helophorus* may be reared on damp blotting paper in pill boxes, in a similar way to some Carabids, but ensuring that the paper is saturated rather than flooded. They should be fed on *Tubifex* worms; when fully grown they are provided with a bank of earth in which to pupate.

Aquatic phytophagous larvae

This includes most Haliplidae, Dryopidae, Elmidae and probably Scirtidae. They are best reared in aquaria-terraria as suggested for the semi-aquatic Hydrophilidae above, but with plant material and fresh-water algae as food. Some, such as *Hydraena*, will survive on encrusting algae found on decaying leaves at the water's edge. The water should be changed regularly, and replaced with water similar to that from where they were collected. An alternative to soil at the edge of the container, is to use a slightly tilted box or tank, and to wedge a few layers of bark in at the shallow end. This provides a damp, but not submerged, pupation site.

A further, interesting group of plant feeding but aquatic larvae, are those of the chrysomelid genera *Donacia* and *Plateumaris*. These may be collected from late summer until mid winter, although the latter season involves dipping one's hands into icy-cold water and the handling of cold slimy mud. By means of a strong hook at the end of a stout handle, or by the use of a grappling iron (if the water is deep) or by the use of the hands alone where it is sufficiently shallow, masses of tangled rhizomes and roots of *Typha, Nuphar, Potamogeton, Phragmites*, etc., may be torn up. Portions should be washed carefully and not too violently in the water and searched over. If the locality is a favourable one, glabrous brown cocoons, oval in form, will probably be found attached by one end to the sides or under surfaces of the rhizomes or occasionally to the roots growing from these, together with small fat white larvae attached by a posterior pair of hollow respiratory spines to the epidermis and cortex of the same plant structures. The cocoons will be found to contain pupae or newly-emerged adult beetles. During the summer and autumn, larvae of *D. vulgaris* and *P. sericea* may be collected easily and if the portions of rhizome are placed in water in shallow receptacles, the mode of feeding adopted by the larva and the making of the cocoon can be observed. The period of pupal life is short and within a period of two or three weeks, adult beetles, prepared for hibernation, can be extracted from the cocoons. In coastal areas *Plateumaris braccata* (Scop.) should be sought for in the rhizomes of *Phragmites*.

Reference

Macan, T.T. 1982. The Study of Stoneflies, Mayflies and Caddis Flies. *The Amateur Entomologist* No. 17.

Dung-inhabiting larvae

Dung-feeding larvae are by no means easy to rear, as the dung is very susceptible to mould, especially when the earth beneath it is damp. If larvae are reared by the following method, however, this complication should be overcome. Fill a medium-sized flower-pot to two-thirds with damp loamy soil and press it well down (Fig. 23B). Next fill the pot up to just below the rim with fresh moist dung and introduce the larvae, making small holes for them if necessary. Then stand the pot in a shallow dish of water and place it in a well-ventilated situation such as a shady corner of the garden. Do not moisten the dung from above. Some dung-living larvae (e.g., many Histeridae, Staphylinidae, etc.) are predaceous on other larvae (especially of Diptera), so the dung should be well stocked with the latter before the former are introduced.

Soil-inhabiting root feeders

This includes many larvae of Elateridae, as well as some chafers (Scarabaeidae) and Byrrhidae, Dascillidae, etc.

Scarabaeid larvae which feed on fibrous roots should be reared in jam-jars filled with earth and root fibres or, better still, in a flower-pot containing a living plant.

An alternative, intensive method which has been used successfully to study the development of such larvae individually, is to make a plaster of Paris block, 20cm square by 5cm deep. The surface of the block is then drilled out with a cork borer or similar tool, making a series of chambers, about 2cm diameter by 1cm deep, in the surface of the block. Larvae are kept separately in each chamber; the whole block is kept moist by standing in a tray of water, and covering the block with a glass sheet. Each larva is fed daily.

Larvae of Elateridae (click beetles) are found in a wide range of soil types and conditions. The breeder must stimulate very different kinds of natural habitats if he is to be successful with click beetles. Since so many species turn cannibal on the least provocation, it is important to keep individuals apart. Corked vials, with a small runnel cut down one side of the cork to allow aeration, filled to at least an inch below the cork with soil or wood, are very suitable. Small jam jars, tins with a few holes in the lid, or any other closed receptacle will do equally well. The wood or soil should be damp but not wet. Larvae will eat their way through a cork or escape through the nick in the cork if they can at all reach it – and they can achieve the seemingly impossible in this respect! For food, living turf or roots of the plants beneath which the larvae were found may be supplied to the soil feeders, or grains of corn may be substituted for nearly all species. This is particularly useful when breeding in vials; each time the soil is changed, fresh grains are provided; it is then simple to see if the larvae have been feeding since the last meal. Mouldy grains must be replaced quickly. In spring, the larvae of many species feed at a great pace, and one larva may consume a dozen sprouting grains in a few days; if more are not supplied, the breeder is asking for trouble in the way of bored-cork escapes. Some species, like *Hypnoidus quadripustulatus* (Fabr.), appear to have very short feeding periods, whereas others, like *Agriotes* spp., feed during many months of the year. *Corymbites cupreus* feeds strongly in spring and summer, but rarely in autumn and winter; *Athous campyloides*, on the other hand, feeds nearly as much in autumn and winter (if mild) as during the other seasons.

Terrestrial phytophagous larvae

This accounts for most Chrysomelidae and Curculionidae. Surface-feeding larvae, found on the leaves of their host plant, include many Chrysomelids (hence their name "leaf-beetles"), but few weevils except the genus *Cionus*, whose larvae are covered with a glutinous layer of slime. Rearing such species seldom presents any difficulty as their larval period is comparatively short. They develop satisfactorily in a glass-topped tin containing a layer of damp blotting paper and a sprig of their food plant, which should be replenished at the first signs of wilting. Sometimes it will be found convenient to introduce a few plants into the garden so that a regular supply of the right foliage may be at hand. Alternatively, the host plant may be kept with its stem in water, but care must be taken to prevent larvae leaving the plant and drowning in the water, especially when they are searching for somewhere to pupate. Rearing phytophagous beetle larvae is in many ways similar to rearing Lepidoptera larvae and methods used by lepidopterists may usefully be copied.

Species whose larvae live internally within plant tissue (most weevils, many Halticinae – flea beetles) are more difficult. Plants thought to be containing larvae can be transplanted into pots in the house and enclosed so as to capture any emergent adults. Leaf-mining species of *Rhynchaenus* can be reared on cut shoots of their host tree stood in water; virtually any beech tree will yield the larval mines of *R. fagi* if young shoots are collected in May. Leaves rolled by Rhynchitine weevils, and seed pods infested by Apionidae, can be collected and maintained until emergence of the adult beetles.

Beetles that breed in flowers, such as the abundant *Meligethes*, and species of *Anthonomus* can be reared similarly by collecting infested plant material.

Wood and bark inhabitants

There is a great variety of beetle habitats associated with wood, from standing trees in varying stages of health, through recently cut logs, and their gradual stages of decay to thoroughly rotted stumps filled with wood mould. The immature beetles associated with these vary accordingly: Cerambycidae, Buprestidae (if lucky), Scolytidae in fresh or relatively-recent dead wood; Lucanidae, some Elateridae and Scarabaeidae (e.g. the Rose Chafer), Melandryidae etc. in well decomposed but dry conditions. The general rule mentioned earlier applies here: always attempt to rear wood-inhabiting larvae in as near natural conditions as is possible. Remember that many larvae in such habitats are predatory, so a variety of prey organisms also need to be present. For sub-cortical species such as Scolytids, if pieces of bark of suitable size are placed with their inner faces clamped towards one another it is usually possible thus to obtain pupae and later the beetles. Owing to the curvature of the strips of bark, it is of course necessary to use some pressure to put the two faces in sufficiently close apposition and therefore the portions used must be of limited size. Desiccation is here as fatal as excess of moisture, the latter condition being conducive to a deadly growth of moulds. Fitting the curving pieces of bark against household fire-logs has not proven successful, but others might try it nevertheless.

In order to rear Cerambycid larvae, a comparatively sound section of the infested branch should be sawn off and brought back with the larvae. Holes of a diameter slightly greater than the maximum breadth of the larva should be drilled longitudinally at one end: they should be of sufficient depth to accommodate individual larvae, which should be inserted head first. When the larvae are collected, it should be observed whether they are feeding in the heart wood or under or in the bark and the holes drilled accordingly. The section should then be inserted to a depth of 2–3" in a tray of moist sand. A day or two later – by which time the larvae will be secure in their burrows – the section should be inverted so that they are feeding head-upwards unless they have been observed to feed head-downwards in the field, as a small number of species do. The wood should be lightly sprayed with slightly saline water periodically, and the sand moistened with a weak aqueous solution of potassium permanganate to prevent the formation of mildew. To prevent the adults subsequently escaping, the wood and the tray should be enclosed in a cylinder of wire gauze or perforated zinc. Alternatively, one may simply keep sections of wood without introducing extra larvae into them, in the hope that larvae already in the wood will complete their development and emerge as adults. Be patient though; drying out of the wood in particular extends the developmental period, and beetles have been known to take years to emerge from cut timber (the maximum noted is 30 years in the case of an American buprestid!).

Larvae such as Lucanidae and Elateridae, living in crumbling, decayed wood, should be placed with sizeable portions of the decayed wood in plastic boxes or similar containers. Spaces between larger fragments of wood should be filled with smaller pieces and wood powder. The lid should be perforated to prevent condensation, but the wood should be kept slightly moistened but not wet.

Detritus and fungal feeders

Very many Staphylinidae, Lathridiidae, Mycetophagidae etc., are found in a range of habitats including fungal fruiting bodies, litter and refuse, birds' nests, straw and compost heaps. A container should be filled with the material from which it is hoped to rear larvae, and treated as for larvae in thoroughly rotten wood (see above). Larvae found by sieving the material can be added if required. If the material is placed in a darkened box, with a small hole on top to which is taped an inverted clear glass or plastic vial, this functions as a crude "emergence trap". The larvae in the material will avoid the light, and remain in the damp conditions of their rearing medium; emergent adults, however, are usually attracted to the light, and will be found crawling around in the vial, trying to escape.

PART V

BIOLOGICAL RECORDING
By Paul T. Harding

Most people acquire knowledge for its own sake, because they are interested, but knowledge should be shared and exchanged because each individual has a slightly different assortment of knowledge to contribute. Biological records are just a very specialised type of knowledge – of the occurrence of species – but one which it is essential should be shared and exchanged. The sources, the co-ordination and the uses of biological recording are described below.

What is a biological record?
At the simplest level a record answers 4 questions:
    What species?
    Where was it found?
    When was it found?
    Who identified it?

Existing record cards produced by the Biological Records Centre (Figs. A, B, C.) aim to record at least this information for modern records. In most cases the "Where was it found?" element is expressed as a locality name and as a grid reference (usually also as a vice-county). The other elements of a record are also capable of expansion.

The sources of biological records
Anyone with some knowledge of a taxonomic group can contribute to biological recording, the main source of records in the British Isles being amateurs (i.e. those who voluntarily study the topic). Other sources are professional scientists in museums, universities and research institutes (many of whom metamorphose into "amateurs" in their spare time), university research students, and conservationists including reserve wardens and survey teams.

Co-ordination of biological recording
Co-ordination is at 3 main levels – local, national and international.

Local – Local recording is rather irregular; some counties have an active local biological records centre, usually based at a museum or at the county trust for nature conservation, a local entomological society or group or a local expert collating records for a county list. In some counties, some or all of these may exist so that the few experienced coleopterists in an area may find themselves in great demand to examine sites and provide lists.

Habitat Hawthorn blossom		Vice-county name E. SUFFOLK
		Date 5 1 9 7 2
Recorder P.T.Harding	Determiner P.T.H.	Compiler P.T.H.
		Altitude — metres/feet
		Source Field ✓ Mus. Lit.
No.	No.	No.

Grid ref: 6 2 3 5 6 5 1 1

VC No.:

LOCALITY: STAVERTON PARK 5555

COLEOPTERA:CERAMBYCIDAE 6455

71801	Prionus coriarius		73101	Trinophylum cribratum
71901	Arhopalus rusticus		73201	Gracilia minuta
71902	tristis		73301	Obrium brunneum
72001	Asemum striatum		73302	cantharinum
72102	Tetropium gabrieli		73401	Nathrius brevipennis
72201	Rhagium bifasciatum		73501	Molorchus minor
72202	inquisitor		73502	umbellatarum
72203	mordax		73601	Aromia moschata
72301	Stenocorus meridianus		73701	Hylotrupes bajulus
72401	Acmaeops collaris		73801	Callidium violaceum
72501	Grammoptera holomelina		73901	Pyrrhidium sanguineum
72502	~~ruficornis~~		74001	Phymatodes alni
72503	ustulata		74002	testaceus
72504	variegata		74101	Clytus arietis
72601	Alosterna tabacicolor		74301	Anaglyptus mysticus
72701	Leptura fulva		74401	Lamia textor
72702	livida		74501	Mesosa nebulosa
72703	rubra		74601	Pogonocherus fasciculatus
72705	sanguinolenta		74602	hispidulus
72706	scutellata		74603	hispidus
72707	sexguttata		74701	Leiopus nebulosus
72801	Judolia cerambyciformis		74801	Acanthocinus aedilis
72802	sexmaculata		74901	Agapanthia villosoviridescens
72902	Strangalia aurulenta		75001	Saperda carcharias
72903	maculata		75002	populnea
72904	melanura		75003	scalaris
72905	nigra		75101	Oberea oculata
72906	quadrifasciata		75201	Stenostola dubia
72907	revestita		75301	Phytoecia cylindrica
			75401	~~Tetrops praeusta~~

Other species and comments

COLEOPTERA: CERAMBYCIDAE 6455

Biological Records Centre September 1981 RA 45

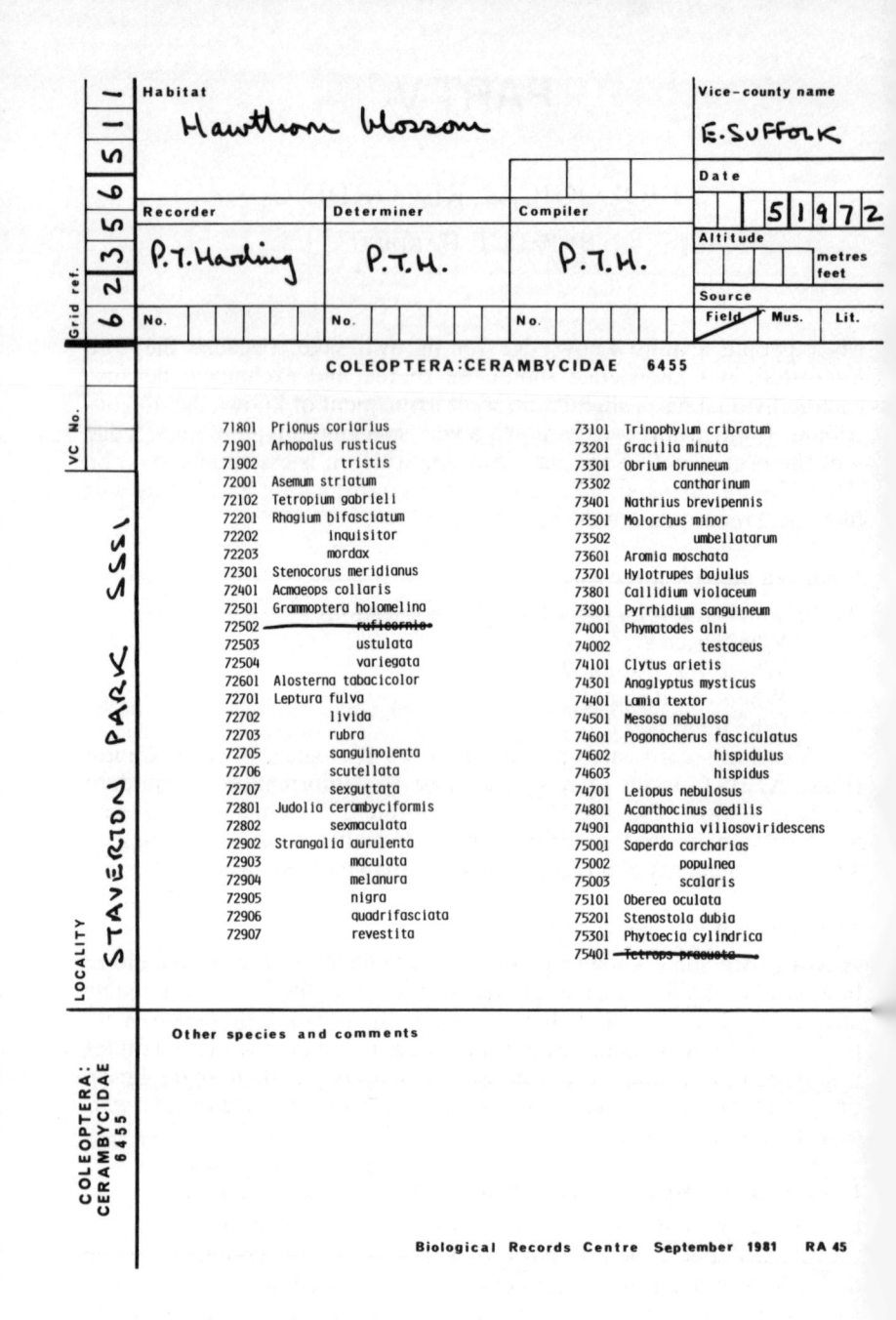

Fig. A. A "species list" or field card for one of the B.R.C. recording schemes.

ORDER COLEOPTERA	GENUS & SPECIES ISCHNOMERA SANGUINICOLLIS	SUB-SPECIES
COMPILER P.T.HARDING	SOURCE (Collection/Reference) P.SKIDMORE & F.A.HUNTER Fld Mus Lit Ent. Mon. Mag. 1980, 116, 129-132	

Grid Reference	V-C	Collector/Recorder	Determiner	Locality	Notes (Habitat, etc)	Date
45/60-82-	62	F.A.HUNTER	F.A.HUNTER	DUNCOMBE PARK		17.6.1979
45/60-82-	62	P.SKIDMORE	P.SKIDMORE	— " —		23.6.1979
45/6 -8-	62	B.A.COOPER	F.A.HUNTER	HELMSLEY		25.6.1946
45/56-84-	62	P.SKIDMORE	P.SKIDMORE	ASHBERRY PASTURES. N.R.		27.6.1971
32/34-42-	36	P.SKIDMORE	P.SKIDMORE	MOCCAS PARK		25.5.1965
			Fig B			

Fig. B. A "one species" card useful for abstracting records from collections or publications, or for entering records when identifying series of specimens collected during a season.

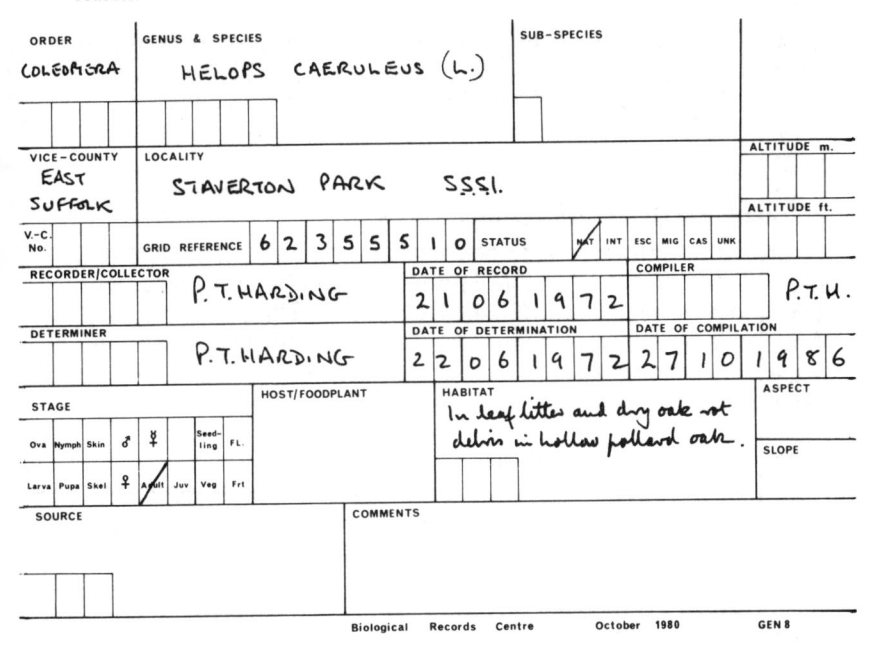

ORDER: COLEOPTERA
GENUS & SPECIES: HELOPS CAERULEUS (L.)
SUB-SPECIES:
VICE-COUNTY: EAST SUFFOLK
LOCALITY: STAVERTON PARK SSSI.
ALTITUDE m.
ALTITUDE ft.
V.-C. No.
GRID REFERENCE: 62 5555 10
STATUS: NAT INT ESC MIG CAS UNK
RECORDER/COLLECTOR: P.T.HARDING
DATE OF RECORD: 21 06 1972
COMPILER: P.T.H.
DETERMINER: P.T.HARDING
DATE OF DETERMINATION: 22 06 1972
DATE OF COMPILATION: 27 10 1986
STAGE: Ova Nymph Skin ♂ ☿ Seedling Fl. Larva Pupa Skel ♀ Adult Juv Veg Frt
HOST/FOODPLANT:
HABITAT: In leaf litter and dry oak-rot debris in hollow pollard oak.
ASPECT:
SLOPE:
SOURCE:
COMMENTS:
Biological Records Centre   October 1980   GEN 8

Fig. C. An individual record card used for rare species and confidential records.

National – The Biological Records Centre (B.R.C.) at Monks Wood was set up to help amateurs and professionals to pool their knowledge of the occurrence of species. The collection of information is organised through various national recording schemes. There is a scheme for most popular groups of Coleoptera e.g. ground beetles, water beetles, ladybirds, longhorns, click beetles. A full list of schemes is available from B.R.C. (address below), together with details of the availability of record cards.

National atlases are produced by B.R.C. from data collected by the national schemes. A full list of atlases is available from B.R.C. However, atlases are not the only use to which data from B.R.C. schemes are put. The B.R.C. data bank is available to conservationists to help evaluate and document important sites and to monitor changes. The data are also available to research workers wishing to analyse data or to study individual species. B.R.C. exercises discretion over the release of data on rare species and confidential records.

International – The co-ordination of international recording requires international funding. Unfortunately funding for such work on invertebrates is very scarce, but despite this a European Invertebrate Survey (E.I.S.) has been set-up. E.I.S. aims to document the occurrence of species in Europe drawing on information from national records centres (such as B.R.C.) wherever they exist. Further details of E.I.S. can be obtained from B.R.C.

Uses of biological recording

The Coleoptera is a large and diverse group of insects. Beetles play an important role in many ecological processes and in almost every habitat. Organised biological recording is one means of making the knowledge of experts more widely available, so that the importance of beetles is recognised.

Information exchange between specialists

Records allow information to be exchanged between individual specialists, but on a longer timescale information can be exchanged between generations through a centrally-held data bank. This role will increase in importance as fewer journals accept papers with basic recording information. By pooling data through recording schemes a more completed picture of the distribution of species can be obtained.

Conservation of species and sites

To be able to conserve species and sites, information on the occurrence of species is essential. There are too few professionals to be able to supply the information, thus conservationists are increasingly dependent on amateur specialists. However, those specialists do not want to spend their time providing lists for a variety of organisations, nor is it a sensible use of their expertise, which is in the field and at the microscope. Better that they submit their records once, to a central organisation designed to store and disseminate information. In many cases the specialist will need to interpret the information once it has been provided.

Research and education

The potential use of biological records is considerable, but again, the availability of the information is better organised through a central organisation to save the specialist additional work.

Summary

All coleopterists collect biological records as well as beetles. There is a demand for the knowledge and expertise of coleopterists, most of which can be answered by their involvement in well organised national and local recording schemes. The contribution which amateur naturalists have made and, it is hoped, will continue to make to the documentation and conservation of our flora and fauna is unrivalled in the rest of the world.

Biological Records Centre
Institute of Terrestrial Ecology
Monks Wood Experimental Station
Abbots Ripton
Huntingdon
Cambs PE17 2LS
Tel: 04873 381

CONSERVATION AND THE COLEOPTERIST
By Dr. Roger S. Key

Introduction – The pace of change

With increased leisure time, mobility and access to the countryside, we as coleopterists are rather more fortunate than our predecessors in that a much wider range of habitats and localities are within easy range for investigation. At the same time, we are, however, much *less* fortunate in that a huge proportion of the natural habitat in that countryside has either been totally destroyed or degraded to the extent that it would be almost unrecognisable to previous generations of entomologists.

It is unncessary to catalogue all of this destruction and degradation of habitat – entomologists will be familiar with the changes they have experienced in their own lifetimes, however long or short those may be. A good example to illustrate what has happened, would be the fate of Britain's lowland woodland, our richest habitat supporting the widest diversity of beetles, during the last half century or so.

It is estimated that 54% of Britain's broadleaved woodland has been *totally cleared* since 1933, and now grows wheat, cows or concrete (N.C.C. 1984). Furthermore, an unknown, but very significant proportion, now supports only introduced conifers, grown in high density plantations, sometimes camouflaged with a deciduous fringe. Much of that which remains has been sadly neglected, particularly so since the last war when the market for coppice timber products dwindled almost to nothing. The all-important open spaces in many woodlands, in the form of rides, glades and young coppice plots, which are so necessary to the adult dispersal phases of so many of our beetles and the early stages of others, have closed in and become dense, dark and inhospitable to insect life.

One result of these dramatic changes in our countryside has been that very many species of beetle have become very rare and difficult to find. Many species have, of course, always been relatively rare, but a grand total of fifty-four appear to have died out altogether since the turn of the century, (Shirt, 1987) and many more have not been seen for a long time and may now also be extinct. Many more also have retreated from much of their former range and are now restricted to those parts of the country where the pace of development has been somewhat less rapid. Others have become rare but remain widely distributed, only occasionally turning up as isolated individuals, at scattered sites throughout their whole range, but now very rarely found in thriving colonies.

Moreover, sites where large numbers of the more interesting species co-exist in natural communities have become very few and far between. We have only to look at the descriptions of forays to classic beetling sites such as Whittlesea Mere in Cambridgeshire, Wheatley Wood in South Yorkshire, Birch Wood in Kent etc. in 19th century publications such as the *Entomologists' Weekly Intelligencier*, *Science Gossip*, the *Zoologist*, etc. to realise what has been lost. All of these were magnificent localities, supporting large numbers of rare species, but all are sadly no longer in existence.

Even where the once-rich sites still exist, today's species lists are often mere ghosts of those recorded in the past, usually because of degradation of the habitat. For example, last century, thirty very rare water beetles were regularly recorded from Askham Bog near York. The site still exists, is a nature reserve and indeed still supports quite a rich insect fauna, but no more than four of those water beetle species appear to have survived into the latter half of this century. The remainder have been lost because of drainage of surrounding land, pollution, drying out and scrubbing over of the remaining habitat (Fitter & Smith, 1979).

With this increasing rarity of individual species, and of good beetle habitat, comes the desirability to take into account the conservation of species and their habitat when we are pursuing our entomological studies. Our interest in beetles can actually become far more rewarding if we try and make our activities contribute towards their conservation and the sites on which we have found them. Also, we should endeavour to ensure that our own activities as coleopterists in no way jeopardise the survival of the species we are studying, or damage their habitat, its flora and the rest of the fauna.

In this chapter I will attempt to outline how the coleopterist can achieve these objectives, at the same time, hopefully enhancing the enjoyment of his or her interest.

Why is conservation of beetles important?

Before addressing these points, however, and attempting to outline how we can conduct our studies with the furtherance of conservation in mind, this question ought first to be answered. It should more realistically be asked: "why is *conservation* important", for it is really impossible to consider the question on the basis of single groups such as "beetles". Arguments as to why conservation is important can be grouped according to three principles – ethical, economic and ecological, which may be very much interlinked.

Ethical issues

This can be argued from two standpoints. First, on behalf of the species under consideration, that it is "wrong" for us to cause the extinction of other species or the destruction of particular ecosystems. This argument potentially has strong philosophical and even religious overtones. It is probably not one that can really be argued scientifically, being, *prima facie*, a subjective judgement. I leave the reader to make up his own mind on the unlikely question of "rights for beetles"!

The other ethical issue, however, is our responsibility to our own species, to those who come after us, and also to those among the current generations who derive either a livelihood, enjoyment, intellectual stimulation or whatever else from the continued existence of species of wildlife and their habitats. We have inherited a biosphere which is already very badly damaged by our antecedents and our everyday activities are continuing to contribute to that destruction. Surely we have a moral duty to pass on to our descendants as much of the natural riches of this planet as is possible, including the maximum number of species still in their habitats, eventually to enable attempts to be made to make good some of our past mistakes.

This applies at all geographic levels, from a global overview ("extinction is forever"), through continent and nationwide considerations, down to the importance of local sites and populations of species for the enjoyment of the local human inhabitants. Perhaps, in this context, people perceive there to be something of the taxonomic hierarchy of importance – more people value attractive birds and nice furry mammals than small obscure invertebrates, pretty butterflies more than beetles, and big pretty beetles more than small black ones. Understanding and concern for whole ecosystems is, however, very much gaining ground and is a far more practical interpretation.

This ethical argument really boils down to – "it is wrong for us to destroy species and their habitat because people now, and in the future, want them to continue to exist".

Ecological issues

Leaving aside such weighty philosophical argument, there are solid ecological and even economic reasons why biological conservation is so important. The very survival of all life on earth, including ourselves, is dependent on innumerable interlinked biological nutrient and energy cycles. It is argued that these have always been very robust primarily because of their complexity and variability and the whole of life on earth has been linked to one huge, self-regulating living organism, that we have now started to debilitate. This fairly new idea, the *Gaia hypothesis*, remains rather controversial. It may be that the reduction in diversity caused by our actions reduces the number of potential pathways for these cycles, that they become less robust and increasingly vulnerable to adverse conditions or the dominance of just a few species and these we interpret as pests. We are only now starting to understand that biological control of pests and potential pests is not a new idea, necessarily involving shipping around the world exotic insect species to feed on ones that we have unfettered from their natural enemies, but is something that has been carried on by the wildlife around us all along.

One example of the contribution made by invertebrates might be the recent realisation of the value of weeds and their associated insects, in particular the beetles, to the survival of young ground living "game" birds, on arable land. This has led to the Game Conservancy's "Conservation Headlands" project, which has shown that it is possible to achieve a dramatic increase in populations, back almost to their former levels, of these birds, at very little cost, simply by not spraying with pesticide a narrow band of crop at the perimeter of each field. A diversity of insect species survive, upon which the young birds are dependent for food.

Economic issues

In a rather more direct sense, invertebrates are set to contribute more and more to our human economy. Much has been made of the contribution of the world's flora to its pharmacopoeia and of the consequent need to conserve botanical diversity.

Historically, only a few invertebrates have contributed such useful substances as shellac and cochineal, materials now superseded by synthetic alternatives. Insects have, however, been used in oriental folk medicines for millennia and recent research is beginning to reveal an equally diverse spectrum of invertebrate chemicals, often defensive substances, ranging from simple organic acids to complex biologically active analogues of mammalian hormones with wide potential in the treatment of human malady.

Current uses of invertebrate-derived medicines range from a highly effective anti-tuberculosis drug derived from the anal secretion of a species of ant, to an equally effective adhesive for corn plasters on sweaty feet, derived from the gummy defensive secretions of a centipede! Invertebrate products are also being widely screened in the search for drugs effective against the current scourge of AIDS. Waterbeetles show particular promise in the synthesis of mammalian hormone analogues (Richter, 1983) and the substance *paederin*, derived from the attractive red and black rove beetles of the genus *Paederus*, apart from being among the most toxic animal substances known, in minute quantities it appears also to promote mammalian tissue growth and is now being used in the treatment of chronic ulceration (Pavan, 1986).

Far from being difficult to justify the conservation of invertebrates on economic grounds, they represent an enormous untapped resource of useful products which we should be safeguarding for the future.

Conservation and you, the coleopterist

All this may seem far removed from our beetling forays into the British countryside, and indeed much of it is. Relatively few of Britain's nearly four thousand species are in danger of extinction on a world basis and there are still remaining examples of most types of our major ecosystems, including a very few that are still reasonably intact. Newly recognised factors such as acid rain, climatic change and perhaps the greenhouse effect may now, however, be posing new threats to once-safe sites. Also, many sites that people value highly are still coming under threat and being destroyed so that many species of beetle continue to decline.

How best then, can we, as coleopterists, do something to assist in the conservation of the species that we find and the sites on which we have found them? Let us see what are the possibilities.

Britain has a network of voluntary and statutory conservation bodies which is perhaps the most comprehensive in the world. With it we now have the opportunity, maybe not to halt completely the continuing degradation of the whole countryside, but at least to conserve those places that we consider to be the "best" bits, either as sites with some degree of statutory protection – the Sites of Special Scientific Interest – S.S.S.I.s – or as nature reserves of one kind or another. It is also important to exert some influence towards conservation in the wider countryside away from the protected sites if at all possible, ideally through education and example.

The need for information to protect sites

To achieve this, we need to be able to justify, sometimes even to Her Majesty's Inspector in a Public Inquiry, either that the particular places that we are trying to conserve have some intrinsic merit, or that they are the "best" ones in some particular way, either nationally, regionally or locally. To do this, we are then forced to define and justify exactly what we mean by "best". Ammunition, in the form of scientific evidence, is needed for these arguments, which, on occasion, can turn into emotionally charged battles.

To win these arguments, conservation bodies need to know exactly what the wildlife interest is on a site they are trying to protect. They need to know where else, if anywhere, similar interest can be found on a local and national basis and, if all goes well and the future of the site is secured, how the wildlife can be maintained by appropriate management. For this, the chief need is for *information* – information as to which species occur at the site, where else they occur, how frequently, and an idea of their biology.

For the plants, birds and mammals at a site, and to some extent, for the butterflies, moths and dragonflies, we are fortunate that a vast body of skilled naturalists exists throughout the country, and a great background of information is already available about the local and national status, distribution and biology of most of the species.

When we try to defend a site purely on the basis of its invertebrate interest, particularly one where the interest does not coincide with high botanical or ornithological quality, the task is *much* more difficult. There are far fewer coleopterists than lepidopterists, and far fewer entomologists as a whole than botanists or ornithologists. Many localities have *never* been investigated entomologically, and yet comparison with sites with similar habitat is invariably demanded in order to justify the claims we make about the quality of a particular locality. Furthermore, the ecology of most species of beetle is at best poorly understood and for many, not understood at all. All too frequently our knowledge stops merely at the ability to identify the right sort of habitat in which we can find adults of a particular species.

If we are to have the satisfaction of seeing successful protection of our favourite sites, we should endeavour to ensure that we use our powers of observation and skills as entomologists as efficiently as possible. Most important, we should also ensure that we communicate our findings so that they can be used by others. There must be a prodigious amount of potentially extremely useful observation locked up in our notebooks and on the data labels in our collections, information which could extend our joint understanding of beetles considerably if it were all put together.

A who's who of users of entomologists' information

There are quite a number of organisations which are now in a position to utilise for conservation purposes the information that we produce as a result of our studies. I will endeavour to outline the functions of each.

The National Biological Records Centre (B.R.C.), part of the Institute of Terrestrial Ecology at Monks Wood, runs quite a large number of Coleoptera recording schemes which are invaluable in giving an overview of a species national distribution and also collate ecological information on the species themselves. The B.R.C. recording schemes are the subject of a special chapter in this volume.

There also exist a large number of Local Biological Records Centres throughout the country, often being funded and housed by local authorities either in local museums or in the county or district planning departments. These are quite variable in the degree of beetle and other entomological information that they store and also in the way that it is stored, some being fully computerised, others relying on site and/or species based index card systems. These records centres play an important role in supplying information to the local authority when it is formulating policy towards wildlife conservation and also making planning decisions concerning sites of wildlife value.

I work for the government's own conservation body, the Nature Conservancy Council,* in its Invertebrate Conservation Branch at its headquarters in Peterborough. We rely, to a very great extent, on information supplied by amateur entomologists. The Branch is part of the advisory, research and scientific policy making section of the organisation, and we commission research and survey of invetebrate species and sites, placing particular emphasis on identifying the management needs of particular species and communities.

The N.C.C. has a regional structure for which they divide Britain into 15 regions, each with a regional and, in many cases, a number of subregional, offices. These are responsible for the nitty-gritty of identification, notification and defence of S.S.S.I.s and the management of National Nature Reserves. It is with these regional offices that an entomologist's first contact with N.C.C. is usually made and they can be located using the local telephone directory.

---

* Since writing this, the government has split the N.C.C. into separate agencies for England, Scotland and Wales. The I.S.R. is now part of a central co-ordinating body which remains based at Peterborough.

Central to the activities to invertebrate conservation is the Invertebrate Site Register (I.S.R.*) which tackles invertebrate conservation in two complementary ways.

First there is a computer database of sites throughout the country that have been found to have been of interest for their invertebrates, together with the records of the more significant species recorded there. We have collated this information from as wide a range of sources as possible, including amateur and professional entomologists, local records centres, the national Biological Records Centre. We also use the huge amount of data available in the literature, using a large computerised entomological bibliography that has been created, called ENTSCAPE.

The site register is used by regional staff of the N.C.C. as a basis for identification of sites to receive statutory protection, either as S.S.S.I.s, or, as on a number of occasions now, as National Nature Reserves. Information is made more widely available to the County Wildlife Trusts and other conservation bodies, to be used in a wider context in local site defence and in the formulation of site management policies.

The other, complementary, side to the I.S.R. is the preparation of reviews on the national status of particular groups of invertebrates. These review information on the scarcer species and include sections on the threats faced by each and, where possible, suggestions on appropriate management of their habitat. Reviews already produced at the time of writing include butterflies, micro and macro moths, grasshoppers, molluscs, caddisflies, stoneflies, mayflies, bugs, lacewings, bees, wasps and some flies. A review of scarce beetles is also nearing completion (Hyman *et. al.*; in prep), produced from a consensus of information and opinion from over 100 coleopterists throughout the country. The I.S.R. was also partly involved in the production of the insect volume of the Red Data Book (Shirt, 1987) detailing the very rarest species, and a Red Data Book covering other invertebrates is soon to follow. Using the reviews and the Red Data Book, species lists can be more easily interpreted as to their conservation significance, even by non-specialists.

The County Wildlife Trusts, of which there are nearly fifty, are co-ordinated through the Royal Society for Nature Conservation, based at Lincoln. The Trusts each cover an individual county or area of Britain and are responsible for the care and management of a very large number of nature reserves including some very important beetle sites. They also act in an independent advisory capacity on matters of wildlife importance in the county, and often run a Local Biological Records Centre. They are obviously particularly interested in information on the insect life of their reserves, and also in a wider context in the county which they cover.

Many other organisations are now starting to realise the importance of invertebrates on their sites and would appreciate information collected by the amateur entomologist. The National Trust, the biggest conservation landowner of them all, now has a biological conservation team, where insect conservation, particularly of the old parkland beetles, is playing an increasingly major role. The Ministry of Defence and the Electricity Industry are also taking pride in the wildlife, including the invertebrates, on the land in their care, and are investing money and manpower into its conservation. There are also the Royal Society for the Protection of Birds, whose literature these days has included articles on the invertebrates on their reserves, the Woodland Trust which is purchasing woodland all over Britain, and there is also now one society devoted entirely to the conservation of one particular group of insects, The British Butterfly Conservation Society. None is as yet dedicated to the conservation of beetles or invertebrates in general.

A problem in information flow and a pending solution

This multiplicity of the conservation organisations requiring the results of our labours can be quite bewildering and may, to some extent, actually impede the flow of information. Each may request of us the information resulting from our studies, each with its own recording form. A day in the field and the subsequent time spent on identification could therefore, if we conscientiously submitted the information to all who request it, lead to a more daunting bout of form filling than the annual tax return! Understandably this has not been wholly welcome to the coleopterist, whose interest lies in the beasts themselves rather than in repetitive form-filling.

At the time of writing all is set to change with the formation of the National Federation for Biological Recording, with one of its specific aims being to simplify the way we collate and use biological information in this country. Modern developments in computing should soon make it possible to insert our records at any point into a national network of records centres, and the various bodies that need our information will have rapid access to it, provided that we have released the information using one of a number of levels of confidentiality.

A start is already under way, and a number of local records centres and Trusts take part in the development of a biological recording database programme called "RECORDER", developed by Dr. Stuart Ball of the N.C.C. It will obviously take time to build up the system but the need for information for conservation purposes is with us *now*. In the meantime it is still vital that the information that we produce becomes available to the conservationists that need it.

All of the bodies described above, are, however, in a fair degree of communication with one another, so a judicious decision as to where to deposit our information at one or more points in the system will ensure that it is put to work. In the meantime – keep the vital information flowing in; things are due to get better soon!

Protection and management of sites

A further way in which information flow from the field entomologist may assist in the protection of sites of conservation value is in our vigilance regarding imminent threats to sites of entomological importance. Conservation bodies are woefully understaffed and their officers are frequently almost desk-bound by the legal niceties of Public Inquiries, correspondence with their members or with other organisations, fund raising and the writing of reports and management plans. Rarely are they able to keep an eye on all developments at all sites which come into their area of concern and often they have to rely on tip-offs from local naturalists that sites of value may be coming under threat.

In the case of sites of entomological importance, this not only means being wary of an advancing bulldozer, chain-saw gang or whatever, but, on occasion, being wary of the over-enthusiastic conservation worker, often wielding the self-same equipment. The general level of understanding is improving considerably, but not always are the needs of the smaller and more obscure denizens of reserves and other conservation areas fully taken into account by those in whose care they fall. This is obviously not through intent, but again from lack of vital information, leading to ignorance, often of the species' very existence and certainly of their habitat requirements. Tactful, but firm, advice from a visiting entomologist as to the great importance of the dead wood that may be being cleared and burnt for "tidiness" sake, or of the flowery grassland that is being grazed very hard to create botanical diversity but to the exclusion of the insects, is usually welcomed by the site's managers.

It is important here not to take too critical an approach – positive suggestions as to ways future management might be improved to take into account the insects' needs are far more acceptable than negative criticism. Of course the most constructive way of exerting more influence on site management is by becoming actively involved with the conservation organisation itself. An entomologist on a council of management, reserve management committee or in a volunteer management team rapidly comes to be regarded as a useful expert and his or her advice sought as decisions are taken.

With this, however, comes the responsibility to make sure that the advice given is correct. Before stepping in with advice that might radically effect the way management is carried out, it is essential that the basic ecological principles of habitat management are understood. It is not possible here to reproduce a manual on all the principles of invertebrate conservation, indeed none currently exists. Articles on invertebrate conservation now occur regularly in journals such as *Biological Conservation*, *Ecological Entomology*, *Field studies* etc., and the N.C.C. produce a number of booklets on the conservation needs of various groups of invertebrates – currently ones on butterflies, dragonflies, bees & wasps and molluscs, and a further one is being prepared on the fauna of dead wood. They also produce more specialist publications on invertebrate conservation management in their "Focus" and "Research & Survey" series (see bibliography).

Perhaps one of the most fundamental principles in habitat management for invertebrates is to recognise the need for continuity of correct conditions. Most beetles and other invertebrates have annual life cycles and often have very limited powers of dispersal. This means that the conditions have to remain suitable *every single year* and so many species are, unlike annual plants which can survive as seeds, very vulnerable to quite short periods of adversity. This makes them very often excellent indicators of continuity of quality in some habitats (Harding & Rose; 1987, Garland; 1983). Many of the most threatened species are those associated only with a particular phase of an ecological succession, often the earliest (bare ground or ruderal plants) or, in particular, the final (hulks of post-mature trees) stages. The maintenance of a desirable *status quo* is usually only possible with a considerable input of management and often we have to fight against natural successional trends in the vegetation. In all but the most mature habitats, an attitude: "it's quite nice now and to interfere would be to spoil it" is likely to lead to an eventual decline in interest.

Insect work on nature reserves

Attitudes of conservation bodies towards entomological pursuits on reserves have varied considerably in the past. They have ranged from, in a few instances, a total ban on collecting on reserves, to permission for controlled collecting dependent upon a number of conditions being complied with. Some of these conditions may have been rather ill thought out. The restriction placed by one organisation, for example, was that a maximum of one male and one female of any species was all that it was permissible to remove for identification purposes, and that these then had to be released afterwards. For those studying the *Aleocharinae*, or many other groups of small, difficult to identify species requiring genitalia dissection, this may have been a rather difficult condition to comply with and the released beetles unlikely to continue to propagate their species!

Fortunately, very good relations have now built up between entomologists and the conservation bodies, so that the only restriction is usually on the removal of specimens of butterflies which are usually easy to identify in the field. The production of a small report or species list giving the results of the visit may be requested as a condition of the permit, and should be submitted even when not specifically asked for.

A point worth mentioning to the officers of conservation bodies when promising to submit species lists, is the need for a degree of patience on their part. They are likely to be used to instant responses from ornithologists, botanists and most lepidopterists and it is often not realised that the identification and confirmation of the more difficult groups may take a considerable amount of time. On a number of occasions, I have been reproached for a tardy response a few weeks after a recording visit to a reserve, when the majority of the specimens had yet to be mounted, let alone identified. A reasonable interval to suggest might be to wait until well after the end of the main autumn field season.

When submitting species lists to organisations whose officers are unlikely to be entomologists, a straight list of six-legged Latin names is usually pretty incomprehensible and, to be of maximum use, such lists need to be annotated as to the significance of what has been found. If some of the species recorded are real "goodies" then it is best to say so, even though this is adding to our burden of writing. Eventually enough experience is gained to tell us which are those species that indicate something significant about the site – ones that are "faithful" to a particular type of habitat and consistently turn up in only the best examples of it. These need not necessarily be the rarer ones, and the significance of particular species may vary considerably in different parts of the country. A number of guides as to which species might be worth annotating already exist: the insect volume of the Red Data Book (Shirt; 1987) and various parts of the N.C.C.'s National Review of Terrestrial Beetles (Hymen *et. al.* in prep), the review of ancient parkland saproxylic species produced by the Institute of Terrestrial Ecology (Harding & Rose; 1987), and an equivalent produced for use only in northern England (Garland; 1983). Finally, a note at the top or bottom of the list indicating that the remaining species are more ubiquitous, and the finished article is much more likely to be put to effective use.

Our behaviour as coleopterists

Having indicated the most constructive ways in which the coleopterist can contribute to the conservation of beetles, I must now turn to the ways in which our own behaviour in the field might adversely affect the beasts that we are studying and their habitat, and what we may do to mitigate this.

Whether or not we are likely to jeopardise their survival or their habitat and other wildlife at that site really depends on the zeal with which we carry out our fieldwork and create a reference collection. How do our field techniques affect the habitat? How many specimens should we take? Is it necessary to take specimens at all? Is it justifiable to go to a known site to collect a rarity?

In 1972 the Joint Committee for the Conservation of British Insects (J.C.C.B.I.) produced a Code for Insect Collecting and all that is contained therein is highly relevant to ourselves as coleopterists. The code dwells largely with subjects of particular relevance to the lepidopterist and a few extra points might be made that are more specific to the coleopterist, whose field methods tend to be rather different from those used by pursuers of the scaly-winged. I have outlined a "Coleopterists' Code of Practice" regarding our field techniques, as an addendum to the J.C.C.B.I. code.

It is difficult to make such a code anything other than a long list of "do nots", although I have tried to be as positive as possible. I have chosen the points that I have made on the basis of bitter experience of the effect that I know I have had on the wildlife habitat I have visited and the relationships that I have developed with landowners and other naturalists. No doubt other coleopterists could add to the code. The basic need is mainly for common sense and restraint.

The collection of rarities

The Victorian literature is scattered with examples of entomologists' extravagant reports as to the length of the series of particular rarities that they had discovered, with offers of specimens to all and sundry. Such behaviour is, mercifully, now rare in this country in the latter part of the 20th century although a rarer worrying development in Britain has been the recent proliferation of entomological dealers' "trade fairs", and, although there is a concentration on the Lepidoptera, rare Coleoptera sometimes also feature.

The problem does still exist abroad, however. Collectors and, in particular, dealers in parts of central Europe have made such depredations on the fauna of some vulnerable habitats, that a number of governments have felt it necessary to afford legal protection to a large number of species of insect, including beetles, and, in some countries, have imposed a total ban on insect collecting in National Parks and in some major ecosystems. It is necessary to exercise self-restraint to avoid the excesses of the previous century, lest others may regard it necessary to call for a ban on some of our entomological pursuits.

In Britain, at the time of writing, fortunately we have but two species of beetle currently deemed sufficiently threatened to warrant legal protection under the Wildlife and Countryside Act of 1981 and its subsequent reviews. These are the leaf beetle *Chrysolina cerealis (L.)* and the click beetle *Limoniscus violaceus (Muller)*, both only known from a small area of a single site, the latter possibly only from one or two individual trees. *These should not be collected.*

There are also some additional ones where the habitat is very restricted and individual populations are likely to be very small and where extreme restraint should be shown when collecting specimens. On the whole, the effects of the coleopterist on natural populations must be miniscule, when compared with the effects of the habitat destruction that is going on all around. The important point to note, however, is that the coleopterist concentrates on populations of already scarce species, and works specifically and systematically upon their habitat.

Collecting from known sites

When we discover that a particular species, especially a rarity, has been found by another coleopterist, there may be the temptation to take the easy option and add the species to our own collection from the same site – in other words, to use the common parlance among the bird-watching fraternity, to "twitch" the rarity.

There *may* be some justification for this. By finding the species at a known locality, our personal knowledge of the species and its habitat increases and so, perhaps, might our ability to find additional localities for the species. On the whole, however, the tendency to "twitch" usually adds little to our knowledge of the species and might significantly jeopardise the continued existence of some populations of species, particularly through continual damage to its habitat. It is far better to marry the knowledge gained from another's fortuitous find to our own knowledge of similar habitat and try to find previously unknown populations of the species. By so trying and failing, we confirm, as much as is possible to prove a negative, that the species is genuinely rare and therefore also confirm the importance of the original find.

Of course another justification for visiting a known site for a species is to make a genuine study of the species' biology. It is far more interesting, satisfying and useful to discover more about a species than simply to add a specimen to a collection. What does it feed on as an adult? and as a larva? Is its larva described? What eats it? What is its reproductive behaviour? Where does it oviposit? What exactly about the habitat differentiates the places in which it can be found from those where it is absent? The possible questions are endless. For most species a few hours lying on one's stomach simply *watching* the beasts is bound to come up with new, eminently publishable, information. The spirit of J.H. Fabre is now all too rare in our high speed world!

Public relations and permission

The remaining consideration of our behaviour as coleopterists concerns our relationship with the owners and other legitimate users of the countryside. Could our activities in any way interfere with any other person's legitimate use or enjoyment of the site, such as the disturbance of crops or livestock, or of birds at a popular birdwatching area, or involve the spoiling of the site's appearance? In other words we are dealing here with our abilities in public relations.

First there is our relationship with the owners of the land on which we are making our investigations, whether it is a nature reserve or private land. In either case it is obvious that permission to collect is required. After all, our intention may be to take away specimens from that land, and, as outlined in the code, our methods may leave some visible effect on the habitat that we are investigating.

On private land it is sometimes difficult even to find out to whom it belongs in order to seek permission, and there is often the temptation to trespass. There is danger in this of souring future relationships between the landowner and all naturalists and conservationists if the entomologist is "caught", or leaves obvious signs of his investigations which might easily be interpreted as vandalism. At the very most, activities should be confined to the margins of public footpaths in areas with no general public access, and even then, one's legal right is only for access and not to hunt for beetles! Common sense indicates what is and is not acceptable behaviour on private land. Permission should be sought wherever possible and the temptation to "nip over the fence" should be resisted.

It is always worthwhile to give a little bit of feedback to the private landowner as to what you have found. Instead of a full species list, which is unlikely to be particularly meaningful, a bit of face-to-face enthusiasm, directed towards some obscure little black beetle, as well as towards the bigger, prettier ones – the cardinal beetles and longhorns – may get over the idea that some people genuinely care about things that live under stones and in bits of dead wood. It is always worthwhile to praise the value of the habitat in the landowner's care.

If there is a chance, get talking about management of the land in relation to the habitat. Foresters and landowners are usually unaware of the value of dead or unhealthy trees, and they take a lot of persuading to change a lifetime's attitude that tidiness is essential when dealing with timber. It is always better to be positive, rather than critical in any recommendations made and it is also important that any recommendations you make are ecologically sound (see above). We are only just starting to get the message about the different requirements of invertebrates across to many conservationists, so we need to be doubly patient with the unconverted.

Relationships with other naturalists

The other group of people with whom we may need to maintain a good understanding are our fellow naturalists. For many of these, the idea that we might need to imprison in specimen tubes, or even, on occasion, to kill, the wildlife that we study, can be very difficult to comprehend. Many students of butterflies and dragonflies now are totally against the taking of specimens, even as vouchers, arguing that the making of a collection is equivalent to the old hobby of collecting birds' eggs, and should similarly be outlawed. This can be an emotive subject, where reason does not always prevail. Justification for the necessity for the retention of specimens for identification and for the keeping of voucher specimens has already been touched upon. It may also be useful to point out the number of specimens taken in relation to the likely overall size of a viable population of beetles and the scale of mortality from other causes. The depradation by the coleopterist is usually negligible by comparison provided our field behaviour, as outlined in the code, has been exemplarly.

Our educational role as coleopterists

There is a definitive need to have some educational contact with the young. Books such as "The Junior Naturalists' Handbook" (Watson, 1962) my own *vade mecum* as a youth, which detailed how to go about making a collection, tend to be rather unfashionable these days. Most children's books and wildlife groups now tend to stress the observant, but "hands off" approach to wildlife, particularly emphasising the importance of releasing all specimens at the end of the day. I have been thoroughly chastised by a number of children for the crime of incarcerating a specimen in a tube, particularly compounded when I had admitted that I thought the species might be rare and that I might have actually to kill it in order to find out. While there is much merit in this – it would hardly do if an entire generation of youngsters suddenly developed the urge to become avid collectors – the real enthusiast, the one with a genuine interest in glueing beetles to small pieces of cardboard, should be given full encouragement and allowed to make a collection. He or she should also be encouraged to get involved with both the entomological and conservation fraternities where the involvement of more young people would be particularly welcome. Most of us had some adult "mentor" as a child, someone who actually encouraged our interest in creepy-crawlies, and it would be nice to think that the next generation of entomologists might get similar encouragement from ourselves.

In conclusion

In this chapter I have tried to put over my own ideas as to how we as coleopterists can contribute to the conservation of the beetles themselves and the places in which they live. Hopefully this will add a further dimension to the satisfaction and enjoyment that we already get from our interest. Conservation problems are, sadly unlikely to go away and the coming years are likely to produce more and more situations where it is vital for entomologists to become practically involved in conservation. There is a growing awareness of the significance of these smaller denizens of the countryside and what they can tell us about the places in which they live, and the more that we can encourage this, the better are the hopes for the survival of Britain's flora, fauna and wild places.

Keep up the good work!

References & Bibliography

Chelmick, D., Hammond, C.O., Moore, N. & Stubbs, A.E., 1980. The Conservation of Dragonflies. Nature Conservancy Council, London. 24pp.

Else, G., Felton, J. & Stubbs, A.E., 1979. The Conservation of bees and wasps. Nature Conservancy Council. London. 14pp.

Fitter, A. & Smith, C., 1979. A Wood in Ascam – A Study in Wetland Conservation. Sessions. York. 164pp.

Fry, R.A. *et al.* (1991). Habitat Conservation for Insects. The Amateur Entomologist. Vol. 21.

Garland, S.P., 1983. Beetles as primary woodland indicators. Sorby Record 21; pp. 3–38.

Harding, P.T. & Rose, F., 1987. Pasture woodlands in lowland Britain – a review of their importance for wildlife conservation. Institute of Terrestrial Ecology. Huntingdon. 89pp.

Hyman, P. *et. al.* (in prep). A National Review of British Terrestrial Beetles. Nature Conservancy Council.

Kerney, M. & Stubbs, A.E., 1980. The Conservation of Snails, Slugs and Freshwater Mussels. Nature Conservancy Council, London. 24pp.

N.C.C., 1981. The Conservation of Butterflies. Nature Conservancy Council, London. 28pp.

N.C.C., 1984. Nature Conservation in Great Britain. Nature Conservancy Council. Peterborough. 111pp.

Pavan, M., 1986. A European Cultural Revolution: The Council of Europe's "Charter on Invertebrates". Council of Europe. 51pp.

Richter, R., 1983. The struggle for existence. Translation of article by F. Hebauer, 1982, "Uberlebenskampf wasserbewohnender Kafer." Nationalpark 36(3): pp. 48–49. The Balfour-Browne Club Newsletter 28: pp. 2–4.

Shirt, D.B. (ed.), 1987. British Red Data Books: 2. Insects. Nature Conservancy Council. Peterborough. 402 pp.

Stubbs, A.E., 1972. Wildlife Conservation and Dead Wood. Journal of the Devon Trust for Nature Conservation 4: pp. 169–182.

Stubbs, A.E., 1982. Conservation and the future for the field entomologist. Proc. Trans. Br. Ent. Nat. Hist. Soc. 15: pp. 55–67.

Watson, G., 1962. The Junior Naturalist's Handbook. A. & C. Black. London. 176pp.

A COLEOPTERIST'S CODE OF PRACTICE

In addition to the J.C.C.B.I. code, a few other points can be made specific to the activities of the Coleopterist, some of which may be of lesser relevance to other entomologists. I have deliberately not set my recommendations out in the form of a discrete, itemised code, seeking to justify each of the statements made.

FIELD TECHNIQUES

Dead wood

Whether it be fallen to the ground, or still part of the living or dead parent tree, invariably dead wood is a scarce and very valuable resource at a site. It is also one that is frequently not appreciated by those in charge of the site, to the extent that it may often be removed and destroyed, sometimes even on nature reserves. It is therefore a habitat under threat. The easiest and most effective way to investigate the fauna is to employ one of various "offensive weapons" such as a sheath knife, axe, tiling hammer, cold chisel, crow-bar, etc. (at the time of writing, legislation is under way restricting the carrying of some such objects in public places).

The effect on the dead wood can be catastrophic, as far as the fauna is concerned. Apart from direct effects of the destruction of habitat, the remainder of the wood may be rendered barkless, open to the effects of drying winds and hungry birds. It can, however, be reasonably easy to minimise our effects on the fauna of this vulnerable habitat by keeping to the following "code" which has been arrived at from a consensus of attitudes of a large number of coleopterists and invertebrate conservationists.

1. It is best to concentrate on trees or branches whose future in providing good quality habitat may be in doubt. For example, the trunks of trees recently split in half exposing moist heart rot or previously enclosed hollows are likely to deteriorate rapidly through drying out. Bear in mind, however, that at least one more generation of insects may emerge from the larvae and pupae in the wood before it becomes unsuitable for insect life, so even here exercise restraint. Don't dig too deeply into the heart rot and always replace material excavated, not as a loose heap of chippings, but somewhat compacted.

    Fallen trees and dead wood in very visible situations are the ones most often removed, either for safety or reasons of "visual amenity". Such wood may often end up on the bonfire or be sawn up as firewood and therefore less damage is done when investigating it. Beware, however, of dismantling someone's legitimately collected firewood stack. At least seek permission before beginning an investigation of it.

2. *Never* endanger the long term survival of living trees which have areas of dead bark or partly exposed dead wood by making "the first incision". This will cause the habitat subsequently to degrade by drying out, or become a more likely target for removal by the hygiene-conscious forester.
3. While bark that has been removed and replaced is never as satisfactory a habitat as undisturbed bark on dead wood, it is still worthwhile to make the attempt to replace it. There are some coleopterists who carry about a tap hammer and supply of small nails specifically for this purpose. For bark on vertical trunks, this is often the only way to replace it without it subsequently falling off again and it is often best to refrain from removing such bark if there is no prospect of being able to replace it.
4. Try to concentrate activities. It is usually better to investigate a small quantity of dead wood thoroughly, even if this means investigating it to destruction, than making superficial studies on a large quantity. In this way the smaller or deeply burrowing species are more likely to turn up while the majority of the habitat is left undisturbed.
5. Large bracket fungi may take years to develop so do not remove all such fungi from dead wood and be sure to replace that which has been investigated somewhere close to its source, in a damp situation.
6. Old nests, leaf litter, fungi and wood mould from hollows and forks should be replaced after being investigated so that larvae and pupae within it have a chance to complete their development.
7. Replace overturned logs. This applies equally to stones, litter, moss, piles of excavated soil, etc.
8. If there is obvious use of beetle burrows by solitary wasps and bees, avoid damage to these in particular. Replace such logs in their *exact* original position and condition. The wasps relocate their burrows by sight and may be unable to return if they cannot recognise their home after it has been removed or damaged.
9. Bear in mind the other potential biological interest of dead wood and bark – interest to the mycologist, lichenologist and bryologist.
10. Old individual trees of known value for particularly rare species should be left alone, rather than visited by a succession of entomologists. To reiterate what is said above, there is little need for collecting readily identifiable rare species from known populations.
11. There is much scope for the development of non-invasive methods of investigating the fauna of many habitats. Emergence traps have hardly ever been used on dead wood and could be profitably tried in places where they are less likely to be interfered with by members of the public. Also, many or most of the dead wood species visit nectar sources during their brief adult dispersal phase. Much about the fauna of a rich dead wood site can be deduced from examination of nearby hawthorn blossom and the flowers of umbellifers and composites.

Pitfall trapping

Beware of excessive damage to the vegetation while inserting and examining the traps, either by trampling, or in the unjudicious siting of the traps themselves. Most pitfall preservative solutions are highly toxic to vegetation and should not simply be poured onto the ground as the trap is examined. While pitfalls only collect a relatively small sample of populations, if ones used for censuses continually capture very large numbers of a rare species, then it is worth considering resiting them. "Lids on stilts" should be used to prevent small mammals, amphibians and reptiles from falling in and drowning.

Grubbing, beating and sweeping

These can be potentially quite damaging to vegetation. Exercise intelligent restraint when faced with orchid meadows, rare plants, etc. – the sweep net may inadvertently double as a scythe. Few things are more embarrassing than the emptying out of the net a forlorn little bunch of flowers in front of a reserve manager at the end of a bout of sweeping. Such antics are unlikely to result in the invitation for a return visit.

The most effective form of beating is a short, sharp shock – a sudden tap on a branch is more likely to dislodge the beetles thereon than a prolonged thrashing which can damage the foliage and branches. Searching vegetation and learning to recognise feeding signs can be much more productive and lead to a far better understanding of the species we are studying.

Species of other groups, in which we are not so interested, are more often than not less robust than the average beetle. It is necessary to examine the contents of the sweep net and release non-target species after a fairly small number of sweeps, otherwise the result is a mangled ball of squashed flies and bugs (often with the more interesting beetles rolled up in the middle!).

Beware of nesting birds when carrying out any of these operations. Excessive disturbances and a prolonged presence near a nest site may cause birds to desert.

Aquatic habitats

Water margins are particularly vulnerable habitats, with soft mud and luxuriant, but easily damaged, species of plant. When collecting water beetles or those living among waterside plants, the most effective methods often involve treading down emergent and submerged vegetation in order to float out the beetles. Not surprisingly, this can have a devastating effect on the vegetation itself and the method should be used with considerable restraint, confining activities to a very small proportion of the available habitat. Although such habitat destruction may be temporary, it can be unsightly in the extreme and justifiably incur the wrath of the landowner or reserve warden.

Ant-nests

These can be investigated in a relatively non-invasive way by the placing of stones or logs on the surface of the nests and then revisiting after a few days when beetles may be found clinging to the underside. Burying a bundle of twigs in the nest for subsequent re-exhumation and investigating is again preferable to digging out the whole nest.

Surplus material

What do we do with surplus specimens that may have been reared from material collected earlier and also what to do with the remaining pabulum after the majority of beetles of interest have emerged?

The temptation may be to release the beetles in the garden and add the pabulum to the compost heap to allow any remaining insects to "take their chance". This is unsatisfactory for a number of reasons.

First, it is a drain on the natural populations of the insects. If it is at all possible, effort should be made to return surplus living specimens and pabulum to the site from which it was taken.

Second, there is the remote possibility that another entomologist may discover the released specimens or ones emerging from discarded material. These need not even be beetles and confusion may be spread among other entomologists, clouding our knowledge of the natural distribution of species and sometimes thwarting our efforts to conserve them by making them appear more widespread than they really are.

If it is *absolutely* not possible to return such material, it is probably best destroyed and surplus livestock killed and distributed to other entomologists.

Introductions

A planned release and introduction is a more serious affair and is inadvisable without informing and consulting either the N.C.C., the Biological Records Centre or the J.C.C.B.I., who have produced a code concerning attempts at introduction or reintroduction of insect species – "Insect re-establishment – a code of conservation of practice". (1986). This can be obtained from the Nature Conservancy Council's Headquarters at Peterborough.

Reference
A Code for Insect Collecting. Joint Committee for the Conservation of British Insects. 1972.

Appendix I

GLOSSARY

By J. Cooter

As the reader will no doubt notice, the majority of terms and definitions listed below have been taken from Fowler (1887), *Col. Brit. Islands*, 1: ix–xv. No attempt has been made to define or list the various anatomical terms the student will encounter when attempting identifications or more formal studies of the Coleoptera. These are fully explained in other works, (especially recommended are Crowson, R.A., 1954 *The Natural Classification of the Families of Coleoptera*, reprinted 1967 with additional papers. Crowson, R.A., 1956 Royal Entomological Society of London Handbooks for the Identification of British Insects, 4 (1), Introduction and Keys to Families. Imms, A.D., 1951 *A General Textbook of Entomology*. Torre-Bueno, J.R., 1937 *Glossary of Entomology*, Brooklyn Entomological Society.

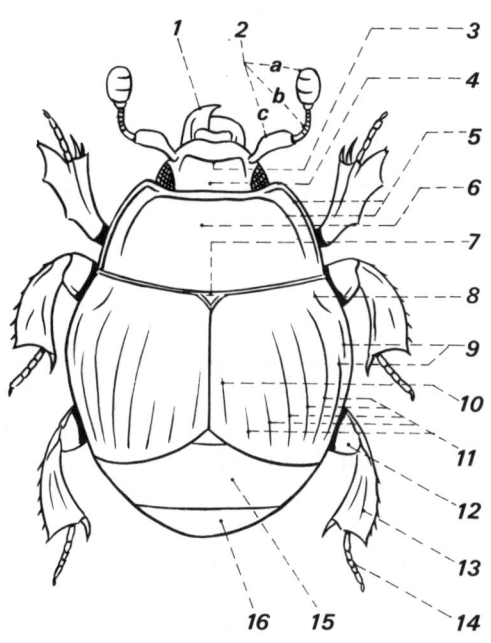

Figure 24. *Hister unicolor* L., dorsal view (After Halstead, D.G.H., 1963 *Handbk. Ident. Br. Insects*, **4(10)**:5.

(1) mandibles; (2) antenna, (a, club; b, funiculus; c, scape); (3) frontal stria; (4) frons; (5) pronotal striae; (6) pronotum; (7) scutellum; (8) humeral stria; (9) sub-humeral striae; (10) sutural stria; (11) dorsal striae; (12) posterior femur; (13) posterior tibia; (14) posterior tarsus; (15) propygidium; (16) pygidium.

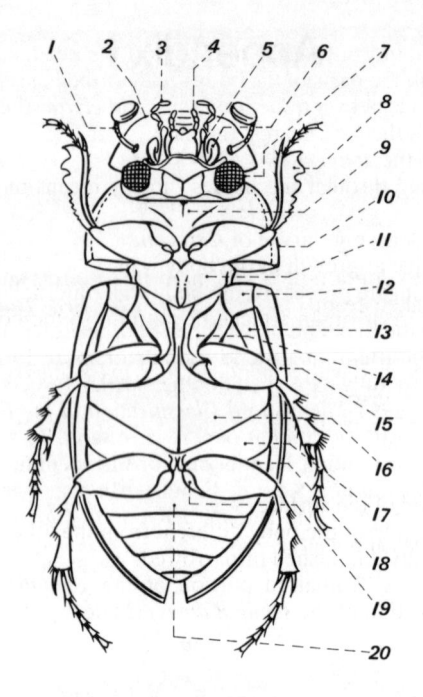

Figure 25. *Aphodius rufipes* (L.), ventral view. (After Britton, E.B., 1956 *Handbk. Ident. Brit. Insects*, **5(11)**:4).

(1) antenna; (2) clypeus; (3) maxillary palp; (4) labial palp; (5) maxilla; (6) mandible; (7) eye; (8) anterior leg; (9) prosternum; (10) pronotum; (11) mesosternum; (12) mesepimeron; (13) metepisternum; (14) middle coxa; (15) elytral epipleuron; (16) metasternum; (17) posterior coxa; (18) first visible abdominal segment; (19) trochanter; (20) second visible abdominal segment.

Both reproduced by permission of the Royal Entomological Society of London.

Abdomen. The hindermost principle division of the body.
Aborted. Incomplete, undeveloped.
Acicular. Terminating in a sharp point like a needle (*acus.*)
Aciculate. Covered with small scratches which appear as if made by a needle.
Aculeate. Produced to a point; as applied to one group of *Hymenoptera* it means furnished with a sting.
Acuminate. Terminated in a point.
Agglutinate. Fastened closely together so as to form one piece.
Aedeagus. (Etymologically correct *aedoeagus*). The *median lobe* plus *tegmen*. The term "penis" should not be used; this is a mammalian organ, the two are not analagous.
Aeneus. Bright brassy or metallic golden-green colour.
Alutaceous. Covered with minute cracks like the human skin.
Ambulatory. Relating to walking, see also *cursorial* and *natatorial*.
Anal. Pertaining to the apex or extreme end of the *abdomen*.
Annulate. Ringed (of colour).
Ante-. A prefix signifying in front of; e.g. *anteocular*, in front of the eye.

Anterior. Foremost; in front; of the part nearest the head.
Apex. In Coleoptera the parts of the body are described in relation to an imaginary central point, between the *pronotum* and *elytra*; the part furthest from this is the *apex*, the part nearest the base.
Apical. Relating to the *apex*.
Apneustic. Breathing through the tissues, not by means of special respiratory organs.
Apodous. Without legs; e.g. larvae of *Cercyon*.
Apodeme. Any cuticular growth of the body wall.
Apophysis. Any elongate process of the body wall, internal or external.
Appendage. Any part or organ attached by a joint to the body or to any other structure.
Appendiculate. Furnished with appendices, e.g. extra lines or furrows at the end of other lines or furrows.
Approximate. Brought near to one another.
Apterous. Wingless.
Articulated. Jointed.
Asperate. Roughness of the surface.
Asymmetrical. One side of the body different from the other side.
Attenuated. Gradually diminished or narrowed.
Azygos. (not *azygous*) Unpaired portion of the *genitalia*, (i.e. *genital tube* beyond the junction of the *seminal ducts*) (Male).

Base. see *apex*.
Basal. Pertaining to the *base*.
Basal Orifice. Basal (proximal) opening of the *aedeagus* through which the *ejaculatory duct* enters. Often displaced to the ventral side.
Basal Piece. The unpaired basal (proximal) part of the *tegmen*, usually *sclerotised* and may form a complete ring or tube around the *median lobe*. Absent in some cases.
Bead. A fine elevated line generally at the perimeter or centre of a *sclerite* or structure. Example: a basal bead of the pronotum.
Bi-. A prefix to signify something which is composed of two parts, e.g. *bifid*, two-pronged; *bilobed*, divided into two lobes, *biforous*, having two apertures.
Border, Bordered. When a margin has a raised edge.
Buccal. Relating to the mouth or sides of the mouth.
Bursa Copulatrix. The proximal blind end of the vagina, with which it is narrowly connected. It receives the *internal sac* during copulation (Female).

Calcar. A spur, strong spine or pointed process.
Callosity, Callus. A slight projection or elevation, usually rounded.
Callose. Furnished with such a projection.
Campodeiform. Of a larva; the three principle divisions of the body are defined, and possessing three pairs of legs.
Canaliculate. Furnished with one or more channelled furrows.
Capillary. Slender and hair-like, applied to antennae.
Capitate. The term used to describe an abruptly clubbed antenna.
Carina. A keel or longitudinal elevated line.
Carinate, Carinated. Furnished with a *carina*.
Castaneous. Chestnut-coloured.
Catenulate. Chain-like.

Caudal marginal line. The line of junction which runs parallel to the hind coxae on the first abdominal sternite, which may either follow the hind coxal edge exactly or diverge from it to a varying degree posteriorly. Useful in specific diagnosis of the genera *Meligethes* and *Carpophilus*.

Cercus. Usually a paired, attenuated – sometimes segmented – process projecting from the ninth abdominal tergite. (Plural, *cerci*). The *cerci* of many coleopterous larvae are long, and directed posteriorly.

Chitin. Chemically, a nitrogeneous polysaccharide of very complex structure, often admixed with other substances of doubtful composition. Pure Chitin is colourless. Chitin forms the essential constituent of the insect exoskeleton.

Cicatrix. A large deep, scar-like impression.

Ciliate. Furnished with cilia or fringes of hairs more or less parallel.

Clavate. Clubbed or club-shaped, especially of the antennae (Fig. 26).

Clypeus. A cranial *sclerite* anterior of and generally fused to the *frons* see fig. 16 (larva).

Common. Extending over two neighbouring portions of the body, e.g. elytra with a common spot.

Compressed. Flattened laterally.

Concolorous. Uniform in colour.

Confluent. Running into one another, applied to coloration and puncturation.

Connate. Soldered together.

Connecting Membranes. Sharp and Muir (1912, p. 485) state that this term cannot be commended. The *genital tube* may exist without *sclerites*, in such a case are these *connecting membranes*? However the term is widely used and is very useful; there are two *connecting membranes*.

First Connecting Membrane – connects the *median lobe* to the *tegmen*.

Second Connecting Membrane – connects the *tegmen* to the *apex* of the *abdomen*. (Male)

Coprophagous. Feeding on excrement.

Cordate, Cordiform. Heart-shaped, usually applied to the *pronotum*.

Coriaceous. Leathery.

Cornea. Lens of *ocellus*.

Corneous. Horny.

Costate. Furnished with elevated longitudinal ribs (*costae*).

Costiform. In the shape of a *costa* or raised rib.

Coxal. Relating to the *coxae*; the *coxal* or *cotyloid cavities* are the cavities in which the *coxae* articulate.

Crenate, Crenulate. Furnished with a series of very blunt teeth which take the form of segments of small circles.

Crepuscular. Occurring at twilight, relates to habits.

Cretaceous. Chalky.

Cribriform. Perforated, like a sieve.

Cruciform. Cross-shaped.

Cupule. Small cup-shaped organs with which the anterior *tarsi* of certain males (especially among the *Dytiscidae*) are furnished; they are used as suckers for adhering.

Cursorial. Adapted for running.

Cuspidate. Sharply pointed.

Cyathiform. Cup-shaped.

Decumbent. Lying down.

Deflexed. Bent down, compare *reflexed*.

Dehiscent. Gaping apart towards the *apex*.
Dentate. Toothed, furnished with small teeth or tooth-like prominences.
Denticulate. With a row of small teeth.
Depressed. Flattened as if by pressure from above. (Compare *compressed*).
Dichotomy. A division into pairs.
Dichotomous Table. A scheme for identification which gives an alternative choice at each step.
Digitate. see *Palmate*.
Dimorphic, Dimorphous. Presenting two distinct types in the *same sex*, e.g. females of *Dytiscus circumcinctus* Ah. may have smooth or sulcate elytra.
Dimorphism, sexual. Differences in form *between the two sexes* of a species.
Disc. The middle, central portion.
Distal. Away from the centre of the body or point of attachment.
Distinct. Of spots, punctures, etc.; not touching or running into each other.
Divaricate. Used of two parts that are contiguous at the base and very strongly *dehiscent* at the apex (forked).
Dorsal. The upper surface.
Ductus ejaculatorius. The ectodermal, mostly median and unpaired, exit tube of the efferent system, opening by the *gonopore* at the tip of the *median lobe*. Its distal part sometimes called the *bulbus ejaculatoris* when enlarged. (Male).
Ductus spermathecae. The primary canal through which the sperms enter the *spermatheca* from the *vagina* or *bursa copulatrix*. (Female).

Ecdysis. Moulting of the external skeleton.
Elytra. The chitinous anterior pair of wings; the wing cases (singular = *elytron*).
Emarginate. Notched, with a piece cut out of the margin.
Entire. Without excision, emargination, or projection.
Entirely. Sometimes used when describing colour, meaning the whole insect or whole part being referred to.
Epigeal. Fungi maturing above the soil.
Eruciform. Of larvae when the three major divisions are not evident, the legs may be degenerate or wanting.
Excised. Cut away.
Exoskeleton. The hard chitinous integument of the Arthropods, fulfills the same function as (endo-) skeleton, but is external.
Explanate. Widened out or expanded. Joy (1932) defines it as "A slight hollowing out, close to margin, most commonly used for the sides of the thorax".
External Lobe. A synonym for *basal piece*.

Facets. The lenses or divisions of the compound eye.
Facies. General aspects of a species, genera or group of insects.
Farinose. Presenting a mealy appearance.
Fascia. A coloured band.
Ferruginous. Rust-red colour.
Filiform. Generally applied to antennae. Thread-like; elongate and of about the same thickness throughout. (Fig. 26).
Fissate. A type of lamelliform antennae, the terminal segments are only poorly laminate. Fig. 26. e.g. *Lucanus cervus*.
Flabellate, Flabelliforme. Fan-shaped; of antennae: having the upper segments prolonged into branches.

Flagellum. The sclerotised terminal prolongation of the *ductus ejaculatoris* usually concealed within the *internal sac* when in repose, but sometimes very long and constantly protruding through the *ostium* (male). In more general terms a flagellum is any long thin process attached to a larger (often globular) organ, e.g. the spermatheca in *Longitarsus* and some *Agathidium* species.
Fossorial. Adapted for digging or burrowing.
Fovea. A large round or elongate depression on the surface.
Foveate. Furnished with such depressions.
Free. Of the head: visible, not hidden by the thorax; of a part of the body: movable, not fused to the adjoining part.
Frontal Suture. Paired suture between the frons and one or other of epicranial halves. See Fig. 16 (*larva*).
Frons. An unpaired cranial sclerite, bears the median ocellus (where present).
Funiculus. The segments of the antennae between the *scape* (first elongate segment) and the club; especially applied to the Curculionidae.
Fuscous. Brown or tawny-brown.
Fusiform. Spindle-shaped, broadest in the middle and gradually narrowed in front and behind to a more or less pronounced point.

Geniculate. Elbowed or kneed, abruptly bent upwards or downwards (Fig. 26).
Genitalia. The reproductive organs as a whole. Both sclerotised and unsclerotised parts, male or female.
Genital Opening. The opening of the *ductus ejaculatorius* into the *ductus communis* on ventral surface.
Genital Segments. Those segments principally, but not exclusively, partaking in the formation of the copulatory organs; in male abdominal segment *ix*, in the female abdominal segments *viii* and *ix*. The segments are usually reduced or modified.
Genital tube. The *median lobe* plus *tegmen* plus connecting membranes.
Gibbous, Gibbose, Hump-backed, very convex.
Glabrous. Smooth, hairless and without punctures or raised sculpture; quite glabrous surfaces in Coleoptera are usually shining.
Gonopore. The external opening of the genital duct (in male and female). In the male, it is often on an intromittent organ. When a genital chamber or an *endophallus* is formed the opening of this cavity may be narrowed so as to form a second *gonopore* in which case the mouth of the *ductus ejaculatoris* (in males) or *oviductus communis* (in females) is called the *primary gonopore*, the latter opening the *secondary gonopore* or *gonotreme*; in males *phallotreme*, in females *vulva* or *oviporus*. Opens between abdominal segments *viii* or *ix*, or on abdominal segment *viii*. This is an ambiguous term and it should be deduced whether primary or secondary gonopore is meant. The term should be reserved for the primary gonopore.
Granulate. With small, rounded-off elevations.
Granulation. Applied to the eyes, the granulation of which is said to be fine or coarse accordingly as the facets are more or less numerous and pronounced.
Gressorial. Adapted for walking.
Gular. Pertaining to the throat.

Habitat. The natural environment where the beetle normally lives.
Halteres. Or balancers: two small knobbed appendages attached to the thorax. They are modified wings which vibrate very rapidly and act as balancers

during flight. They are characteristic of Diptera, though also occur in the Strepsiptera.

Heteromerous. With the posterior tarsi composed of less joints than the anterior and intermediate ones.

Hirsute. Set with thick long hairs.

Hispid. Set with short erect bristles.

Humeral. Relating to the shoulder, (*humerus*).

Hybrid. The offspring of two different species.

Hypermetamorphosis. A metamorphosis in which the insect passes through two or more markedly different larval instars, usually accompanied by marked change in larval life, e.g. Meloidae.

Hypogeal. Subterranean fungi, developing and becoming mature in the soil; this includes the truffles (*Tuber* spp.).

Hypha. The filament of a fungus.

Imago. (Plural *imagines*). The perfect, completed state of an insect; the adult.

Imbricate. Overlapping one another like tiles on a roof.

Impunctate. Without punctuation.

Incrassate. Thickened.

Infuscate. Darkened; more or less fuscous in colour.

Inquiline. Living within another organism or in its nest but not as a parasite.

Insertion. Point of attachment of movable parts, e.g. antennae.

Instar. The progressive stages in the history of an insect constitute complete and distinct periods each of which is known as an *instar*; the new stage after each ecdysis.

Integument. The body wall.

Internal sac. Invaginated cavity at distal end of the median lobe of the *aedeagus* into which opens the *ductus ejaculatoris*. Everted during copulation.

Interspace. The space between punctures.

Interstices. The spaces between *striae* or rows of punctures: the term is properly applied to the *elytra* only, the interspaces on the *pronotum* etc. being called intervals. However the term is often loosely used.

Iridescent. Exhibiting prismatic colours, changing in different lights.

Juxta. In composition means near, as juxta-ocular, situated near the eye.

Keel. A fine raised line.

Labrun. An unpaired cranial sclerite, the "upper lip" covers the base of the mandibles. See Fig. 16 (*larva*).

Lamina. A flat plate or scale.

Lamellate, Lamelliform. Of antennae: having the apical segments like leaves of a book; as in the Lamellicornia (see Fig. 26.).

Lateral. Pertaining to the sides.

Lateral Lobes. Of the *aedeagus*. Often used in a purely descriptive manner; a *paramere*, (not in all cases are *parameres* lateral or paired).

Lignicolous. Dwelling in wood.

Lignivorous. Wood-feeding.

Linear. Narrow, elongate, parallel sided; applied to the whole insect or a particular portion.

Lineated, Lineate. With longitudinal stripes.

Lobes. Parts of an organ separated from one another by a more or less deep division.
Lunulate. Crescent-shaped.

Maculate. Spotted.
Mandibles. The biting jaws. (See Fig. 16. *larva*).
Margin. The outer edge. Margined – furnished with a more or less distinctly produced outer edge.
Maxillae. The lower jaws, always smaller than the *mandibles*.
Median. Central.
Median Suture. (Median epicranial suture), the suture between the frons and the two epicranial halves. (See Fig. 16 *larva*).
Membraneous. Of the consistancy of a membrane.
Mesonotum. The upper surface of the *mesothorax*.
Mesothorax. The middle segment of the *thorax*.
Metamorphosis. The different transitions undergone by an insect during its development from the egg, through the larval stages to the imago or perfect adult form.
Metanotum. The upper surface of the *metathorax*.
Metasternum. The under surface of the *metathorax*. (Fig. 25.)
Metathorax. The third segment of the *thorax*.
Molar. The grinding surface of the mandible.
Moniliform. Of antennae; as if formed of beads. (Fig. 26.).
Mucronate. Terminating in a sharp point.
Mutic. Without point or spine.
Myrmecophilous. Living with ants.
Myxomycete. The "slime moulds". Not true fungi as they never form hyphae but spend most of their life as a naked mass of protoplasm (*plasmodium*).

Natatorial. Adapted for swimming.
Necrophagous. Feeding on dead and decaying matter.
Normal. Usual or natural; this term is used very loosely, but it is often very useful, and its meaning in comparison is always easily understood from the context.

Obconical. A reversed cone, with the thickest part in front often used to describe antennal segments which become thicker towards their apex.
Obsolete. Almost effaced or only slightly marked.
Ocellated, Ocellate. Furnished with round spots surrounding by a ring of a lighter colour.
Ocelli. Small extra simple eyes usually situated on the top of the head. Ocelli (as distinct from compound eyes) are the only ones present in the larval stages (except in third instar larvae of some non-British Strepsiptera).
Ochraceous. Brownish-yellow.
Onisciform. Shaped like a wood-louse.
Onychium. The last segment of the tarsi which bears two *onychia* or claws.
Operculum. A lid.
Orbital. Relating to the upper border of the eye, as supra-orbital, situated above this upper border.
Ostium. Opening through which the internal sac is everted during copulation. Usually situated dorsally and distally on the *median lobe* of the *aedeagus*.
Oval, Ovate, Ovoid. Egg-shaped or eliptical – these are technically different geometric shapes, but in entomology the term is loosely applied.

Ovipositor. The female organ by means of which eggs (*ova*) are deposited. Only one type is to be found in the Coleoptera, this is often called the oviscapt type. Formed by the prolongation or modification of the posterior abdominal segments. Wanting in some Coleoptera.

Palmate. Widened and divided like a hand: if the divisions are long and slender, the term *digitate* is used.
Palpus. (Plural *palpi*) Auxiliary organ of the mouth-parts.
Palillae. Small rounded tubercles.
Paramere. A pair of appendages (sometimes coalescent or even completely joined) forming the distal (apical) part of the *tegmen* and usually protruding on either side of the *median lobe* of the *aedeagus*.
Parthenogenesis. The development of an embryo from an egg without fertilisation.
Patella. A little bowl or cup.
Patelliform. Cup or bowl shaped.
Pectinate. Toothed like a comb. (Fig. 26).
Peduncle. A piece supporting an organ, or joining one organ to another like a neck.
Penis. Originally meaning the intromittent organ, the term has been used interchangeably with *aedeagus* or with *phallus*. Thus it is not homologous throughout the insect class and should preferably be replaced by *aedeagus* or *phallus* respectively.
Pentamerous. With five joints.
Penultimate. The last but one.
Perfoliate. Formed of joints separated as it were strung together by a common thread or narrow support running through them (Fig. 26).
Phytophagous. Feeding on plants.
Piceous. A somewhat dark to very dark colour with green or yellow tint.
Pilose. Hairy, covered with hair-like pubescence. *Verticillate-pilose*, of antennae, with hairs set round the vortex of each segment.
Plasmodia. In *myxomycetes*. the multinucleate motile mass of protoplasm, characteristic of the growth phase.
Pitchy. Blackish-brown: a somewhat loose colour term (see *piceous*).
Pleural. The lateral surfaces of the segments.
Pleurite. Lateral plates of the segments.
Plicate. Furnished with a fold or folds.
Polymorphous. Of various forms.
Polyphagous. Feeding on many kinds of food.
Pores. Large isolated punctures.
Process. A projection: any prolongation.
Productile. Capable of being lengthened out or produced.
Pronotum. The upper surface of the *thorax*.
Propygidium. Penultimate dorsal segment of the abdomen (visible in certain Histeridae, etc., to which the term is applied: it is not used of the Staphylinidae).
Prosternum. The under surface of the *thorax*.
Prothorax. The first segment of the *thorax*.
Pseudopod. A "false foot" – a fleshy protuberance on the ventral surface of the terminal segment of the larval abdomen.
Pubescence. Shiny hairiness or down.
Puncture. A small depression on the surface, usually round.
Punctate (*Punctuation, Puncturation, Punctation*). Covered with punctures.

Punctate-striate. With rows of punctures imitating and taking the place of *striae*, opposed to *striate-punctate*, with punctured *striae*; however, the terms have been loosely used often interchangeably.
Pupa. In Coleoptera, the stage preceding the imago.
Puparium. The *integument* or "skin" of the pupa.
Pygidium. Last dorsal segment of the *abdomen*.
Pygopodium. Terminal flat sclerite of abdomen (characteristic of *Elateridae*).
Pyriform. Pear-shaped.

Quadrate. Square.
Quadri-. In composition indicates four times, e.g. *quadrimaculate*, with four spots.

Raptorial. Adapted for seizing and devouring prey.
Receptaculum seminalis. The *spermatheca*, often including the *ductus spermathecae*.
Reflexed. Bent up.
Remiform. Oar-shaped.
Reniform. Kidney-shaped.
Reticulate. Covered with a network of scratches or cross *striae*.
Retinaculum. A produced tooth-like process, usually arising at the inner margin of the mandible.
Rostrum. Prolongation of the head between the eyes, especially applied to the Curculionidae. *Rostrate* – in the form of a beak or *rostrum*.
Rufous. Reddish.
Rugose. Wrinkled, roughened.
Rugulose. Slightly wrinkled.

Salient. Extended, jutting out.
Saltatorial. Adapted for leaping.
Scansorial. Adapted for climbing.
Scape. The term applied to the first segment of the antennae when it is considerably developed. Chiefly applied to the Curculionidae.
Sclerites. The chitinous plates which collectively make up the exoskeleton. They may be very hard, or quite soft.
Scrobe. Lateral furrow of the *rostrum*, holding the base of the antennae when at rest; chiefly applied to the Curculionidae. The scape fits into this.
Sculpture. Modifications of the surface in the way of punctuation, striae, elevations, etc., as opposed to structure, which has reference to the shape and construction of the various parts of the body.
Scutellum. A dorsal plate of the *mesonotum*. (Fig. 24.).
Securiform. Hatchet-shaped.
Secretion. Matter produced by glands of the body.
Sensillae. A simple sense organ; a body hair connected to the nervous system.
Serrate, Serriform. With teeth like a saw. (Fig. 26).
Seta. A long outstanding bristle or stiff hair.
Setaceous. Gradually tapering to the tip, like a bristle.
Setiform. Shaped like a bristle.
Setose, (Setigerous, Setiferous). Set with or bearing *setae*.
Shagreened. A surface divided into microscopically equal areas, very fine sculpture with no punctuation (like shark skin).

Simple. With no unusual addition or modification, e.g. without spines, dilation, emargination, etc.
Sinuate. Slightly waved.
Spatulate. Narrowed at base and enlarged towards extremity.
Spermatheca. The receptable of the sperms during coition. An ectodermal invagination ventrally and posteriorly at the end of the abdominal segment *viii.* In Coleoptera sometimes connected with the *bursa copulatrix*, sometimes opening by *seminal canal* (female).
Spiracle. Respiratory openings on the surface of the body, the external orifice of the *tracheae.*
Sporophore. The spore-producing or supporting structure, the "fruiting-body" in myxomycetes (the powdery fungi of Fowler and other authors).
Spur. Spike-like projection, occurs on the legs of many beetles and is a useful character in diagnosing, for example, some *Longitarsus* species.
Squamose. Covered with larger or smaller scales (*squamae*).
Stadium. The interval between *ecdyses*.
Sternite. The ventral plates of a segment.
Sternum. The ventral surface.
Striae. Impressed lines.
Stridulation. The sound produced by the friction or scraping of one surface against another.
Strigose. Scratched.
Striole. An abridged or rudimentary *stria. Striolate* – furnished with such small striae.
Style. A pointed process.
Sub-. In composition, indicates almost or slightly, as sublinear, subquadrate, etc.
Subulate. Tapering, terminating in a fine and sharp point, like an awl.
Sulcate. Furrowed.
Sulciform. Shaped like a furrow.
Sutural. Pertaining to the *suture*.
Suture. The line on which the elytra join (= elytral suture); the line of junction of any two adjacent parts.
Synanthropic. Living with humans.

Tegmen. The single divided sclerite situated basally (proximally) of the *median lobe* and often surrounding it when in repose. Usually divided into *basal piece* and *parameres*. The *tegmenite* is an isolated *basal sclerite* of the *tegmen* situated on the second *connecting membrane*.
Temple. The part of the head behind the eye.
Tergite. The dorsal plate or *sclerite* of a segment.
Terminal. The last, the end of a series of segments.
Testaceous. Yellowish, usually with a dusky tinge; not a bright yellow, although the term is loosely used, it is applied to almost all yellowish or reddish-yellow shades. Also having a hard outer cover.
Tetramerous. With four joints.
Tomentose. Cottony.
Transverse. Broader than long.
Trivial name. The name of the species.
Trophi. Parts of the mouth used when feeding.
Truncate. Abruptly cut right across in a straight line.
Tubercle. A small abrupt elevation of varying form.

Unicolorous. Of the same colour throughout.
Unisetose. Bearing one *seta*.

Variolose. Covered with small impressions or pits, a pock-marked appearance.
Vellum. The thin membrane forming part of the apical and marginal portions of a *paramere*.
Vellum aedeagus. The thin membraneous covering of the intromittent organ.
Vermiculate. Covered with irregular, sinuate, worm-like *striae*.
Versicolorous. Of various colours.
Vertex. Upper flattened surface of the head (site of ocelli when present).
Vesicant. Raising a blister, (as applied to *Cantharis* spp. etc.).
Villose. Covered with long loosely-set hairs.

Xylophagous. Wood-feeding.

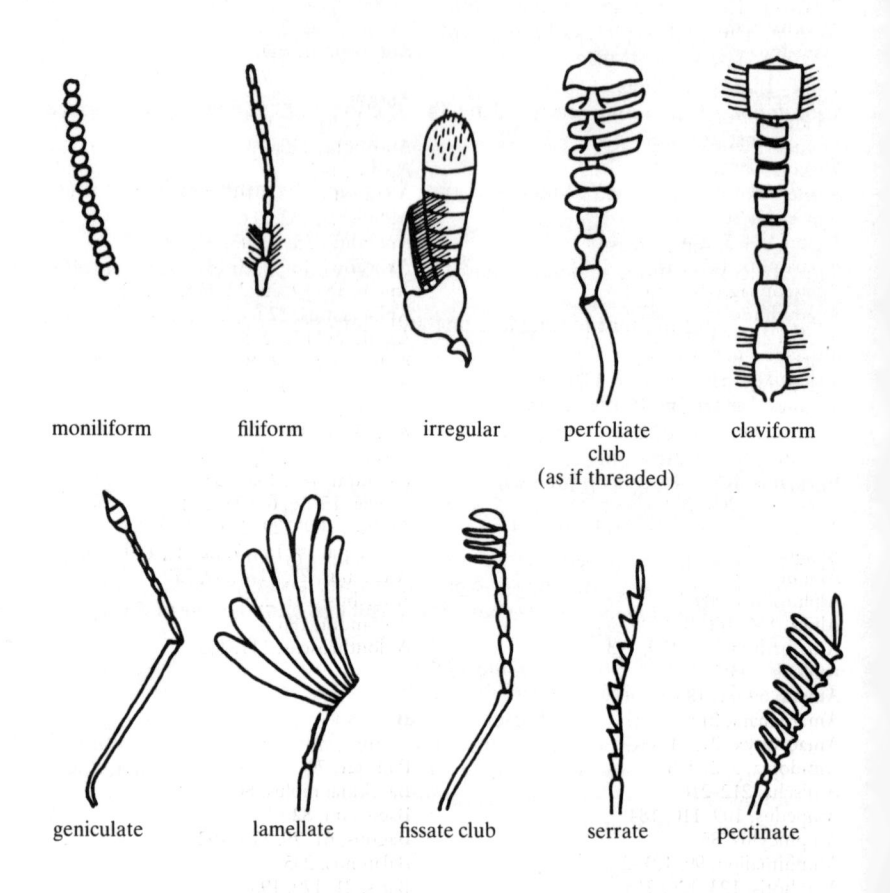

moniliform   filiform   irregular   perfoliate club (as if threaded)   claviform

geniculate   lamellate   fissate club   serrate   pectinate

Fig. 26: Various forms of antennae found in the Coleoptera.

Index of Genera referred to in text

A

Abax, 70
Abdera, 187
Abraeus, 80
Acalyptus, 180, 194
Acanthocinus, 145–6, 188
Acilius, 245
Aclypea, 89
Acmaeops, 188
Acrantus, 119, 198
Acritus, 25, 80
Acrotrichis, 14, 81, 83, 212, 215–6
Acrulia, 21
Actocharis, 23
Acupalpus, 69
Adalia, 133, 136
Adelocera (= Agrypnus)
Aderus, 20, 143, 187
Adonia, 133–4
Aegialia, 99, 102
Aeletes, 80
Aepus, 17, 22, 23
Agabus, 74–5, 244
Agapanthia, 147, 234
Agaricophagus, 84, 85
Agathidium, 15, 25, 85–6
Agelastica, 164–166
Aglenus, 139
Aglyptinus, 84
Agonum, 25, 69–71
Agrilus, 106–7, 184, 228
Agriotes, 20, 109, 248
Agrypnus, 108
Ahasverus, 200, 209
Aleochara, 17, 23, 92–3, 212–215, 224
Alianta, 15
Alophus, 22
Alphitobius, 209
Altica, 156, 166–8, 189
Amalorrhynchus, 179, 193
Amalus, 193
Amara, 69–71, 183
Amarochara, 213
Amauronyx, 213–4
Amidobia, 212, 216
Amischa, 212–216
Ampedus, 109, 110, 184
Amphicyllis, 85
Amphimallon, 99, 101–2
Amphotis, 123, 126, 213
Anacaena, 77, 245
Anaglyptus, 188
Anaspis, 20, 140–1, 187
Anatis, 186

Anchonidium, 177
Ancistronycha, 115–6
Anisodactylus, 27
Anisoplia, 102
Anisotoma, 25, 85
Anisoxya, 187
Anitys, 117, 185
Anobium, 117, 119, 185, 229
Anomala, 99
Anommatus, 25
Anoplus, 192
Anotylus, 23, 92
Anthaxia, 106, 184
Antherophagus, 28, 130
Anthicus, 22–3, 27, 142
Anthocomus, 120
Anthonomus, 15, 21, 179, 193–4, 248
Anthophagus, 25
Anthrenus, 27, 117, 209, 229
Anthribus, 172, 191
Apalus, 142
Aphanisticus, 18, 107, 184, 228
Aphidecta, 133, 186
Aphodius, 25, 29, 45, 97, 99–102
Aphthona, 156, 167, 189
Apion, 18, 19, 22, 27, 173, 191, 235
Aplocnemus, 120
Apoderus, 172, 235
Apteropeda, 166–8
Araecerus, 172
Arena, 23
Arhopalus, 147, 234
Aromia, 147, 188, 234
Asemum, 147, 188, 234
Atheta, 15, 16, 61, 93, 212–5
Athous, 109, 229, 249
Atomaria, 14, 16, 17, 22, 27, 130
Attagenus, 27, 117, 209, 229
Attelabus, 191, 235
Aulonium, 186
Aulonthroscus, 111, 112

B

Badister, 70
Baeckmanniolus, 80
Baeocrara, 82
Bagous, 61, 79, 178, 192
Balaninus, 235
Baris, 21, 179, 193
Barynotus, 22, 175
Batophila, 167, 189
Batrisodes, 96, 212–5
Bembidion, 17, 23, 24, 26, 69–72

Berosus, 76
Bibloplectus, 95
Biphyllus, 25, 186
Bitoma, 139
Blaps, 71, 231
Blastophagus (= Tomicus)
Bledius, 23, 24, 26, 70, 93
Bolitophagus, 186
Boreophilia, 61
Brachonyx, 179, 194
Brachygluta, 96
Brachypterolus, 121, 122, 185
Brachypterus, 121, 185
Brachytarsus, 171, 172, 191
Bradycellus, 17, 70, 71, 183
Bromius, 162, 188
Broscus, 23, 70
Bruchela, 171
Bruchidius, 188
Bruchus, 188
Bryaxis, 96
Buprestis, 106
Byctiscus, 172, 191
Byrrhus, 17, 104, 105
Bythinus, 96
Byturus, 186

C

Caenocara, 117, 231
Caenopsis, 174
Caenoscelis, 17, 130
Cafius, 92
Calathus, 25, 70
Callidium, 147, 234
Calomicrus, 164–6, 189
Calosoma, 183
Calvia, 133
Cantharis, 23, 114–6, 229
Carabus, 23, 37, 46, 68, 70–2, 221
Carcinops, 17, 80
Cardiophorus, 108–9, 228
Carpalimus, 24, 26, 27
Carpophilus, 15, 123–6, 204
Cartodere, 25, 137
Cassida, 23, 119, 156, 169, 190, 235
Cathormiocerus, 174–5
Catopidius, 86
Catops, 61
Cephennium, 90
Cerambyx, 146
Cerapheles, 120
Cercyon, 23, 27, 76–7, 223
Cerophytum, 113
Cerylon, 15, 186
Cetonia, 95, 99, 102, 212, 216, 226
Ceuthorhynchidius, 179, 193

Ceutorhynchus, 18, 19, 178–9, 193
Chaetocnema, 156, 167–8, 190
Chaetophora, 104
Chalcoides, 166–7, 190
Chilocorus, 186
Chlaenius, 23, 70–1
Choleva, 27, 61, 86–7
Chrysanthia, 142, 187
Chrysolina, 162–4, 188–9, 267
Chrysomela, 162–3, 189
Cicindela, 18, 69, 223, 241
Cicones, 139, 186
Cidnopus, 109, 228
Cidnorhinus, 179, 193
Cimberis, 171
Cionus, 177, 192, 248
Cis, 15, 138
Claviger, 96, 214–5
Cleonus, 176, 192
Cleopus, 177, 192
Clitostethus, 134, 186
Clivina, 24
Clytra, 159, 160, 211–2, 235
Clytus, 147, 234
Coccidula, 186
Coccinella, 133–4, 186, 212
Coelambus, 74
Coeliodes, 179, 193
Coelostoma, 76
Colenis, 84
Colon, 20, 27, 87
Coniocleonus, 192
Conopalpus, 140, 233
Copris, 100, 102
Corticaria, 23, 137
Corticarina, 23, 137
Corticeus, 186–7
Corylophus, 23, 132
Corymbites, 248
Creophilus, 224
Crepidodera, 167, 190
Criocephalus (= Arhopalus)
Crioceris, 158–9, 188
Cryphalus, 151, 198
Cryptarcha, 123, 126
Crypticus, 27
Cryptocephalus, 160–1, 188
Cryptolestes, 205, 207, 209
Cryptophagus, 14, 22, 28, 130, 200
Cryptorhynchus, 192, 235
Crypturgus, 151
Ctenicera, 109
Ctesias, 25, 27, 117, 229
Curculio, 180, 194
Curimopsis, 104–5
Cyanostolus, 27
Cychramus, 123, 126
Cylindronotus, 231

Cymbiodyta, 245
Cypha, 15
Cyphon, 65, 103
Cyrtusa, 84
Cytilus, 105

D

Dascillus, 226
Dasytes, 120
Demetrias, 71, 183
Dendroctonus, 148–9, 153–4
Dendrophilus, 17, 80, 212–6
Dendroxena, 89, 184
Denticollis, 109, 228
Deporaus, 172, 191, 204
Dermestes, 27, 117, 205, 209, 229
Derocrepis, 167
Deubelia, 17
Dianous, 18, 23, 92
Diasticus, 99, 102
Dibolia, 190
Dicheirotrichus, 27, 70
Dicronychus, 18, 184
Dienerella, 15
Diglotta, 23
Dinarda, 212, 214, 216
Diplocoelus, 186
Dirhagus, 113, 184
Dolichosoma, 20
Donacia, 21, 157, 188, 246
Dorcatoma, 117–8
Dorcus, 80, 89, 102, 130, 184, 224
Dorytomus, 17, 178, 192–3, 236
Dromius, 15, 25
Drupenatus, 179, 193
Drusilla, 212–16
Dryocoetes, 151, 198
Dryocoetinus, 151, 198
Dryophilus, 117, 185
Dryophthorus, 215
Dryops, 24, 78, 108
Dycronychus, 108, 109
Dyschirius, 23, 24, 26, 70
Dytiscus, 44, 45, 242, 245

E

Elaphrus, 24
Elater, 108–9, 184, 229
Eledona, 231
Ellescus, 17, 180, 194
Elodes, 103
Enicmus, 25, 137
Enochrus, 76, 77, 245
Epiphanis, 113

Epitrix, 190
Epuraea, 26, 123, 125–7, 231
Eremotes, 192
Ernobius, 21, 25, 117, 185
Ernoporus, 151, 198
Eubria, 78
Eubrychius, 179, 193
Eucnemis, 113, 184
Euconnus, 27, 90, 215
Euheptaulacus, 101, 102
Euophryum, 178
Euplectus, 14, 27, 95
Eurycolon, 87
Euryptilium, 82
Euryusa, 215
Eutheia, 90, 215
Exapion, 171
Exochomus, 135, 186

F

Fleutiauxellus, 109
Furcipus, 179, 194

G

Galeruca, 164–6
Galerucella, 156, 164–6, 188, 235
Gastrallus, 117, 185
Gastrophysa, 162, 189
Geodromicus, 25
Georissus, 77
Geotrupes, 100–102
Glischrochilus, 15, 123, 126–7, 231
Gnathoncus, 16, 17
Gnatocerus, 205, 209
Gnorimus, 98, 102, 184
Gracilia, 25
Grammostethus, 80
Graphoderus, 75
Gronops, 192
Grynobius, 229
Grypus, 178, 193
Gymnetron, 180, 193
Gynandrophthalma, 159, 160, 188
Gyrinus, 75, 78
Gyrohypnus, 212–3
Gyrophaena, 92

H

Hadrobregmus, 120, 185
Halacritus, 80
Haliplus, 73
Hallomenus, 233
Halobrecta, 23, 92

Haploglossa, 16, 212–3, 215, 224
Harminius, 184
Harmonia, 134
Harpalus, 70, 71, 183, 222
Helichus, 27, 78
Helochares, 76, 77, 245
Helodes (= Elodes), 226
Helophorus, 76–7, 183, 223, 245–6
Helops, 26
Hemicoelus, 117
Heptaulacus, 101, 102
Hermaeophaga, 167, 189
Hetaerius, 81, 212, 214, 216
Heterocerus, 24, 27, 106
Heterothops, 23, 212–3
Hippodamia, 134
Hippuriphila, 166, 180
Hister, 80, 224
Homoeusa, 213–5
Hoplia, 102
Hydaticus, 45, 75, 245
Hydnobius, 29, 84–5
Hydraena, 77, 246
Hydrochara, 245
Hydrochus, 77, 245
Hydrocyphon, 103
Hydronomus, 192
Hydrophilus, 62, 76, 245
Hydroporus, 74, 75, 245
Hydrosmecta, 92
Hydrothassa, 162–4, 189
Hygrobia, 74, 223
Hylastes, 15, 151, 198
Hylastinus, 198
Hylecoetus, 185
Hylesinus, 151, 198
Hylis, 113, 184
Hylobius, 20, 22, 177, 192, 235
Hylurgops, 151, 198
Hypebaeus, 20, 120, 185
Hypera, 18, 176–7, 192, 235
Hypnoidus, 248
Hypocaccus, 18, 26, 80
Hypocoleus (= Hylis)
Hypulus, 187

I

Ilybius, 74, 223, 242
Ilyobates, 213, 215
Ips, 148, 151, 198
Ischnodes, 109, 184

J

Judolia, 188

K

Kateretes, 121–2, 185
Kissister, 80, 183
Kissophagus, 152, 198
Korynetes, 27, 119

L

Labidostomis, 159, 160, 188
Laccobius, 77
Laccophilus, 74
Lacon, 108–9, 184
Laemostenus, 29, 71
Lamia, 188
Lamprinodes, 214, 216
Lamprohiza, 116
Lamprosoma, 161
Lampyris, 116, 229
Langelandia, 25
Laricobius, 184
Larinus, 176, 192
Lasioderma, 15, 117, 199, 209
Latheticus, 209
Lebia, 71
Leiodes, 20, 27, 29, 84–5
Leiopus, 147, 188, 234
Leiosoma, 177, 192
Leistus, 25, 70
Lema, 158, 188
Leperisinus, 181, 198
Leptacinus, 212, 215–6
Leptideella (= Nathrius), 25
Leptinus, 16, 17, 28, 213
Leptura, 148, 188
Lesteva, 23
Licinus, 18, 70
Lilioceris, 158, 159
Limnebius, 77, 245–6
Limnichus, 78, 105
Limnobaris, 179, 193
Limobius, 177, 192
Limoniscus, 109, 184, 267
Limonius (= Cidnopus)
Liogluta, 61, 213
Liophloeus, 175, 192
Liparus, 177, 192
Lithocharis, 27
Litodactylus, 179, 193
Lixus, 176, 192
Lochmaea, 165–6, 189
Lomechusa, 214
Lomechusoides, 26
Longitarsus, 19, 156, 167–8, 189
Lophocateres, 119
Lucanus, 37, 98, 184, 224
Luperus, 164, 165, 189

Lycoperdina, 28
Lymantor, 198
Lymexylon, 185
Lyprocorrhe, 212
Lythraria, 168, 190
Lytta, 142, 187

M

Macronychus, 27, 77
Macroplea, 79, 157, 188
Magdalis, 177, 192, 235
Malachius, 120
Malthinus, 115
Malthodes, 114–5
Manda, 93
Mantura, 166–7, 190
Mecinus, 180, 194, 235
Megapenthes, 109, 184
Megatoma, 27, 117, 229
Melandrya, 140, 187, 233
Melanimon, 27
Melanophila, 25, 106–7, 184
Melanotus, 109, 229
Melasis, 113, 184
Meligethes, 21, 123–4, 126–7, 171, 185, 231, 248
Meloe, 28, 142
Melolontha, 99, 101–2, 226
Mesites, 178
Mesosa, 147, 188, 234
Metabletus, 70
Metoecus, 28, 142–3, 233
Miarus, 180, 194
Miccotrogus, 194
Micralymma, 17, 70, 92
Micrambe, 131
Micrapsis, 133
Micrelus, 179, 193
Micridium, 82
Microcara, 103
Microglotta (= Haploglossa)
Microlomalus (= Paromalus)
Micropeplus, 13, 25
Microptilium, 82
Microscydmus, 96
Mniophila, 14, 167
Molorchus, 147, 234
Monochamus, 153
Mononychus, 29, 179, 193
Monotoma, 13, 22, 27, 129, 212, 216
Mordellistena, 141, 187
Morychus, 105
Mycetaea, 17
Mycetochara, 187
Mycetophagus, 139
Mycterus, 140

Myllaena, 94
Myrmechixenus, 139
Myrmetes, 80
Myzia, 133

N

Nacerdes, 23, 142
Nanophyes, 173–4
Nargus, 86
Nebria, 22–3, 25, 46, 70–2, 221
Necrobia, 27, 119
Necrodes, 89
Necrophorus, 27, 37, 86, 89, 224
Negastrius, 109
Nemadus, 86
Nemozoma, 119, 185
Neomysia, 186
Nephanes, 82
Nephus, 134, 186
Neuraphes, 90, 91, 214
Nitidula, 123, 126, 231
Nossidium, 81
Notaris, 178, 193
Noterus, 74
Notiophilus, 69–72, 221
Notothecta, 212–3, 216
Notoxus, 142

O

Oberia, 20, 188
Obrium, 148, 188
Ochina, 117, 185
Ochthebius, 77, 245–6
Ocyusa, 17
Odacantha, 73, 183
Odontaeus, 27, 99, 102
Oeceoptoma, 17, 89
Olibrus, 131, 186
Oligella, 81
Oligota, 93
Omalium, 23, 92
Omophron, 69
Omosita, 123, 126, 231
Oncomera, 142
Ontholestes, 224
Onthophagus, 100–2
Onthophilus, 80
Oodes, 70, 73
Opatrum, 27, 214, 215
Ophonus, 71
Opilo, 120, 185
Orchesia, 233
Orectochilus, 75
Oreodytes, 24

291

Orobitis, 26, 193
Orsodacne, 157, 188
Orthocerus, 139
Orthochaetes, 178
Orthoperus, 132
Orthotomicus, 151, 198
Oryzaephilus, 199, 203, 205–6, 209
Osmosita, 27
Ostoma, 119, 185
Othius, 212–216
Otiorhynchus, 17, 174, 176, 192
Oulema, 158, 159, 199
Oxylaemus, 139
Oxyomus, 99
Oxypoda, 25, 212–3, 215–6
Oxytelus, 27

P

Pachytychius, 193
Palorus, 209, 231
Parabathyscia, 25
Paracymus, 245
Paromalus, 80, 183
Peranus, 16
Phaedon, 156, 162–4, 188
Phalacrus, 21, 131, 186
Phaleria, 23, 27
Philonthus, 22–27, 46, 93–4, 224
Philopedon, 175
Phloeonomus, 26
Phloeophthorus, 181, 198
Phloeopora, 93
Phloesinus, 151, 198
Phloetribus, 151
Phloiophilus, 15, 119
Phloiotrya, 187
Phosphaenus, 116
Phylan, 27
Phyllobius, 62, 175, 192, 235
Phyllobrotica, 164–66, 189
Phyllodecta, 162
Phyllopertha, 99, 102
Phyllotreta, 156, 166–8, 189
Phymatodes, 147, 188, 234
Phytobius, 179, 193
Phytodecta, 156, 162, 164, 188
Phytoecia, 147, 234
Phytonomus (= Hypera)
Phytosus, 18, 23, 92
Pilemostoma, 169, 190
Pissodes, 22, 192
Pityogenes, 151, 198
Pityophagus, 15, 123, 186, 231
Pityophthorus, 151, 198
Placusa, 92
Plagiodera, 162–3

Platambus, 24
Plataraea, 213
Plateumaris, 79, 157, 188, 246–7
Platycerus, 102
Platydema, 27
Platydracus, 212, 214–5
Platypus, 150, 181, 198, 236
Platyrhinus (= Anthribus)
Platystomos, 191
Plectophloeus, 95
Plegaderus, 79, 183
Pleurophorus, 105
Pocadius, 28, 123, 126
Podabrus, 115
Podagrica, 167, 190
Pogonocherus, 22, 188, 234
Pogonus, 26, 27, 70
Polistichus, 27
Polydrusus, 175, 192
Polygraphus, 151, 198
Pomatinus, 78
Poophagus, 179, 193
Porcinolus, 105
Prasocuris, 162, 163, 189
Pria, 123–4, 185
Prionocyphon, 103
Prionus, 147, 188, 234, 236
Prionychus, 187
Pristonychus, 29
Procraerus, 109, 110, 184
Psammobius, 18, 102
Pselactus, 23, 178
Pselaphaulax, 96
Pselaphlus, 96
Pseudocistela, 187
Pseudopsis, 13
Pseudostyphlus, 193
Psilothrix, 120
Psylliodes, 156, 166–8, 190
Psyllobora, 133
Pteleobius, 151
Ptenidium, 23, 81, 212–3, 215–6
Pterostichus, 23, 25, 69–71, 221
Pteryx, 82
Ptilinus, 119, 231
Ptiliola, 82
Ptiliolum, 82
Ptilium, 82
Ptinella, 82
Ptinus, 16, 28, 118, 209, 215
Ptomophagus, 86
Pycnomerus, 139
Pycnota, 93
Pyrochroa, 140, 187, 233
Pyronema, 80
Pyrrhalta, 164–5, 189, 235
Pyrrhidium, 188
Pytho, 140, 187

Q

Quedius, 16, 18, 23, 92–4, 211–16

R

Reichenbachia, 96
Rhagium, 26, 147, 188, 229, 234
Rhagonycha, 114–5
Rhamphus, 180, 194
Rhinocyllus, 192
Rhinomacer, 21, 191
Rhinoncus, 179, 193
Rhinosimus, 15
Rhizophagus, 15, 26, 129, 231
Rhizotrogus (= Amphimallon)
Rhopalodontus, 138
Rhynchaenus, 180, 194, 248
Rhynchites, 172, 191, 235
Rhyncolus, 192
Rhynocyllus, 176
Rhyssemus, 102
Rhyzopertha, 185, 199, 209
Riolus, 77
Rutidosoma, 179, 193
Rybaxis, 96
Rypobius, 132

S

Salpingus, 22, 25, 187
Saperda, 147, 153, 188, 234
Saprinus, 80
Saprosites, 98, 102, 226
Scaphidema, 186
Scaphidium, 91
Scaphisoma, 91
Scaphium, 91
Schizotus, 140, 187
Sciodrepoides, 86
Scirtes, 103
Scolytus, 149, 151, 154, 181, 198, 236
Scraptia, 20, 140
Scydmaenus, 91
Scydmoraphes, 91
Scymnus, 134, 186
Selatosomus, 108–9, 184
Sepedophilus, 14, 15
Serica, 99, 102
Sericoderus, 132
Sermylassa, 26, 164–6, 189
Siagonium, 224
Sibinia, 194
Silpha, 89, 224
Silusa, 26
Simplocaria, 105

Sinodendron, 98, 102, 184
Sitona, 175–6, 192, 200, 235
Sitophilus, 178, 199, 204, 206
Smicronyx, 193
Smicrus, 27, 82
Soronia, 26, 123, 126, 231
Spercheus, 77, 223
Sphaeridium, 77
Sphaerites, 79
Sphaeroderma, 166–7, 190
Sphindus, 25
Sphinginus, 185
Sphodrus, 29, 71
Staphylinus, 26, 46
Stegobium, 117, 209
Stenagostus, 109, 110
Stenelmis, 77, 78
Stenichnus, 90, 212, 215
Stenocarus, 179, 188, 193
Stenolophus, 69
Stenopelmus, 178, 192
Stenostolus, 188
Stenus, 18, 20, 23, 24, 94
Stethorus, 133
Stilbus, 131, 186
Strangalia, 145, 147–8, 234
Strophosomus, 192
Subcoccinella, 133
Sunius, 214
Syagrius, 192
Symbiotes, 25, 215
Syncalypta, 104
Synchita, 139

T

Tachyporus, 14, 15
Tachys, 24, 71
Tachyusa, 24
Tachyusida, 215
Tanysphyrus, 178, 192
Taphrorychus, 152, 198
Tapinotus, 179, 193
Tarsostenus, 119
Telmatophilus, 15, 129, 186
Tenebrio, 62, 204–5, 209, 231
Tenebriodes, 119, 205
Teredus, 139
Teretrius, 80
Tetratoma, 15, 140
Tetropium, 147, 188, 234
Tetrops, 188
Thalycra, 26, 123, 126
Thanasimus, 119, 185, 231
Thaneroclerus, 119, 120
Thiasophila, 212, 213
Throscus (= Trixagus)

Thryogenes, 178, 193
Thymalus, 119, 185
Tilloidea, 119
Tillus, 119, 185
Timarcha, 162, 163
Tipnus, 118
Tomicus, 149, 151, 154, 198
Trachodes, 192
Trachyphloeus, 174
Trachys, 107, 184, 228
Trechus, 26, 71
Triarthron, 84
Tribolium, 199, 205–9, 231
Trichius, 98, 102
Trichocellus, 17, 183
Trichodes, 119
Trichohydnobius, 84
Trichonyx, 95
Trichophya, 93
Tricocellus, 70
Trinodes, 27
Triplax, 15
Trixagus, 20, 111, 112
Tropideres, 191
Tropiphorus, 175
Trox, 99, 102, 226
Trypodendron (= Xyloterus)
Trypophloeus, 151, 198
Tychius, 180, 194
Tychobythinus, 96, 214
Tychus, 96
Typhaea, 139, 200, 203
Typhaeus, 100, 101, 226

V

Velleius, 12, 28, 93, 224

X

Xantholinus, 215
Xestobium, 117, 120, 185, 229, 231
Xyleborinus, 151
Xyleborus, 152, 181, 198
Xylechinus, 198
Xyletinus, 120, 185
Xylocleptes, 152, 181, 198
Xylodrepa, 224
Xyloterus, 15, 151, 198

Z

Zabrus, 71, 183
Zacladus, 179, 193
Zeugophora, 158, 188
Zilora, 187
Zorochros, 23, 109
Zyras, 212–216

Notes

Notes

Notes

For details of membership of the
Amateur Entomologists' Society and
a full list of current publications please write to:-
THE A.E.S., P.O. BOX 8774, LONDON SW7 5ZG.

THE AMATEUR ENTOMOLOGISTS' SOCIETY

THE SOCIETY was founded in 1935 to promote the study of entomology, particularly among the amateur or younger generation.

The bi-monthly BULLETIN, issued free to members, contains articles on all insect orders in a style suitable for the amateur, as well as observations by members. The Society relies upon members to contribute articles as much as possible. A WANTS & EXCHANGES LIST, issued with the Bulletin, enables members to buy, sell or exchange entomological material, etc.

The MEMBERSHIP LIST, issued free to members, is revised periodically and enables members to contact other entomologists.

STUDY GROUPS exist within the Society for members interested in specific entomological matters or Orders. An ADVISORY PANEL of experts in most Orders provides help with insect identification and other problem areas.

An ANNUAL EXHIBITION is held in London. Field meetings are held by Study Groups and by local groups of members. The Society holds its Annual General Meeting every Spring in London.

PUBLICATIONS cover a wide range and new titles are added at intervals. Details from THE A.E.S. PUBLICATIONS AGENT
The Hawthorns, Frating Road,
Gt. Bromley, Colchester, CO7 7JN

AFFILIATE MEMBERSHIP is available to schools, societies, libraries and other institutions. The subscription is the same as for Ordinary members.

MEMBERSHIP APPLICATION FORMS available from:–
A.E.S. Registrar
22 Salisbury Road, Feltham, Middlesex TW13 5DP.
(Please enclose stamped/addressed envelope)